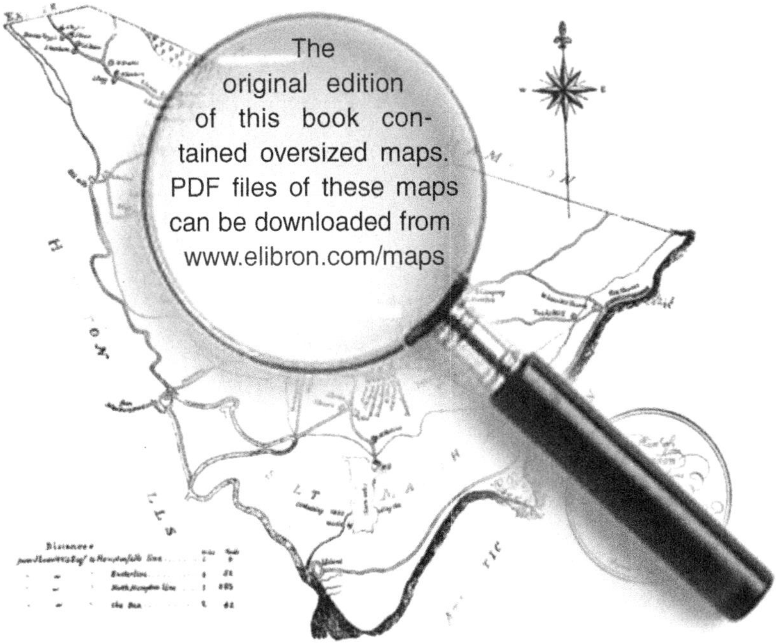
The original edition of this book contained oversized maps. PDF files of these maps can be downloaded from www.elibron.com/maps

ROBIN JOHN TILLYARD

THE BIOLOGY
OF
DRAGONFLIES
(ODONATA OR PARANEUROPTERA)

Elibron Classics
www.elibron.com

Elibron Classics series.

ISBN 1-4212-4898-0 (paperback)
ISBN 1-4212-4897-2 (hardcover)

This Elibron Classics Replica Edition is an unabridged facsimile of the edition published in 1917 by the University Press, Cambridge.

CAMBRIDGE ZOOLOGICAL SERIES

GENERAL EDITOR:—ARTHUR E. SHIPLEY, Sc.D., F.R.S.

MASTER OF CHRIST'S COLLEGE, CAMBRIDGE

THE BIOLOGY

OF

DRAGONFLIES

(ODONATA OR *PARANEUROPTERA)*

CAMBRIDGE UNIVERSITY PRESS
C. F. CLAY, MANAGER
London: FETTER LANE, E.C.
Edinburgh: 100 PRINCES STREET

London: H. K. LEWIS AND CO., LTD., 136 GOWER STREET, W.C.
New York: G. P. PUTNAM'S SONS
Bombay, Calcutta and Madras: MACMILLAN AND CO., LTD.
Toronto: J. M. DENT AND SONS, LTD.
Melbourne and Sydney: THOMAS C. LOTHIAN, PROPRIETARY, LTD.
Tokyo: THE MARUZEN-KABUSHIKI-KAISHA

Plate 1

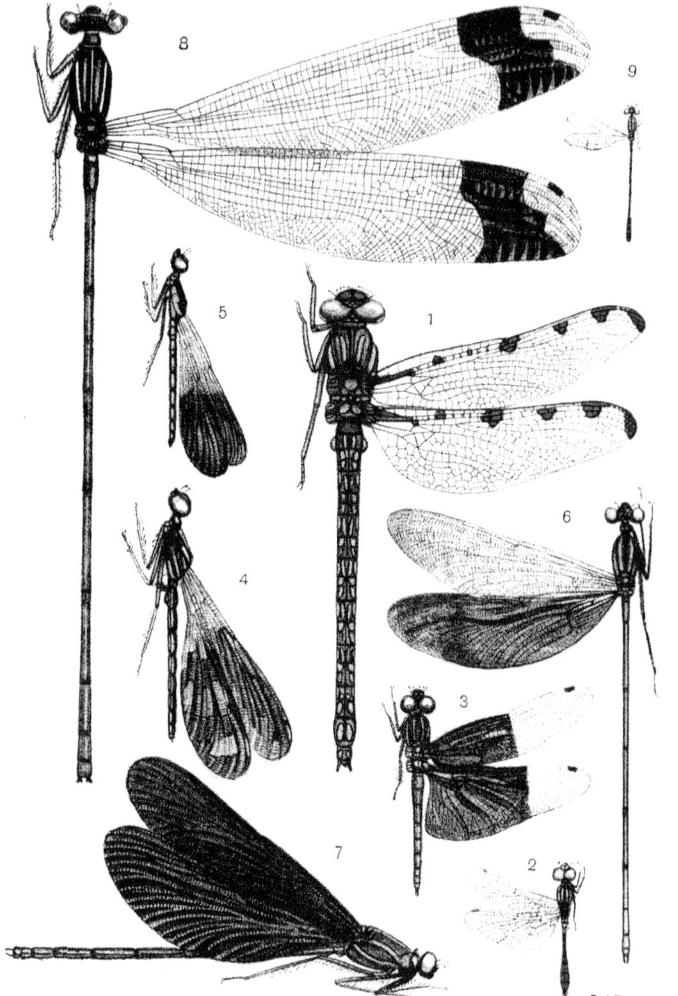

Fig. 1. *Austropetalia patricia* Tillyard, ♀ (*Aeschnidae*, New South Wales).
Fig. 2 *Nannophya australis* Br., ♂ (*Libellulidae*, S. Queensland).
Fig. 3. *Rhyothemis resplendens* Selys, ♂ (*Libellulidae*, N. Queensland).
Fig. 4. *Rhinocypha quadrimaculata* Selys, ♂ (*Calopterygidae*, New Britain).
Fig. 5. *Rhinocypha tincta* Ramb., ♂ (*Calopterygidae*, Papua).
Fig. 6. *Neurobasis chinensis* Linn., ♂ (*Calopterygidae*, Burmah).
Fig. 7. *Matronoides cyaneipennis* Foerster, ♂ (*Calopterygidae*, Borneo).
Fig. 8. *Megaloprepus coerulatus* Drury, ♂ (*Agrionidae*, Central America).
Fig. 9. *Agriocnemis rubricauda* Tillyard, ♂ (*Agrionidae*, N. Queensland)
(All figures natural size.) (Original water-colour drawings.)

THE BIOLOGY OF DRAGONFLIES

(ODONATA OR *PARANEUROPTERA)*

BY

R. J. TILLYARD, M.A. (Cantab.)
B.Sc. (Sydney), F.L.S., F.E.S.

Macleay Fellow in Zoology to the Linnean Society of New South Wales

Late N.S.W. Government Research Student in Zoology in the University of Sydney; formerly Foundation Scholar of Queens' College, Cambridge; and Assistant Mathematical and Science Master at Sydney Grammar School

Cambridge:
at the University Press
1917

Haec est mei

LIBELLULA libri:

τῶν κορδυλῶν φροντίζω τῶν τυπτουσῶν

τῶν γόμφων τῶν πηγνύντων

τῶν αἰχμῶν τῶν λαμπομένων

πάντων καλλιπτέρων

πάντων ἀγρίων.

A living flash of light."

TENNYSON, *The Two Voices.*

TO MY
GOOD FRIENDS AND
MENTORS IN ODONATOLOGY
M. RENÉ MARTIN AND DR F. RIS,
I DEDICATE THIS BOOK, IN MEMORY OF
MANY KINDNESSES RECEIVED
FROM THEM

PREFACE

THE principal object with which this volume has been written is to present as full and complete an account of the Biology of the Odonata as it is possible to offer in the present state of our knowledge of these insects. Thus the book is written primarily for biologists rather than systematists. More than ninety per cent. of the papers so far published on the Odonata have been chiefly systematic in their aim. The remainder are principally morphological, and form a valuable though somewhat small collection of facts on which a book of this kind must necessarily be based. During the last three years, I have fortunately been able to undertake a considerable amount of work on the internal anatomy of the Order, thereby not only confirming many points which were more or less in doubt, but also adding some new discoveries, and here and there rectifying errors. Most of this work has not yet been published, and will first see the light in the condensed form in which it appears in the present volume. It is hoped that the method of treatment followed in this book, by which the morphological, phylogenetic and physiological view-points have been correlated, in so far as our present knowledge allows, will enable students of the Odonata to take up any line of research on this interesting Order with a full knowledge of what has already been achieved. To this end, the Bibliography has been practically confined to those publications bearing on the biological study of the Order. It would have been impossible to include a complete list of the enormous number of purely systematic publications on these Insects. Only those have been included which contain material of sufficient biological interest to require a reference in the text. All papers, however, dealing with the larvae, life-histories or faunal studies, have been included. With the exception of two or three which it has been quite impossible to obtain, all the works listed have been carefully read and studied by the author. References to the Bibliography are given within square brackets.

A complete scheme of classification down to *tribes* is given in Chapter XIV. It is hoped that this, combined with a very full account of those external parts which are of use in classification, will make the book of value to systematists. For the use of British entomologists, a chapter is included dealing exclusively with the British species (Chapter XVIII). At the request of several entomological friends, a short glossary of biological terms has been introduced, for the use of readers who are not biologists.

In a previous volume of this series[1], the author, Dr O. H. Latter, in dealing with the Insecta, selects three types of this enormous Class for study. These are the Cockroach, the Dragonfly and the Wasp. The Cockroach has for many years served the purpose of a "type" for the Insecta, in all general courses on Zoology. Now there are three essentials which a "type" for general use ought to fulfil. Firstly, it must be obtainable without undue difficulty, and easily preserved for use. Secondly, it should illustrate as generally as possible the morphology of its Class. Thirdly, it should present a fairly typical life-history for its Class. The Cockroach completely satisfies the first two conditions. But it fails in the third, since it offers no illustration of the very important phenomenon of metamorphosis, which is associated with the great majority of the Orders of Insecta. It

[1] *The Natural History of Some Common Animals*, by O. H. Latter, M.A. Cambridge, 1904.

seems, therefore, to be in the highest degree advisable, that Dr Latter's lead should now be followed in all Zoological Laboratories, by the selection of two other Insect "types," which should illustrate respectively the phenomena of incomplete and complete metamorphosis. Of these, one should undoubtedly be the Dragonfly, the other either the Wasp, Bee, Blow-fly or Butterfly.

The metamorphosis of the Dragonfly is easily observed and easily understood, yet it is as striking as that of any other Insect. If we add to this the further advantages that both larva and imago (but particularly the former) provide dissections of comparative ease and unrivalled beauty: that both closed and open tracheal systems can be studied in the one insect: and that the adaptations to the aquatic way of living in the larva are of an interest not to be equalled anywhere in the Insecta, it will be seen that a very strong case exists for the recognition of the Dragonfly as a second biological "type" for the Class. If this book helps towards the general acceptance of this view, by providing a text-book suitable for the use of zoological students, the author will feel more than repaid for his labours.

Little need be said as to the design of the book. I have made original text-figures in every case except where the material was not available. In such cases, the authorship of the figure is gratefully acknowledged, but the lettering has been altered to suit the text of the book. Of the original drawings, figs. 23, 30, 33, 35, 37, 38, 125, 128, 139, 146 and 154 have already appeared in my contributions to the *Proceedings of the Linnean Society of New South Wales* [167, 169, 171, 172, 176], while figs. 79–83 are taken from a paper contributed to the *Journal of the Linnean Society of London* [179].

Almost all the specimens studied and figured have been collected around Sydney. Readers will therefore not fail to discover a very strong Australian "flavour" in the book. Those who know how rich and varied the Australian Dragonfly-fauna is, will look upon this rather as an advantage than otherwise. While care has been exercised in selecting, for general description, well-known genera of wide distribution, yet there has been no hesitation in making use of the more archaic Australian forms, when these latter could throw new light on the phylogeny of the Order.

Throughout the book, the names *Gomphus*, *Aeschna*, *Calopteryx* and *Agrion* are used in the Selysian sense. In common with practically all other students of the Order, I regard these names as *nomina conservanda*. Their alteration would not only lead to immense confusion, but would in a sense vitiate the whole of the classical or Selysian literature. Odonatologists are more concerned than anyone else in the retention of these names, and they will never consent to their arbitrary removal by committees formed of men who know little, and care less, about the requirements of this very special branch of entomology.

The MS. of this book was completed on March 19th, 1915, and was forwarded to Cambridge in July, 1915, with the Bibliography brought up to that date. During the first correcting of proofs, opportunity was taken to bring the work on the Respiratory System (Chapter IX) into line with the recent researches of Calvert and myself; but it was not considered practicable to add any new records to the Census and lists of genera. The Additional Bibliography on p. 373 was added in July, 1916, when the first proofs were returned to Cambridge. The sending of the original MS., figures and plates, three sets of galley-proofs and three sets of page-proofs (involving in all a total of some fifty packets and letters), to and fro between England and Australia in these troublous times, without the loss of a single item, is an achievement which I cannot let pass without offering my grateful thanks to all who were responsible for it.

In conclusion, I desire to acknowledge with very sincere thanks the help received in the preparation of this work. My thanks are due to my wife, for the execution of figs. 25, 167, 184, 185, 187, 188, and for her valuable help in indexing; to Mr F. W. Carpenter, M.A., of Sydney Grammar School, for much arduous work in the field, for advice and help in the preparation of wing-photographs, and for the two beautiful photographs reproduced in Plate IV; to Dr F. Ris, of Rheinau, Switzerland, for the loan of publications from his library, and for much useful help and criticism; to M. René Martin, of Paris, for an account of the Ethiopian fauna; to Professor P. P. Calvert, Ph.D., of Philadelphia, for an account of the fauna of the New World, and for other information; to Dr F. F. Laidlaw, of Uffculme, Devon, for an account of the Oriental fauna; to Mr R. A. Muttkowski, B.Sc., of Madison, Wisconsin, for information on the Nearctic fauna; to Mr K. J. Morton, of Edinburgh, and Mr Herbert Campion, of Ealing, W., for specimens of, and information concerning the British and European fauna; to Dr S. J. Johnston, Lecturer in Zoology, University of Sydney, for assistance in drawing up the glossary; to Mr B. G. Engelhardt, of Hornsby, for geological information; and finally to those many other correspondents, whose exchanges of publications and specimens have indirectly helped me in the completion of this work.

R. J. TILLYARD.

"KURANDA,"
MOUNT ERRINGTON,
HORNSBY, N.S.W.
September 30th, 1916.

CONTENTS

PLATES

ERRATA

Plate I, Fig. 4. *For* New Britain *read* India.
Fig. 5. *For* Papua *read* New Britain.

CHAPTER I

INTRODUCTION

HISTORICAL.

In his subdivision of the Class Insecta, Linnaeus placed all the Dragonflies known to him in the single genus *Libellula*, forming the family *Libellulidae* of the very heterogeneous Order Neuroptera. His pupil Fabricius, in re-arranging the Orders of Insects by the structure of their mouth-parts, constituted the Dragonflies as a separate Order under the name Odonata[1], because of the form of their mandibles. His general scheme quickly fell into disuse. But the name Odonata persisted, and has been adopted by all subsequent writers. Latreille merged the Odonata with the May-flies (*Ephemeridae*) in his Order Subulicornia, based on the form of the antennae. To these Gerstaecker added the Stone-flies (*Perlidae*) forming a group Orthoptera Amphibiotica, characterized by the possession of aquatic larvae. Erichson, in splitting up the old Order Neuroptera, formed an Order Pseudoneuroptera to comprise the Odonata, *Ephemeridae*, *Psocidae* and *Termitidae*. Later workers have almost unanimously rejected these unnatural groupings, and have regarded the Odonata as a distinct Order. The name Paraneuroptera has been suggested by Dr Shipley as a substitute for Odonata, with a view to maintaining the termination "-ptera" for all Orders of Winged Insects.

Recent work on the morphology and phylogeny of the Dragonflies has not only tended more and more to confirm the view that they form a distinct Order of Insects, but has even gone so far as to increase our sense of their isolation. Thus Handlirsch [67], on

[1] Greek ὀδούς a tooth, stem ὀδόντ-. The word should therefore more correctly be Odontata.

phylogenetic grounds, raises them to the rank of a Sub-Class (Libelluloidea). This seems to be going rather too far, particularly now that the very important character of the radial sector has been shewn to hold only for the Anisoptera, and not for the Zygoptera. We shall therefore be content to treat the Dragonflies as a very distinct and isolated Order of Insects, under the name Odonata or Paraneuroptera.

Little attention was paid to the Odonata during the first half of the nineteenth century, and the total of described species remained small, despite the excellent work of Rambur [127] and others. The darkness was dispelled by the appearance on the scene of the great Belgian master, rightly called the "Father of Odonatology," Baron Edmond de Selys-Longchamps. His work on the Order began just before the middle of the century, and lasted until his death in 1890. His whole life was devoted to the study of Natural History. But the Dragonflies soon became his favourite group, and remained so to the end. He not only built up a magnificent collection of these insects, but published monographs of every subfamily except the *Libellulinae.* His collections, which contain an enormous number of types, were placed at his death in the Natural History Museum at Brussels, where, we most sincerely trust, they have remained untouched by the greedy hands of the Despoilers.

The life of de Selys marks the Classical Period of the study of Odonata. It was he who first saw the importance of the wing-venation as the basis of classification, and who started the study of the Order along the right lines. To him we owe the facts that the Odonata now stand as one of the best-understood of all Orders of Insects, and that their classification is based on a natural system supported by an ever-increasing weight of morphological and phylogenetic evidence.

After his death, the work of de Selys was continued by his sons, who arranged for the publication of a series of catalogues of the various groups of animals in their father's collection. These catalogues became expanded into complete and magnificent monographs, dealing with every known species. In the case of the Odonata, the *Corduliinae*, *Aeschninae* and *Calopterygidae* were entrusted to M. René Martin of Paris. He has faithfully completed his task, though the work on the last-named family had unfortunately not left the printer's hands by August 1914. The immense task of dealing with the *Libellulinae* was entrusted to Dr F. Ris, of Rheinau, Switzerland. As a result of more than ten years' labour, he has now completed his work, in a series of brilliantly conceived and executed fascicules which mark the high-water level of systematic achievement in Entomology. The *Gomphinae, Lestidae* and *Agrionidae* were unfortunately entrusted to three German authors. Not one of these has so far made any serious attempt to deal with his obligations, accepted now over thirteen years ago.

The personal influence of de Selys reacted directly upon the next generation of students of the Odonata, and shewed itself in an era of great activity and advancement in our knowledge of the Order. There were many, including Martin and Ris, who could proudly claim that they had sat at the feet of the master, and had been directly fired with his spirit of enthusiasm. The younger generation of Odonatologists, amongst whom I am proud to include myself, owe to these men what they in their time owed to de Selys. For the past twenty years, with unfailing courtesy and industry, they have placed their unrivalled knowledge at the disposal of all who sought it, and have lit afresh the flame of devotion which we trust will now be passed on to succeeding generations. Mainly owing to their endeavours, the systematic study of the Odonata has been brought up to a level of excellence not exceeded in any other Order; and thus the way is now paved for the biologists of the present and the future to pass on to the deeper studies of the morphology and physiology of the group, not by selecting, at a chance, any common type that happens to be convenient, but with full knowledge of the phylogenetic position of their regional fauna, and an added certainty as to the degree of specialization to be expected of the types they choose to study.

Characters of the Order.

Habits. *Carnivorous* in all stages. *Amphibiotic*: the larva or nymph being aquatic, breathing by means of tracheal gills; the imago being aërial and breathing air directly through spiracles. *Hemimetabolous*: the larva active, campodeiform, with no true pupal or resting-stage, but with a considerable change of structure at metamorphosis.

Imago. *Head* with biting mouth-parts, the mandibles thick and strongly toothed; antennae reduced, filiform; three ocelli, two large compound eyes. *Thorax* with small movable prothorax, and large obliquely-placed synthorax (formed by fusion of meso- and metathorax); the legs placed well forward and of little use for walking; the wings placed far backward; two pairs of thoracic spiracles. *Wings*, two equal or sub-equal pairs, unfolded, richly veined; the six typical longitudinal veins all present, but $R + M$ and also $Cu + A$ fused basally for a short distance; a secondary anal vein (A') developed; Sc stopping far short of the apex, at a thickened cross-vein, forming a joint, the *nodus*; a thickened membranous patch or *pterostigma* nearly always present between C and R, near the apex; R either simple or once-branched; M branched either three or four times; an *arculus* and *discoidal cell* always present. *Abdomen* with ten complete segments, and terminal appendages; in shape, narrow, elongated, the pleurites membranous, the large tergites enfolding the reduced sternites; eight pairs of spiracles. *External Genitalia*:—♂ with genital pore on ninth sternite; complicated copulatory apparatus developed from the second and anterior part of the third sternites. ♀ with genital pore in the suture between eighth and ninth sternites; ovipositor either complete or reduced. *Mid-gut* without caeca. Numerous *Malpighian tubules*.

Larva. Similar to imago in general features, but with less elongated and often very broadly rounded abdomen. *Compound eyes* present from birth; *ocelli* absent during most of larval life. *Antennae* larger than in imago. *Labium* very highly specialized, forming an elongated jointed *mask* for the capture of prey. *Legs* strong, longer than in imago. *Wing-sheaths* developed externally and gradually, the hind-wing coming to overlie the fore-wing; developing wings richly supplied with tracheae, around which the main venational scheme is subsequently formed. *Tracheal System* secondarily closed; the spiracles present, but either closed or only partly functional. *Gills* developed in the rectum, or on the anal appendages; rarely also as lateral appendages on some of the abdominal segments.

Position of the Order in the Class Insecta.

The Dragonflies occupy such an isolated position within the Class, that it is a difficult task to indicate any other groups as having any real relationship with them. Morphologically, they have been considered to shew relationships with (*a*) the Stone-flies or Perlaria, (*b*) the May-flies or Plectoptera.

The points of resemblance with the Perlaria are: (i) the presence of three ocelli, (ii) the large number of Malpighian tubules, (iii) the aquatic habit of the larvae. and the fact that both breathe by means of tracheal gills, (iv) the retention of the cerci alone of all the abdominal appendages in the larva (except in a few genera). It seems very probable that not a single one of these characters has been inherited from a common ancestor. As regards the ocelli, fusion of two out of an original four to form a single large median ocellus takes place in a number of widely separated groups of Insects. The Malpighian tubules of the Dragonfly are arranged in groups of five or six together. The young larva has only three tubules. Thus the large number of these organs is again a secondarily developed character. The larvae of the Perlaria have no stigmata, and their tracheal system appears to be a *primarily closed one*. The larvae of Odonata, on the other hand, possess stigmata, and their tracheal system is only *secondarily* closed. This points to the supposition that the Perlaria were aquatic long before the Odonata, and that the aquatic habit is not to be traced back to a common ancestor. Lastly, as regards the cerci, these appendages in the Perlaria remain as undifferentiated filiform organs, and are so carried over into the imago. In the Odonata, they are either reduced, or else developed in the form of caudal gills. Only in the Zygoptera are they carried over into the imago. Combining this with the facts that a few Odonate larvae still possess their lateral abdominal appendages, and that the gill-systems in the two Orders are absolutely different, we cannot be far wrong in assuming that the retention of the cerci has been brought about by gradual and incomplete parallel reductions of the original complete set of abdominal appendages once possessed by the ancestors of *all* Insecta, and does not indicate any special affinity.

While the true affinities of the Perlaria remain a matter of much doubt, we must maintain that no true relationships can be proved between them and the Odonata, on the evidence available to us.

Turning next to the May-flies, the position is somewhat different. Their points of resemblance with the Odonata are: (i) the presence of three ocelli, (ii) the large number of Malpighian tubules, (iii) the aquatic habit of the larvae, and the fact that both breathe by means of tracheal gills, (iv) the retention, in the larvae of the Plectoptera, and in *a few* archaic larvae of Odonata, of the lateral abdominal appendages, in the form of gills, (v) the retention of the cerci in the larvae of both groups, and in the imagines of the Plectoptera and the Zygoptera, (vi) the reduced awl-like antennae, (vii) the presence of only one wing-axillary, and the inability to fold the wings downwards either in a flat or roof-like manner over the body. With regard to the first three points, the same argument may be applied as in the case of the Perlaria, especially as the tracheal system of the Plectoptera is as little like that of the Odonata as is that of the Perlaria. Both the latter and the Plectoptera, in fact, appear to have been aquatic in their larval stages long before the Odonata became so. The fourth character mentioned becomes rather striking when we compare the larva of *Pseudophaea* (fig. 87) or *Cora* with that of an Ephemerid. It may well be that we have here some evidence of a very ancient connection between the two groups. As regards the retention of the cerci, we must repeat the argument used above. With respect to the form of the antennae, the reduction almost certainly rests upon convergence, and cannot be used as an argument for relationship. Similar antennae appear quite independently in other groups of Insects. The last character, (vii), seems undoubtedly to have been derived from a common ancestor. Thus we have in this case a certain amount of evidence for a connection between the two Orders, very long ago.

The palaeontological evidence here helps us considerably. The Protodonata of the Upper Carboniferous were admittedly the ancestors of our present-day Odonata, although an unfortunate gap in the Trias hides the exact line of descent from our view. The Protodonata were large insects with well-developed sub-equal wings. They were undoubtedly Dragonflies in facies and habits, though lacking those high specializations of the wing-venation which mark off the Odonata of to-day. Now we can also trace back the Plectoptera to the Jurassic, where we find forms differing little from those of the present day except in the possession of *less reduced* hind-wings. The fossil record undoubtedly points to a gradual reduction taking place in the hind-wings of the Plectoptera, from an ancestral form in which fore and hind-wings were sub-equal. A study of Permian fossils also bears this out. Coming at last to the Carboniferous period, we find evidence of the existence of a group of moderately delicate insects with four almost equal wings. These, under the name Protephemeroidea, are generally admitted to have been the ancestors of our May-flies. It will thus be seen that, in the Upper Carboniferous, the two Orders Plectoptera and Odonata were already differentiated, but far more closely related than they are to-day. It seems therefore not unreasonable to recognize a very archaic connection between the two, of a far more definite character than we can point to in the case of the Perlaria. We need, however, only remember the numerous and overwhelming points of difference between the two Orders to realize how very slight the relationship really is. We have only to recall the differences in the structure of the head and thorax, the atrophy of the mouth-parts in the May-flies, the very characteristic arrangement of interpolated sectors in their wing-venation, together with numerous points of difference in the internal anatomy, to feel the force of this. We have not mentioned the sub-imago of the Plectoptera, since a somewhat analogous process is evident in the case of the pronymph of the Odonata.

But for the fact that the Planipennia have a complete metamorphosis, we might be tempted to see a considerable resemblance between the very ancient Protascalaphine genus *Stilbopteryx* and the Odonata. That *Stilbopteryx* is remarkably like a Dragonfly in its flight and appearance, nobody who has seen this insect alive will deny. Moreover, the fact that it is admittedly the most archaic of all the *Ascalaphidae* makes it impossible to accept this resemblance as one purely due to convergence. The more highly specialized Ant-lions depart widely from the type shewn by *Stilbopteryx*, and do not resemble Dragonflies in the least. It seems allowable to postulate an exceedingly remote ancestor common to these two groups—an ancestor with an incomplete metamorphosis, and only such structure of wing-venation as underlies both types. From that ancestor, the Odonata broke away by the adoption of the aquatic larval habit. The consequent change in the source and direction of the oxygen-supply to the developing wings may have been the determining factor for those wonderful specializations in wing-venation in which the Odonata are unique. On the other hand, a more direct line of descent would lead to *Stilbopteryx* by the intercalation of a pupal stage, and high specialization in the habits and form of the larva. Palaeontology cannot help us here, since any connection of the kind supposed must have existed before the Protodonata became differentiated out.

To conclude this discussion, we find in the Odonata a singularly isolated group, marked by very high specializations of structure, superimposed upon an exceedingly archaic foundation. No near relatives of these extraordinary insects exist to-day, nor have any such existed since Palaeozoic times. We may, however, most certainly point to the Plectoptera as being the group which has the closest affinity with them, slight though it be. In a serial arrangement of the Orders of the Insecta, the Odonata are best placed next to, and just after, the Plectoptera.

CHAPTER II

THE IMAGO. EXTERNAL FEATURES AND SKELETON

In this chapter we shall deal with the general external features of the Dragonfly, the exoskeleton of the head, thorax, and abdomen, its internal prolongations or endoskeleton, and all the appendages except the wings, which are treated separately in chap. III.

THE HEAD (fig. 1).

The head of a Dragonfly, like that of other insects, shews no clear traces of metameric segmentation. Embryology, however, tells us that it consists of a number of segments, most probably *six*, which can be classified according to their position with regard to the mouth, and the nature of the nerve-centres supplying their appendages. The following table shews us at a glance the usually accepted theory of head-segmentation:

	No. of Segment	Position	Appendages, etc.	Central Nervous System	
Pro-cephalon	First	Preoral	Compound Eyes[1], Lateral Ocelli[1]	Protocerebrum	Brain, or Supra-oesophageal Ganglion
	Second	,,	Antennae, Median Ocellus[1]	Deutocerebrum	
	Third	,,	Intercalary segment, with no appendages	Tritocerebrum	
Gnatho-cephalon	Fourth	Postoral	Mandibles	First Pair of Ventral Ganglia	Sub-oesophageal Ganglion
	Fifth	,,	First Maxillae	Second Pair of Ventral Ganglia	
	Sixth	,,	Second Maxillae (= Labium)	Third Pair of Ventral Ganglia	

[1] Generally agreed upon as not being true appendages.

It seems fairly certain that, originally, only the first of these segments was preoral. Thus it corresponded with the prostomium of Annulate Worms, and is not strictly to be considered as an *appendage-bearing segment*. The mouth itself has moved from an original ventral position, to take up a place at the extreme anterior end of the head. Hence the original ventral surfaces of segments

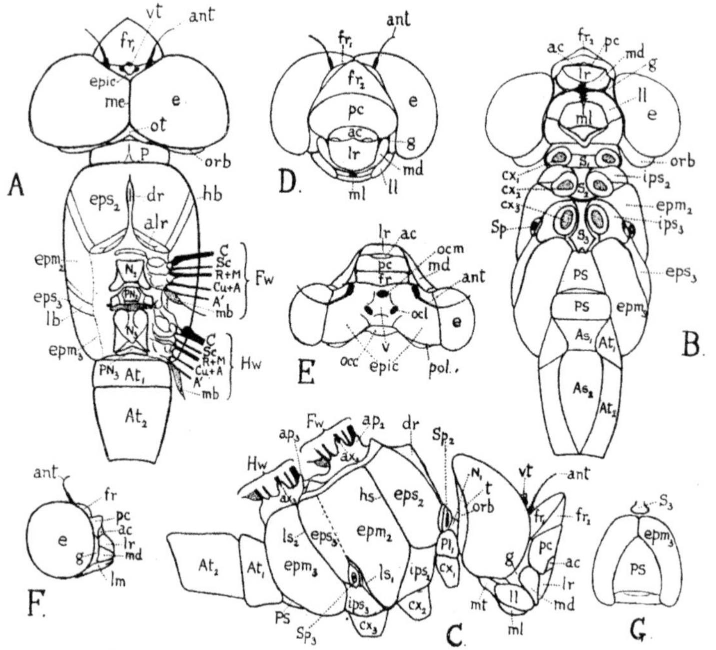

Fig. 1. Exoskeleton of head, thorax, and first two abdominal segments. A. *Austrophlebia costalis* Tillyard, ♀, dorsal view. B. The same, ventral view. C. The same, lateral view. D. Head of the same, front view. E. Head of *Lestes*, dorsal view. F. The same, lateral view. G. Metathorax of *Orthetrum*, ventral view. *A* analis; *A'* secondary analis; *ac* anteclypeus; *alr* ante-alar ridge; *ant* antenna; *ap* costal process; *As* urosternite; *At* urotergite; *ax* axillary; *C* costa; *Cu* cubitus; *cx* coxa; *dr* dorsal carina; *e* eye; *epic* epicranium; *epm* epimerum; *eps* episternum; *fr* frons; fr_1 its superior, fr_2 its anterior portion; *Fw* fore-wing; *g* gena; *hs* humeral suture; *Hw* hind-wing; *ips* infra-episternum; *ll* lateral lobe of labium; *lm* labium; *lr* labrum; ls_1 first, ls_2 second lateral suture; *M* media; *mb* membranule; *md* mandible; *me* median eye-line; *ml* median lobe of labium; *mt* mentum; *N* notum; *occ* occiput; *ocl* lateral, *ocm* median ocellus; *orb* orbit; *ot* occipital triangle; *P* prothorax; *pc* postclypeus; *Pl* pleurum; *PN* postnotum; *pol* postocular lobe; *PS* post-sternum; *R* radius; *S* sternum; *Sc* subcosta; *Sp* spiracle; *t* temple; *v* vertex; *vt* vertical tubercle. (A–D and G $\times 2$; E, F $\times 5\frac{1}{3}$.) Original.

1–3, bearing the clypeus and labrum, have become pushed dorsally upwards into their present positions.

The Compound Eyes (see chap. VII). In the Dragonflies, as is the case also in a somewhat less degree with their Vertebrate analogues the Birds, the shape of the head, and the sizes and relationships of its parts, have become profoundly modified in correlation with the extraordinary development of the sense of sight. The orbits have become expanded to an enormous diameter, in order to accommodate the huge compound eyes (*e*), which, in many forms, occupy by far the largest part of the whole head. In the *Aeschninae* and *Libellulidae* this development culminates in the meeting of the two eyes mid-dorsally on the head in a long line of contact, the *median eye-line* (A, *me*), the true nature of which is explained on p. 146. The stages leading up to this result are preserved for us in a long line of existing Anisopterid forms. Of these, the *Gomphinae* and *Petalurinae* lie nearest to the primitive type. Their eyes are rounded, not excessively large, and separated by a width of epicranium less than the dorsal diameter of the eye. The eyes of *Meganeura* (fig. 156) and other early fossil forms seem to have been very similar to this. In the *Chlorogomphinae* the eyes become somewhat more transversely elongated, and take on a more oval shape. They just fail to touch one another mid-dorsally. In most of the *Cordulegastrinae* a further slight advance is seen, the eyes just meeting in a point. In the *Petaliini* there is another slight advance, the eyes meeting for a very short distance. In the *Brachytronini* the eye-line becomes of moderate length; while in most of the *Aeschnini* and *Libellulidae* it is so long as to obliterate the greater part of the epicranium, and approaches in length the diameter of the eye itself.

In the Zygoptera (E, F) the eye has remained rounded and comparatively small. The tendency has been towards improving the power of sight by a progressive movement of the eyes laterally outwards. Thus the highest forms in this suborder have the head transversely elongated to form a kind of cross-stalk carrying the button-like eyes at its two ends. The distance between the eyes then becomes much greater than the dorsal diameter of a single eye. In the older forms (*Calopterygidae*) the transverse elongation of the head is not so noticeable.

With the above facts borne in mind, we may now profitably study the conformation of the head itself. The various pieces or *sclerites* of the head do not appear to be closely connected with the original segmentation. With the exception, perhaps, of the labrum, clypeus and gula, they cannot be assigned to definite segments. Thus they are chiefly named from their position in relation to the head and mouth.

The Epicranium (*epic*). This is by far the largest part of the head-skeleton. If we compare the head to a hollow rectangular box with a lid on, the epicranium represents not only the lid but the upper portions of the front, back, and sides of the box as well. It is bounded laterally on either side by the eyes, which are set in two large hollows with thickened rims, the *orbits* (A, B, *orb*). Postero-laterally, each orbit becomes somewhat swollen at a place where the eye itself is slightly concave when seen in profile. This is the *temple* (C, *t*). Posteriorly, the epicranium ends in a sharp shelf, the *occiput* (E, *occ*), which varies greatly in size and shape, according to the extent of the eyes. It is broadest in the Zygoptera, where it usually has a wavy outline, and may be divided into a median portion, the occiput proper, and two lateral *postocular lobes* (E, *pol*), corresponding with the more prominent lobes of the same name in the larva. These serve for the attachment of the strong muscles of the mandibles and maxillae. In the *Gomphinae* the occiput is an almost straight shelf, often hairy, and modified in the female, to assist in copulation, by the presence of hooks or tubercles which help to hold the complicated anal appendages of the male in position. In the *Aeschninae* and *Libellulidae*, the occiput becomes reduced to a small *occipital triangle* (A, *ot*). This carries, in some forms, a prominent horn or tubercle. A little in front of the middle of the dorsal surface of the epicranium is a triangular area called the *vertex* (E, *v*), marked by the presence of the three *ocelli*, one of which is median and anterior (*ocm*), the other two lateral, smaller and posterior (*ocl*). Originally flat, as in most Zygoptera, we find that in Anisoptera it tends to become raised up, in proportion to the restriction in the area of the epicranium caused by the increase in the size of the eyes. In the *Gomphinae* the vertex may be a small raised plateau, or a strong transverse ridge with a posterior sunken portion, or a prominent ridge with projecting lateral tubercles. In all these cases, the effect is to force the lateral ocelli forward almost in line with the median one. In the *Aeschninae* and *Libellulidae*, the vertex, like the occiput, becomes excessively reduced, forming a small *vertical tubercle* or *vesicle* (A, *vt*) round which the ocelli are grouped close together.

Just in front of the antennae (p. 15) there runs a transverse groove, the *frontal furrow* or *suture*. This separates the vertex from the projecting anterior portion of the epicranium, known as the *frons* (E, F, *fr*). In the Zygoptera, the frons is a sloping shelf, usually rather narrow, and generally somewhat convex. Below it, on either side, lies a further small piece of the epicranium called the *gena* or cheek (*g*), bordering the eye. In the Anisoptera, and in the genus *Ceriagrion* amongst Zygoptera, the frons becomes raised up into a more or less prominent shelf, having a posterior or superior horizontal portion (fr_1) and an anterior or vertical portion (fr_2), forming a definite forehead. In some *Aeschninae* (e.g. *Austrophlebia*, A, C, fr_1) the frons projects in the form of an anvil, with the superior portion slanting upwards to a median point.

The Clypeus. This is a separate piece inserted below the frons. It forms the chief part of the face of the insect. In all Dragonflies it is divided into an upper *postclypeus* or *nasus* (*pc*) and a lower and smaller *anteclypeus* or *rhinarium* (*ac*). The former embraces and often overhangs the latter (D). In many Zygoptera the post-clypeus forms a kind of shelf comparable with that of the frons in Anisoptera (F, *pc*). In *Rhinocypha* (Plate I, figs. 4, 5) the clypeus projects forwards as a kind of snout.

The Labrum. Below the clypeus, and bounding the mouth from above, lies a strong median piece with a more or less rounded lower border. This is the *labrum* or upper lip (*lr*). It is narrower than the clypeus, and is bounded on either side by the strong pear-shaped bases of the mandibles (*md*). In the Anisoptera it continues the plane of the face vertically downwards; in the Zygoptera it projects obliquely forwards and downwards. Together with the *epipharynx*, which it carries on its inner surface, the labrum forms an organ of taste (chap. VII).

The Mouth. When at rest, the *mouth* is completely closed by the labrum above, and by its appendages below and at the sides. These appendages are the *mandibles* (*md*), the *first maxillae* (mx_1) and the *second maxillae* or *labium* (*lm*); the first two placed laterally, the last ventrally. Hence the mouth is capable of great extension, the gape being as wide as the face itself. The labrum and labium act as upper and lower lips respectively, while the mandibles and first maxillae constitute two pairs of laterally-working jaws.

The Gula. The only remaining portion of the head-skeleton proper is a small membranous piece called the *gula*. Referring again to our simile of the box, this would occupy the lower part of the back of the box, below the occiput. In the higher Anisoptera the large orbits unite in a median line which runs downwards behind the head. Below this line, or, in the other forms, below the lower portion of the occiput, the head joins the neck. If, now, the head be removed, the rounded opening of the *posterior foramen* will be seen, through which the cavities of the head and thorax are in communication with one another. The small *gula* closes this foramen from below. Although so insignificant in size, it is of interest in being generally considered to represent the ventral plate or *sternum* of the sixth or last head-segment. It forms a base of attachment for the labium.

The Tentorium. We have already seen that the frons is really the most anterior portion of the epicranium. The anterior edge of the epicranium, bordering the clypeus, is turned inwards to form an internal chitinous shelf, separating off an upper *frontal cavity* from a lower *clypeal cavity*. Again, the two ends of the frontal furrow also send a pair of ingrowths into the head-cavity. These run downwards and backwards close together, and meet the above-mentioned shelf at its posterior end. They then run on a little further to end up on either side of the posterior foramen. These two rods are in their turn connected by a short transverse rod which divides the posterior foramen into upper and lower halves. The lower passage carries the nerve-cord, the upper the oesophagus, salivary ducts and tracheae. Thus we see that the head has a definite *endoskeleton*, formed of four separate pieces or *endosternites*. The whole structure is called the *tentorium*. The passage between the parts of the tentorium is called the *tentorial* *foramen*. It slants upwards and backwards. Hence the oesophagus, in running forward towards the mouth, passes *downwards* along this foramen. The brain lies above the oesophagus, between the supporting endosternites of the tentorium. From it the two circum-oesophageal commissures pass downwards, encircling the oesophagus, to join the sub-oesophageal ganglion, which lies below the tentorial foramen.

The Appendages of the Head.

The Antennae. These are placed antero-laterally on the epicranium (fig. 1, *ant*; fig. 2). Each has its base inserted into a small pit or hollow, the *torulus*. The antenna itself is always short, and is formed of from three to seven joints. The latter is the usual number. In the *Gomphinae* the number of joints does not exceed four. The first or basal joint is called the *scape*; it is short and thick. The second joint or *pedicel* is longer, and only slightly thickened. The remaining joints or *distalia* are very slender. The comparative lengths of the joints vary in different genera. In *Calopteryx* and *Synlestes*, the pedicel becomes hypertrophied, so as to equal or even exceed in length all the other joints put together. Fig. 2 shews a number of different types of Odonate antennae. The sensory function of these organs is very poorly developed in all Dragonflies (see chap. VII).

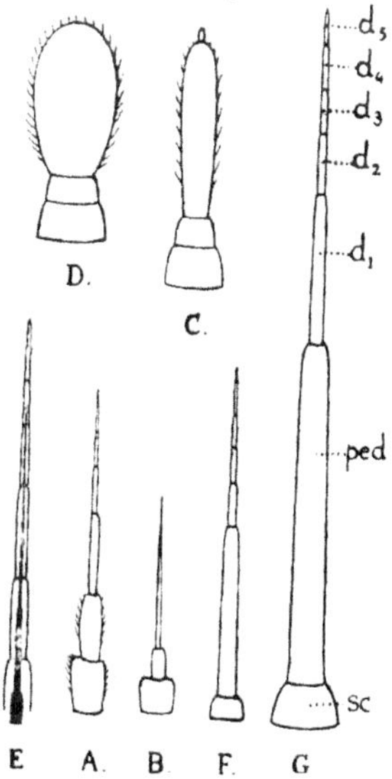

Fig. 2. Various forms of antennae. A. *Aeschna brevistyla* Ramb., imago. B. *Austrogomphus ochraceus* Selys, imago. C. The same, larva. D. *Hemigomphus heteroclitus* Selys, larva. E. *Neosticta canescens* Tillyard, at metamorphosis. F. *Synlestes weyersi* Selys, imago. G. The same, larva. d_1 to d_5 distalia; *ped* pedicel; *sc* scape. (All figures × 11.) Original.

The Mandibles (fig. 3). These are very strong, thick, unjointed appendages, hinged to the head-capsule by a true ginglymus articulation. Viewed externally they are somewhat pear-shaped, the pointed distal end representing the stalk of the pear. The basal portion is shiny, slightly convex, and often brightly coloured, forming a kind of *free* or *mandibular cheek*, in front of the true gena.

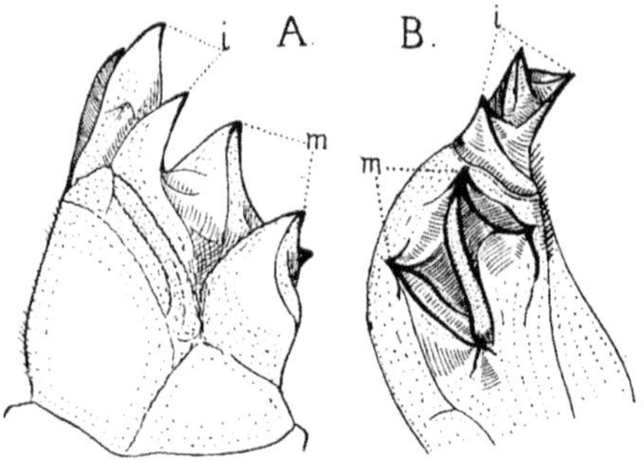

Fig. 3. Mandible of *Aeschna brevistyla* Ramb. A. External view. B. Internal view. *i* incisors; *m* molar. ($\times$13.) Original.

The teeth of the mandible are exceedingly strong. They lie from the distal end to near the base. They may be divided into two groups: (*a*) a distal group of three *incisors* (*i*), and (*b*) a large basal *molar* (*m*). The incisors are sharply pointed and slightly curved. The largest incisor forms the apex of the mandible. The molar has four *cusps* (originally, no doubt, separate teeth), set on a broad raised base in the shape of a **Z**. The cusps are placed one at each end and one at each angle of the **Z**. The structure of the mandibles is very constant throughout the Order. Biramous mandibles occur in a few archaic larvae.

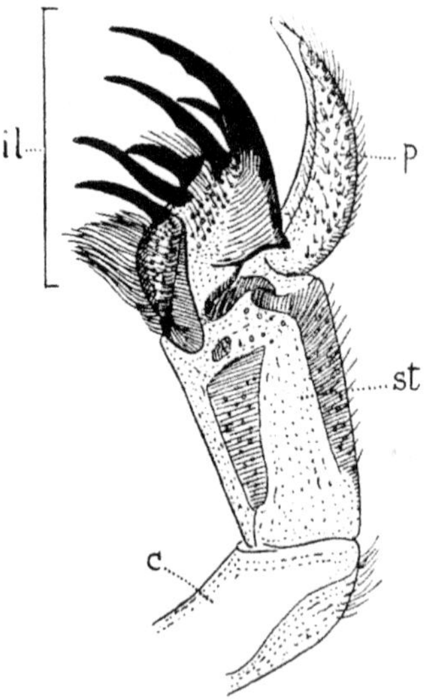

Fig. 4. First maxilla of *Aeschna brevistyla* Ramb. ($\times$15). *c* cardo; *il* inner lobe; *p* palp; *st* stipes. Original.

The First Maxillae (fig. 4). These are jointed appendages, slenderer, longer, and much less powerful than the mandibles. They lie between the latter and the lateral lobes of the labium, and are barely visible when the mouth is closed. The basal joint or *cardo* (*c*) is somewhat twisted. The distal joint or *stipes* (*st*) is longer and wider. It carries on its distal end two freely movable processes or *lobes*. Of these, the *outer lobe* or *palp* (*p*) is curved, hairy, and only moderately pointed. It is obviously an organ of touch. The *inner lobe* (*il*) represents the *lacinia* and *galea* of more typical mandibulate insects, fused together. It has a rather flattened base, from which project six large, sharply pointed and somewhat curved teeth, occupying the outer two-thirds. The inner third forms a small lobe or tubercle clothed with long sensory hairs. The inner lobe is probably both prehensile and tactile in function. Mastication is performed almost entirely by the mandibles.

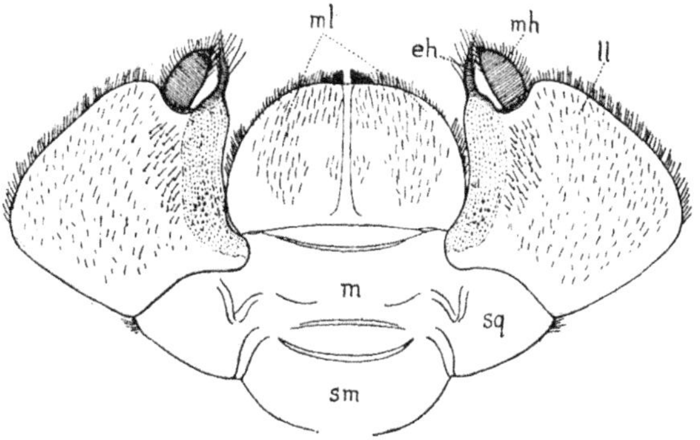

Fig. 5. Labium of *Aeschna brevistyla* Ramb. (×10). *eh* end-hook; *ll* lateral lobe; *m* mentum; *mh* movable hook; *ml* median lobe; *sm* sub-mentum; *sq* squame. Original.

The Labium (figs. 5, 6). In all insects this organ is formed by the fusion of the second maxillae. It is attached basally to the gula, and represents the paired appendages of the sixth or last head-segment. In the Odonata, both larvae and imagines, it is of a highly specialized form. The question of the homologies of its parts has given rise to much controversy. The conflicting views are dealt with in the account of the larval mask on p. 80.

In the imago, a considerable variation of form is seen in the labia of different groups. In all cases, however, we can recognize the same principal parts. There is a broad short base with a projecting middle portion, at the sides of which are articulated two lateral pieces or lobes. The broad base is considered to be formed by the fusion of the *sub-mentum* (*sm*) and *mentum* (*m*). Fused with the mentum distally is the projecting *median lobe* (*ml*), probably representing the *ligula* of other mandibulate insects. The form of this lobe varies greatly. In the Zygoptera and the older Anisoptera it is very prominent and definitely bifid (fig. 6 A). In the *Aeschninae* (fig. 5) the median cleft has disappeared, but

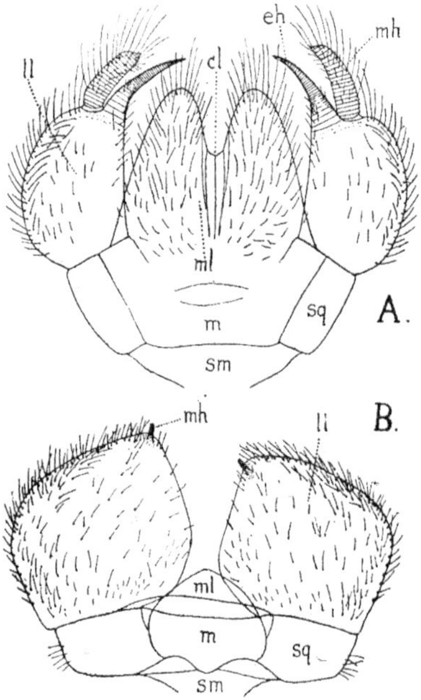

Fig. 6. A. Labium of *Synlestes weyersi* Selys (×25). B. Labium of *Diplacodes haematodes* Burm. (×15). *cl* cleft of median lobe; other references as in fig. 5. Original.

a slight median groove or line marks the place of its disappearance. In addition, one or two hard black tubercles may project on either side of this median groove. In the *Libellulidae* (fig. 6 B) the median lobe is reduced to a small triangular piece, over which the huge lateral lobes fold so as to meet one another in a long median line.

On either side of the mentum, the base is widened by the development of two side-pieces or *squames* (*sq*) which support the *lateral lobes* (*ll*). The changes in form of these lobes are correlated with those of the median lobe. Normally, each lateral lobe has its outer border more or less strongly convex, its inner border either slightly concave, waved or nearly straight. The inner border ends distally in a sharp *end-hook* (*eh*) which is fixed. Just external to this hook is a small articulated piece, called the *movable hook* (*mh*), much resembling a small palpal joint, curved, hairy, and only moderately pointed. The whole lobe is more or less hairy. The end-hook varies in shape. It is most sharply pointed in the Zygoptera (fig. 6 A), and in the older *Aeschninae*. In *Austrophlebia* it is strongly bifid. In *Cordulegaster* it has a set of supplementary teeth. In the higher *Aeschninae* it tends to become reduced in size. The movable hook also tends to become reduced in size along with the end-hook. In the *Libellulidae* (fig. 6 B) the end-hook has disappeared, while the movable hook has lost its articulation, and is only represented either by a small spine or a thickened chitinous spot. The lateral lobes themselves are greatly enlarged, with both inner and outer borders convex. As already described, they meet together above the reduced median lobe. Thus the closed mouth of the *Libellulidae* shews the very characteristic appearance of having a large T-shaped slit, the stem of the T being formed by the line of meeting of these two lateral lobes.

The labium carries internally, projecting from its base, a prominent tongue-like organ projecting into the mouth. This is the *hypopharynx* or *lingua*. This is dealt with in chap. VII. The salivary ducts open close to its base.

THE NECK, OR MICROTHORAX (fig. 7).

In the Dragonfly the neck is exceedingly small and slender, pivoting the head almost, as it were, on a point, and allowing of turning movements to some degree in every direction. Although the neck is firmly fixed to the prothorax, yet it is morphologically part of the head itself. The plates of the neck, or *cervical sclerites* (1–4 *mi*), together with the gula, represent the skeleton of the last head-segment. The neck is supported by four pairs of these plates. Two of these (1 *mi*, 2 *mi*) are rather large and irregular in shape. Posteriorly they are closely connected with the sides of the prothorax. Two others (3 *mi*, 4 *mi*) are very small, and lie freely in the neck membrane [157].

The orifice of the neck, which is the posterior continuation of the posterior foramen of the head, is so minute that it is difficult to believe that the ventral nerve-cord, oesophagus, salivary ducts and tracheal trunks could possibly all pass through it. Yet such is actually the case. It is interesting to note that, if the head of a Dragonfly be removed, either accidentally when striking with the net, or purposely with fine scissors, the body will continue to live for as much as two days longer, while the wings vibrate vigorously whenever the thorax is touched.

The movements of the head on the neck are brought about by a number of pairs of small muscles arising from the prothorax and anterior part of the mesothorax. They are placed dorsally, ventrally, and laterally, so that the head can turn freely in all directions.

THE THORAX (figs. 1, 7).

The thorax of the Dragonfly is very highly specialized, in accordance with the requirements of the very powerful wings and the altered functions of the legs. On the one hand, the prothorax, though remaining a distinct segment, becomes greatly reduced in size. On the other, the meso- and metathorax are large and powerful, and become fused together into one solid mass, which may conveniently be termed the *synthorax*.

In general, three regions can be distinguished in each of the three thoracic segments. These are the *tergum* or true dorsal region, the *sternum* or true ventral region, and the *pleura* or lateral portions. By the addition of the prefix *pro-*, *meso-*, or *meta-*, we indicate to which segment the part in question belongs. The same prefixes may be attached to the name of each particular sclerite.

The Prothorax (fig. 7). This is a small segment supporting the first pair of legs, and the neck or microthorax. It varies greatly in shape. As a general rule, it shews the more normal proportion of parts in the most archaic groups, while it becomes highly reduced or otherwise altered in the most advanced forms.

In all Dragonflies the *tergum* is occupied by a strong sclerite, the *pronotum* (*N*), divided into anterior, median and posterior transverse lobes. The median lobe is generally the largest, and is divided into two by a median longitudinal depression. The posterior lobe may send out a slender strip descending behind the epimerum on either side, and known as the *postepimeral strip*. The *propleurum* is complex in the older forms, such as the *Gomphinae*, where it consists of a large *epimerum* (*epm*) behind, and a narrow elongated *episternum* (*eps*) in front. This latter is constricted into an upper and a lower portion; the lower piece has been named the *proinfraepisternum*. In the *Libellulidae* (A) both parts are reduced, and the episternum is undivided. The *prosternum* (s_1) in the older forms is a single ventral plate with expanded

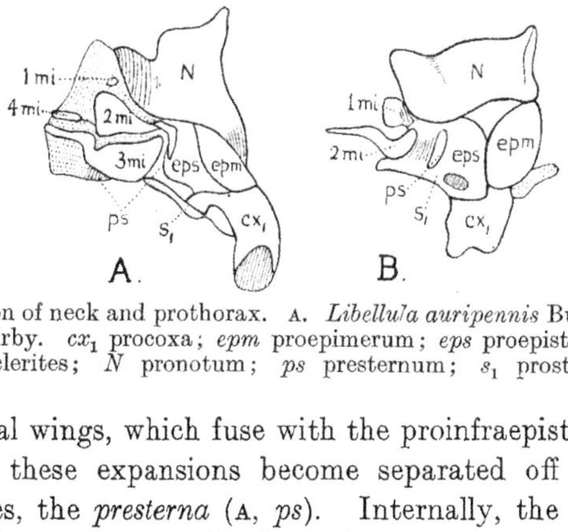

Fig. 7. Skeleton of neck and prothorax. A. *Libellula auripennis* Burm. B. *Lestes uncatus* Kirby. cx_1 procoxa; *epm* proepimerum; *eps* proepisternum; 1–4 *mi* cervical sclerites; *N* pronotum; *ps* presternum; s_1 prosternum. After Snodgrass.

antero-lateral wings, which fuse with the proinfraepisternum. In *Libellulidae* these expansions become separated off as distinct lateral pieces, the *presterna* (A, *ps*). Internally, the prosternum carries a bifurcated piece, the *furca*, forming the endoskeleton of the prothorax. Two slender rods connect the prosternum with the mesosternum, and serve to hold the prothorax in position. They also restrict its mobility upon the synthorax to small dimensions.

In the *Agrionidae* many interesting modifications of structure are to be found in the pronotum. This may take the form of ornamentation—as in the males of the Oriental genus *Amphicnemis*, where the posterior lobe projects backwards and upwards in a huge spine—or it may be concerned with the process of copulation. In such a case, the prothorax of the female is modified to receive the reduced and highly specialized anal appendages of the male, in such a way that a firm interlocking of parts may be secured (see p. 37).

The Synthorax (fig. 1). The remarkable form of the synthorax in Dragonflies is correlated with special conditions of the legs and wings. The former have moved forwards towards the mouth, carrying the sterna with them. The wings, however, have moved backwards away from the head, carrying the terga with them. The reduction in size of both terga and sterna is more than counterbalanced by an enormous development of the pleura. To understand the significance of these changes fully, the student should trace the development of the thorax in the larva, from the time that the wing-buds appear (p. 74).

The well-known *obliquity* or "skewness" of the Dragonfly thorax has been ingeniously measured by Needham and Anthony [107]. If the wings be spread out into a single plane, as if for flight, that plane practically represents the *notal plane* of the synthorax, and its obliquity to the long axis of the body may be measured by the slant of a line drawn through the two axillaries (C, ax_2, ax_3) or other suitable points. This *angle of obliquity* is much smaller in the Anisoptera than in the Zygoptera, the approximate ranges being from 19° to 38° in the former, and from 35° to 62° in the latter. Thus we see that, in this character, the Anisoptera have remained in much the more primitive condition. The ability of the Zygoptera to fold their wings back above the abdomen is of course correlated with the increase in the angle of obliquity.

The obliquity of the synthorax is brought about chiefly by the great development of the *mesepisterna* (eps_2), which grow *forwards and upwards* in front of the tergum, so as to meet in a long middle line, forming the *dorsal carina* or *ridge* (*dr*). Thus they form, in their new position, not only the humeral portion of the thorax, but also all the dorsal part in front of the *ante-alar ridge* (*alr*). The true terga are pushed backwards and lie between the wing-bases. Further, there is a correlated growth of the *metepimera* (epm_3) *downwards and backwards*, which pushes the sterna and the legs forwards, so that the latter come into position close behind the mouth, and are used for holding the prey during flight. In most Dragonflies (G) the metepimera actually meet and fuse together just behind the metasternum (S_3). A median gap left between the line of union and the base of the abdomen is filled in by a thinner plate or plates, the *post-sternum* (*ps*). In the older *Aeschninae*, the fusion is not accomplished, and we have the condition shewn in B.

The Terga. The *mesotergum* is the reduced area lying behind the ante-alar ridge (*alr*) and between the bases of the fore-wings. It bears two prominent plates, often brightly coloured. The upper of these, or *mesonotum* proper (A, N_2), consists of a median shield-shaped *scutum* and a pair of smaller lateral *scutella*. The lower plate, known as the *postnotum* or *pseudonotum* (PN_2), is a short wide piece, likewise divided into three parts, a median *post-scutum* and two lateral *postscutella*.

The *metatergum* lies between the bases of the hind-wings. It carries a large upper plate, the *metanotum* proper (N_3), bearing two convex bosses, and a short, wide, raised *postnotum* (PN_3) in the form of an obtuse triangular ridge.

The Pleura. The suture between the meso- and metapleurum is partly obliterated, except in the *Calopterygidae*, where it runs from between the meso- and metacoxa upwards and backwards to end below and behind the base of the fore-wing. In the other families only the lower half, or less, can be distinguished, running from the coxal region to a point just in front of or above the metastigma. This dividing suture is called the *first lateral suture* (C, ls_1). The *mesopleurum*, lying in front of it, is divided into two pieces by the very distinct *humeral suture* (C, *hs*). These pieces are an anterior *mesepisternum* (eps_2) and a posterior *mesepimerum* (epm_2). From the former a small portion is divided off above the mesocoxa (cx_2) and is called the *mesinfraepisternum*[1] (ips_2).

The *metapleurum*, lying behind the first lateral suture, is likewise divided into two pieces by the very distinct *second lateral suture* (C, ls_2). Of these, the anterior piece or *metepisternum* (eps_3) is rather narrow. It carries the posterior spiracle or *metastigma* (Sp_3). A short cross-suture at the level of the spiracle separates off from it a small lower portion, the *metinfraepisternum*[1] (ips_3),

[1] The arrangement of the thoracic apodemes or endoskeleton lends support to the view held by some students of Odonata that these pieces are really a part of the sternum.

just above the metacoxa (cx_3). The posterior piece or *metepimerum* (epm_3) is broad and subtriangular.

The Wing-Bases. These are outgrowths from the thoracic integument between tergum and pleurum. They are flexible, and merge into the edge of the notum. Each wing-base bears two hard swellings, the *anterior and posterior calli*, the former lying at the base of the costa, the latter at the base of the fused radius + media, with a weaker extension to the cubitus. Each callus is supported by a stiff process from the pleurum. Of these, the anterior or *costal process* (C, ap_2, ap_3) closely supports the costal or anterior callus, while the larger posterior process or *axillary* (ax_2, ax_3) is freely articulated with the posterior callus. The axillary forms the chief fulcrum for the action of the wing-muscles (p. 210). It will be seen that the wing of the Dragonfly agrees with that of the May-fly, but differs from those of all other Insects, in having only the one true axillary, which corresponds with the *second axillary* of other insect-wings [157].

The Sterna. These are very much reduced (B). The *mesosternum* (S_2) is a small plate lying between the two mesocoxae; the *metasternum* (S_3) is a similar plate between the metacoxae. The post-sternum has been already mentioned.

The Thoracic Spiracles. There are two pairs of spiracles on the thorax, the anterior pair being the larger. The anterior spiracles or *mesostigmata* (C, Sp_2) are transversely elongated slits, lying between the sides of the posterior lobe of the prothorax and the mesepisternum, a little above the mesinfraepisternum, in such a position as to be more or less completely hidden from view. These spiracles are developed on special plates, the *mesostigmatic laminae*, which extend so as to fuse with one another across the mid-dorsal line. Thus they form a single *spiracular dorsum* which fuses behind with the steep vertical front of the mesepisterna. The posterior spiracles or *metastigmata* (Sp_3) are smaller, broadly oval, and very distinct. They lie in the metepisterna, at the junction of the cross-suture with the first lateral suture.

The Synthoracic Endoskeleton (fig. 8). Along all the pleural and sternal sutures the thoracic integument is folded inwards to form an internal ridge or *apodeme*. The pleural apodemes of the synthorax number five pairs, named, from in front backwards, the *precostal* (*pc*), the *mesopleural* (pl_2), the *interpleural* (*ip*), the *meta-pleural* (pl_3), and the *posterior epimeral* (*ep*). The mesopleural lies

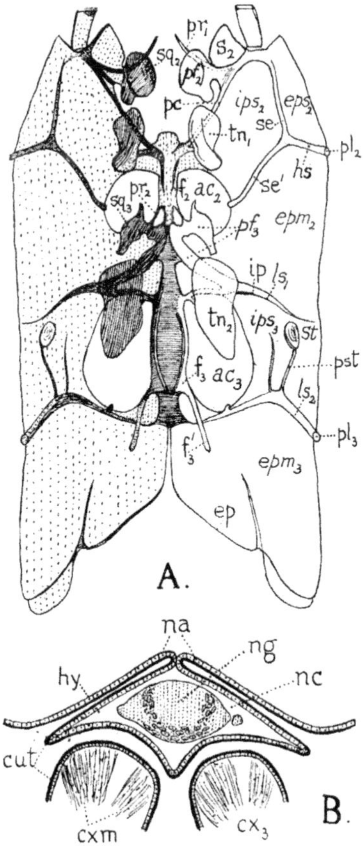

Fig. 8 A. Internal ventral view of skeleton of a Dragonfly's thorax, to shew endoskeleton. Adapted from Berlèse. B. *T.S.* through metathoracic neural canal in *Isosticta simplex* Martin (×15). Original. *ac* acetabulum; *cut* cuticle; cx_3 metacoxa; *cxm* coxal muscles; *hy* hypoderm; *na* sternal apodemes; *nc* neural canal; *ng* metathoracic ganglion; *St* metastigma. For references to parts of endoskeleton, see text; to exoskeleton, legend of fig. 1.

along the humeral suture (*ns*), and is continued ventrally into the *sterno-episternal* (*se*) directed forwards and downwards, and the *sterno-epimeral* (*se'*) directed backwards and downwards, to the acetabulum. In the case of the metapleural (which lies along the second lateral suture, ls_2), there is an additional branch, the *peristigmatic* (*pst*), running to the metastigma (*St*). The interpleural (*ip*) is weakly developed, lying along the first lateral suture (ls_1). The sternal apodemes are known as the *furca* (f_2, f_3) and *prefurca* (pf_2, pf_3). The furca develops a slender posterior process (pr_2, f_3') which in the case of the mesothoracic apodeme serves to connect the mesosternum with the metasternum. A similar process (pr_1) connects the mesosternum with the pro sternum. Anteriorly the furca gives off a large flattened process or *squame* (tn_1, tn_2) which becomes highly specialized to serve as the tendon of attachment for the great elevator wing-muscle. The prefurca likewise develops a squame (sq_2, sq_3) which is less highly specialized in form, but serves as the tendon of attachment for the anterior depressor wing-muscle. The sternal apodemes are so bent inwards towards the middle line that they almost meet along their free edges, isolating below them a small canal of flattened triangular section, the *neural canal* (fig. 8 B). The ventral nerve cord runs through this canal, which serves as a protection to it, and isolates it from all the rest of the thoracic organs.

The Appendages of the Thorax.

The Wings are dealt with in chap. III.

The Legs (fig. 9). All Dragonflies possess the full number of legs, viz. three pairs, one pair to each thoracic segment. Though short in comparison with the size of the body, as well as with their size in the larva, the legs of the Dragonfly are of considerable strength, and are strongly armed. Their position close behind the mouth renders them unfit for walking, though they serve very well for climbing, when that mode of progress is required. By the use of its legs, a Dragonfly can quickly orientate its position, so as to interpose the stem of a reed, or a small twig, between itself and an approaching enemy. During flight, all three pairs of legs are carried folded close up under the thorax. The first two pairs at least are able to hold the prey during flight, while the mouth-parts examine and chew it up. It seems very probable that the fore-legs at any rate are used in the act of capturing prey on the wing. A difference of opinion prevails on this point. The difficulty of making direct observations is obvious, and has not so far been satisfactorily overcome.

Each leg consists of five segments or joints, called (from the base outwards) *coxa*, *trochanter*, *femur*, *tibia* and *tarsus*. There is no trochantine. The *coxa* (*cx*) is a short truncated cone, with its rounded base inserted into a large hollow, the *acetabulum*, situated in the side of the sternum, close to the pleurum. The bases of insertion of the legs of the Dragonfly are very closely approximated, particularly in the cases of the middle and hind-legs (fig. 1 B). The *trochanter* (*tr*) is a short and rather slender joint, having a very short upper surface, so that its distal end runs obliquely downwards, and the articulation with the femur is correspondingly modified. A transverse constriction divides the trochanter into a shorter basal and a longer distal portion. An armature of short thick blunt teeth or tubercles may be present. The *femur* (*fm*) is the strongest joint in the leg. It is a cylindrical, fairly stout, long joint, strongly armed with at least two rows of spines or tubercles. The *tibia* (*tb*) is generally somewhat longer, and much slenderer than the femur; it also is armed with two rows of spines, varying greatly in number, size, and arrangement. The *tarsus* (*ts*) is the short slender terminal segment. In all living Dragonflies, it consists of three joints, of which the basal is the shortest, the distal the longest. In the fossil genus *Tarsophlebia* (fig. 158) there were four joints, the basal being the longest. The joints of the tarsus are armed with small, closely-set spines continuing the lines of the tibial armature. The tarsus ends distally in a pair of strong diverging *claws* (*cl*), each of which usually bears a well-developed inferior tooth. Between the bases of the claws there can be

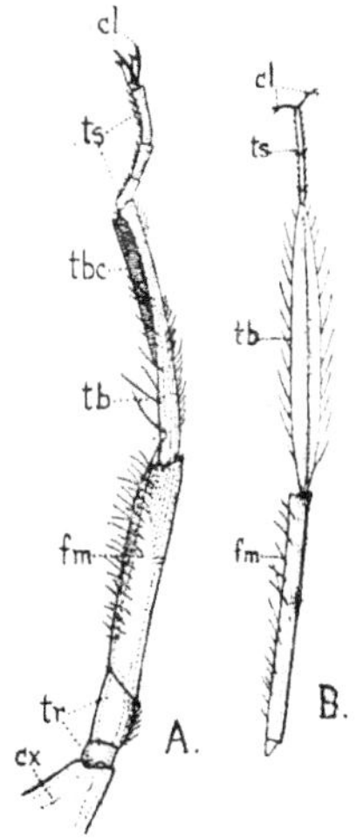

Fig. 9. A. Fore-leg of *Hemicordulia tau* Selys, ♂ (× 5), shewing tibial keel (*tbc*). B. Part of hind-leg of *Platycnemis pennipes* Pallas, ♂ (× 10), shewing dilated tibia. *cl* claws; *cx* coxa; *fm* femur; *tb* tibia; *tr* trochanter; *ts* tarsus. Original.

made out a tiny process, the vestige of a *plantula*, or terminal pad.

Specializations in the Armature of the Legs. So numerous are the variations in this armature that we can only here mention the most outstanding of them:

(1) In the *Corduliinae* and in *Orogomphus*, a *tibial keel* (*tbc*) is present on the under side of the tibia in the male. It is usually only present in the fore-legs. It may be short, and placed at the distal end of the segment, or long, occupying the distal half or more. It seems to be formed by a fusion of the original row of spines which it replaces.

(2) In some *Libellulinae* (e.g. the *Tetrathemini*) the males have a row of curved thorn-like spines on the second and third tibiae.

(3) In the *Cordulegastrinae*, each of the original rows of spines on the femur has become double or triple. The femoral spines are very short in this and some other subfamilies, and may be reduced to knobs.

(4) In many *Platycneminae*, the males have the tibiae of the middle and hind-legs strongly dilated, with the spines along the flattened edges (fig. 9 B).

(5) In the *Calopterygid* genus *Rhinoneura*, the hind femur is thickly clothed with short dense hairs.

(6) In the *Teinobasini*, the inferior tooth of the tarsal claws is absent or vestigial.

THE ABDOMEN (figs. 1, 10).

The abdomen of the Dragonfly consists of ten complete and very distinct segments or *uromeres*, while, according to Heymons [72], the reduced remnants of an eleventh and generally also of a twelfth segment can also be recognized. At its extreme end, the abdomen carries a set of *anal appendages*. In shape, it is always narrow, and much elongated in comparison with its breadth. Probably the most primitive form was that of a fairly narrow elongated cylinder, without either constrictions or expansions. Such a form is approximately retained to-day in the *Petalurinae* and many *Calopterygidae*. In the *Pseudostigmatinae* (Plate I, fig. 8), the abdomen has become excessively slender, and of enormous length. This is a secondary development, correlated with the habit of laying eggs in the water collected between the bases of the leaves of epiphytic Bromeliads. Many of the *Protoneurinae* and *Teinobasini* have the abdomen so slender that it is barely possible to thread it with a fine hair. As a contrast to this, the abdomen in most *Libellulinae* is comparatively short and broad, often strongly depressed, and laterally expanded. In such a form of abdomen, the movements of respiration are seen to the best advantage. In many forms, the third segment of the abdomen is strongly constricted, especially in the males. Many *Corduliinae* have the body shaped like an Indian club, the base of the abdomen representing the handle (*corduliform*). In most *Gomphinae* the abdomen is very strongly clubbed, the sides of segments 7–9 being much expanded, forming a wedge-shaped tip. In the females of most Zygoptera, segments 8–9 are considerably swollen. In most *Aeschninae* and some *Libellulinae*, segments 1–2 and part of 3 are greatly distended, particularly in the female. As a general rule, the body of the female is considerably stouter than that of the male, owing to the presence of the elongated ovaries with their enormous number of eggs.

Each *abdominal segment* (fig. 10) is formed of four pieces or sclerites, viz. a dorsal *tergite* (*tg*), a ventral *sternite* (*st* + *stl*) and two lateral *pleurites* or *pleura* (*plm*). In the Dragonfly, the pleura do not become strongly chitinized, but remain membranous. They are very narrow, and serve merely as a movable connection between tergite and sternite. Their flexibility allows free respiratory movements to take place. They also bear the abdominal spiracles (*sp*). The abdominal sclerites are distinguished from those of the thorax by affixing the prefix *uro-* (e.g. urotergite, urosternite).

The Tergites. In each segment the tergite is wide and convexly arched. It occupies not only the dorsal, but also the whole of the lateral region of the segment, together with the outer portions of the ventral region. We can represent the transverse section of a segment by drawing a segment of a circle somewhat greater than a semicircle. If we then cut off two small pieces from the chord, one near each end, the remaining middle portion of the chord will represent the sternite, while the small outer pieces, together with the whole of the large arc, represent the tergite. Moreover, the ventral pieces of the tergite are so turned in as to overlap and partly hide the sternite from view, especially in the dead insect.

Owing to its great extent and convexity, the tergite is strengthened by the formation of both longitudinal and transverse ridges of hard chitin, called *carinae*. The following carinae may be present:

(i) *Mid-dorsal carina* (*dc*, longitudinal). This occurs very generally, particularly in the larger forms.

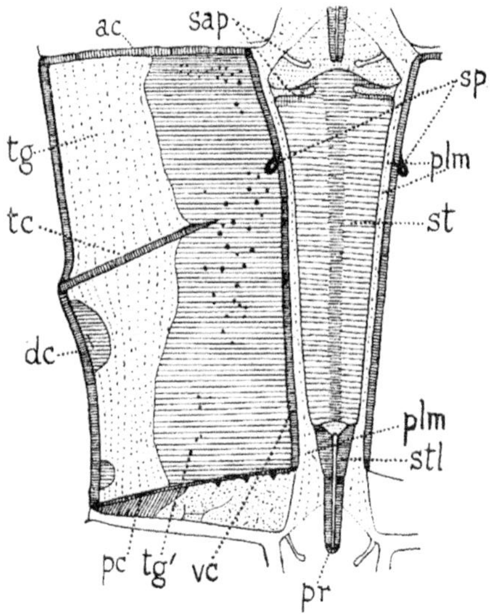

Fig. 10. Urosternite and right half of urotergite from *Petalura gigantea* Leach ($\times 5\frac{1}{2}$). *ac* anterior carina; *dc* dorsal carina; *pc* posterior carina; *plm* pleural membrane; *pr* posterior prolongation of sternellum; *sap* sternal apodemes; *sp* abdominal spiracles; *st* sternum; *stl* sternellum; *tc* supplementary transverse carina; *tg, tg′* urotergite; *vc* ventral carina. Original.

(ii) *Ventral carinae* (*vc*, longitudinal, paired). These border the pleura, and are nearly always present.

(iii) *Supplementary lateral carinae* (longitudinal). They occur along the angle separating the ventral from the lateral portion of the tergite. They are only found in *Aeschninae* and *Libellulidae*.

(iv) *Anterior and posterior transverse carinae* (*ac*, *pc*). These are nearly always present. They serve to separate successive segments from one another, leaving between them a narrow belt of softer chitin forming the *suture*. The sutures are usually blackish, but sometimes brightly coloured.

(v) *Supplementary transverse carina* (*tc*). This lies usually a little basad from the middle of the segment. It occurs in *Petalurinae*, *Aeschninae* and a number of the larger *Libellulinae*, but only on one or more of the most anterior segments. It is incomplete ventrally, and is usually placed slightly obliquely.

Some or all of these carinae may be more or less denticulate. This is particularly noticeable in the *Libellulidae* along (ii), (iii), and the parts of (iv) intercepted between them.

The Sternites. The sternite is usually an elongated subtriangular piece (*st*) narrowing posteriorly, and ending in a small pointed piece or *sternellum* (*stl*) which may be more or less separated off from the rest. Close to its anterior end, the sternite gives off internally and laterally two pairs of *sternal processes* or *apodemes* (*sap*), for the attachment of the segmental muscles. The sternite of the first segment is short, often quite flat, and of a trapezoidal or even rectangular form. It is more closely united with the tergite than in the case of the other segments. The first segment is, indeed, often so short as to become almost completely hidden beneath the overhanging synthorax (fig. 1 A, B).

The Spiracles. Eight pairs of abdominal spiracles or stigmata (*sp*) occur in the abdomen of all Dragonflies, viz. one pair in the pleural membrane of each of segments 1–8. They lie a little distad from the second pair of sternal apodemes, close up to (and often overlapped by) the ventral carinae. They increase in size from before backwards, the eighth pair being usually much the largest. In the older forms they lie slantwise to the body axis; in the higher forms, they become elongated parallel to that axis. Their structure is dealt with on p. 168.

Appendages and Outgrowths of the Abdomen.

The Auricles (fig. 11). These are outgrowths from the sides of the second tergites, found in the males of all those Anisoptera which have angulated hind-wings. They are especially highly developed in the *Gomphinae* and *Petaliini*, where smaller auricles also occur in the females. They are ear-like, rounded, convex above, hollowed out beneath, and often denticulated distally. Their function is unknown; but it seems fairly clear, from their method of occurrence, that they must act in conjunction with the anal angle of the hind-wing in controlling certain phases of flight. Otherwise, it would be hard to account for their absence in *Anax*, *Hemianax* and *Hemicordulia*, the only genera of the *Aeschninae* and *Corduliinae* which have rounded hind-wings in the male. In the *Petalurinae*, the auricles are peculiar and of a primitive form, being merely oblique convex swellings in front of a small depression.

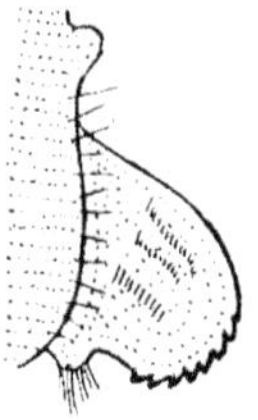

Fig. 11. Right auricle of *Austropetalia patricia* Tillyard, ♂ (× 10). Original.

The Accessory Genitalia of the Male, situated on the sternites of segments 2–3, are dealt with in chap. XI. **The Ovipositor of the Female,** developed from the sternites of segments 8–9, is likewise dealt with there.

The Anal Appendages (figs. 12–16). As these differ considerably in the two suborders, we shall deal with them separately.

1. *Anisoptera.* The males have two *superior appendages* (*sup*) developed from the posterior end of the tenth tergite. They also have, below and between these, a so-called median *inferior appendage* (*inf*) representing the appendix dorsalis of the larva (p. 91). This "inferior" appendage is badly named, since it lies directly above the anus, and is developed from the reduced tergite of the eleventh segment. In the females, the inferior appendage is absent. The superior appendages are, however, present, though apparently functionless. They are usually smaller than in the males, and are called simply the *anal appendages*, since they are the only ones present.

The anus itself (*a*) is surrounded by three small processes or valves, which are the vestiges of an original twelfth segment. Of these, the median dorsal *lamina supra-analis* (*la*) represents the reduced tergite, the two latero-ventral *laminae sub-anales* (*la'*) the reduced bipartite sternite of this lost twelfth segment. A soft paired tubercle below the anus (st_{11}) represents the vestiges of the similar but larger sternite of the reduced eleventh segment. The positions of appendages and vestigial structures are indicated in the diagram (fig. 12 A).

The function of the anal appendages in the male is to grasp the female firmly during pairing. In the Anisoptera, the inferior appendage is always pressed down upon the occiput of the female,

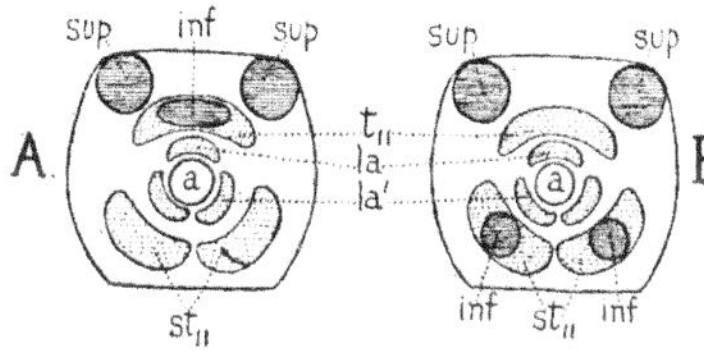

Fig. 12. Diagram to shew the position of the anal appendages of the male, viewed from behind. A in Anisoptera, B in Zygoptera. *a* anus; *inf* inferior appendage; *la* lamina supra-analis; *la'* laminae sub-anales; st_{11} bipartite eleventh sternite; *sup* superior appendages; t_{11} eleventh tergite. Original.

while the superior appendages become engaged between the prothorax and the back of the head, and most usually press forward against the latter. Thus the head is held as it were between the jaws of a vice (fig. 13 A). In the *Petalurinae*, however, the broad leaf-like superior appendages (fig. 13 B) fit closely upon the mesepisterna, while the wide inferior appendage, placed upon the occiput, presses the head back so as to hold the superiors in position. In a number of forms with long or forcipate superior appendages, a similar kind of grip seems to be the rule, though undoubtedly the chief pressure is on the head. In the *Gomphinae* the inferior appendage is strongly bifid, engaging the broad occiput in two depressions; while the superiors, which are too short to be of much direct use, are generally held in place by tubercles or other outgrowths from the back of the occiput. In the *Petaliini* (fig. 14) the inferior appendage is trifid and very long; the superiors very short. The method of pairing has not yet been observed.

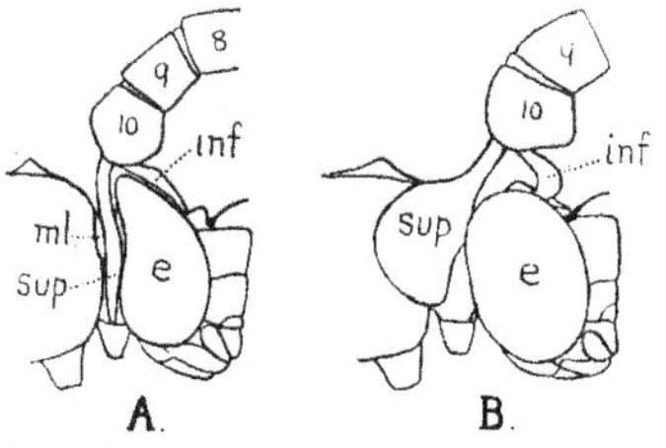

Fig. 13. Position of the anal appendages of the male in Anisoptera during pairing. A in *Aeschna*, B in *Petalura*. *e* eye; *inf* inferior appendage; *ml* mesostigmatic lamina; *sup* superior appendages; *8–10* abdominal segments. A from E. M. Walker, B original.

The diversity of form in these appendages is immense. Except in the *Libellulinae*, and a few *Gomphinae*, practically every species is at once determinable by the form of its appendages. Hence hybridization is a very rare occurrence in the Odonata. In the

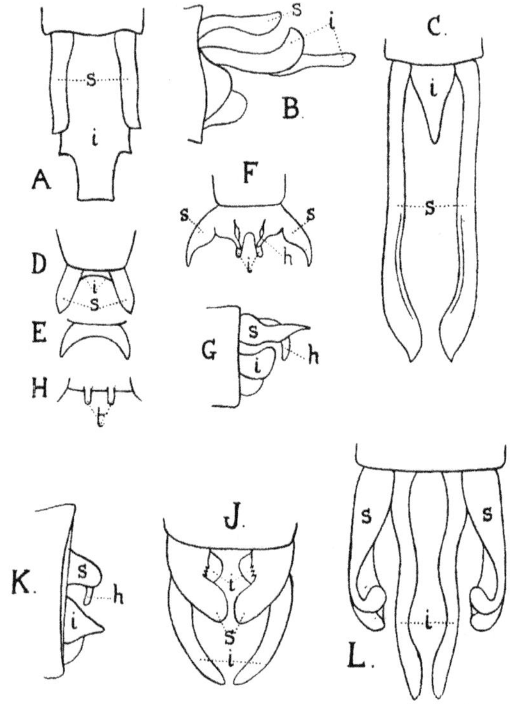

Fig. 14. Anal appendages of male Dragonflies. A–H Anisoptera, J–L Zygoptera. A. *Austropetalia patricia* Tillyard, dorsal view. B. The same, lateral view. C. *Gynacantha rosenberghi* Selys, dorsal view. D. *Gomphus vulgatissimus* Linn., dorsal view. E. Bifid inferior appendage of same (hidden beneath *s* in D). F. *Austrogomphus melaleucae* Tillyard, dorsal view, shewing hooks (*h*). G. The same, lateral view. H. The same, occiput of ♀ shewing tubercles (*t*) which engage the appendages of the male. J. *Isosticta simplex* Martin, dorsal view. K. *Ischnura heterosticta* Burm., dorsal view. L. *Hemiphlebia mirabilis* Selys, dorsal view, shewing white ribbon-like inferior appendages. *i* inferior; *s* superior appendages. (A, B. D–H × 8, C × 5, J–L × 15.) Original.

case of those groups where the appendages are sufficiently similar to admit of the possibility, we find very dissimilar developments of the accessory genitalia of segment 2 in the male. Here, then, a second almost insuperable barrier is set up.

As hundreds of figures of the appendages are available in systematic works, we have contented ourselves with figuring a few types in fig. 14 (see also figs. 170, 174, 178). As regards the superior appendages, the narrow leaf-like form is perhaps the most primitive, and remains extant in most females still[1]. An expansion of this gave rise to the broad *Petalura*-form, while various simple modifications led in the opposite direction to the form seen in *Aeschna* (fig. 170). Developments of a forcipate nature are very frequent. In the *Libellulinae* a slightly reduced pointed type obtains almost throughout, and may be conceived of as lying somewhere between the lanceolate and strongly forcipate types. The much reduced divergent appendages of the *Gomphinae* (fig. 14 D–G) appear to be the most highly specialized of all.

In the case of the inferior appendage (fig. 15) the primitive type was certainly correlated with a wide occiput, and hence may be conceived of as not very far removed from the form found in *Petalura* (a, b_1). A simple change in this form (by median cleavage) leads us directly to the *Gomphine* type (c_1, d_1). With the increase in the size of the eyes and the decrease in the size of the occiput in *Aeschninae* and *Libellulidae*, the appendages tended to lengthen, the inferior becoming sub-triangular and curved, so as to fit over the occipital tubercle (b_2, c_2, d_2). The form in *Petaliini* (c_2') perhaps preserves for us the ancient truncated form, while developing a larger median process to reach forward to the point of union of the eyes.

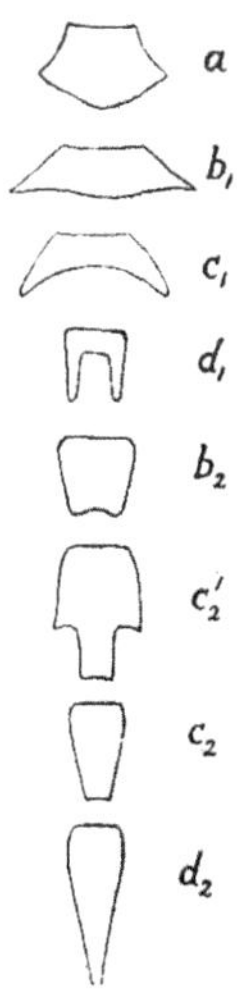

Fig. 15. Assumed phylogenetic series of inferior appendages in male Anisoptera. a *Petalura ingentissima* Tillyard. b_1 *P. gigantea* Leach. c_1 *Gomphus vulgatissimus* Linn. d_1 *Austrogomphus melaleucae* Tillyard. b_2, c_2, d_2, three forms commonly found in *Aeschninae* and *Libellulidae*. c_2' *Austropetalia patricia* Tillyard. The series are (i) a, b_1, c_1, d_1; (ii) a, b_2, c_2, d_2; (iii) a, b_2, c_2'. Original.

[1] Dimorphic appendages occur in the Palaearctic *Boyeria irene*, some females having them long and some short.

2. *Zygoptera*. The males have a pair of superior appendages (*sup*) homologous with those of the Anisoptera. Likewise, these also form the *anal appendages* of the female, which are usually extremely reduced cylindrical or conical stumps. The *inferior appendages* of the Zygoptera are, however, two in number, and have nothing in common with the so-called "inferior appendage" of the Anisoptera. They lie latero-ventrally below the anus, one on either side, and represent the cerci of the larva; that is, they are outgrowths from the reduced bipartite sternite of the eleventh segment. The *laminae anales* are placed as in the Anisoptera. The female, as in the Anisoptera, has no inferior appendages, though Heymons has shewn the existence of their rudiments in *Calopteryx*. The diagram, fig. 12 B, shews the positions of the various parts.

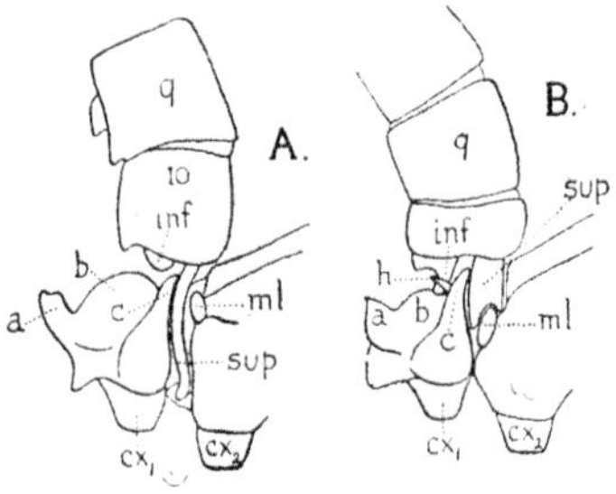

Fig. 16. Position of the anal appendages of the male in Zygoptera during pairing. A in *Austrolestes analis* Ramb., B in *Pseudagrion aureofrons* Tillyard. *a* anterior; *b* median; *c* posterior lobe of prothorax; cx_1 procoxa; cx_2 mesocoxa; *h* hook; *inf* inferior appendages; *ml* mesostigmatic lamina; *sup* superior appendages; *9–10* abdominal segments. Original.

In the Zygoptera, during the process of pairing (fig. 16), the appendages of the male engage the prothorax and the back of the head of the female, in quite a different manner from that described for the Anisoptera. In this case it is the hind lobe of the pronotum which becomes engaged between the jaws of the vice, the head only being touched when the inferior appendages are long enough to pass forward and press against it from behind (as in some *Lestinae*). If the superior appendages are forcipate, a simple transverse lock-grip (fig. 16 A) behind the pronotum appears sufficient to hold the female, bearing in mind the fact that the latter always assists by pressing the pronotum backwards against the spiracular dorsum.

In such a case the inferior appendages rest in front of the hind-lobe of the pronotum, and may exert very little pressure. If, however, the superior appendages be reduced, as in most *Agrionidae*, then a high specialization of both pairs is usually found. This is of such a nature that the inferior appendages, with the aid of various devices of sculpture on the hind-lobe of the female pronotum (grooves, ridges, tubercles, or even strongly curved hooks), are able to press strongly downwards on part of the hind-lobe, while the superior appendages press on it from behind. Thus the hind-lobe is held in a double longitudinal lock-grip, as between the jaws of two vices arranged side by side (fig. 16 B). In some cases the superior appendages press down over, or even behind, the mesostigmatic laminae.

Observations on the actual positions of the appendages during pairing are difficult to make, since the insects nearly always disengage when captured. A good summary of our present knowledge on the subject is given by Williamson and Calvert [201]. (See also Walker [190].)

The most primitive form of Zygopterid appendages is undoubtedly that in which the superiors were short and strongly forcipate, the inferiors still shorter and straight. This form still exists, with minor modifications, throughout the *Calopterygidae* and *Lestidae*. The highly reduced and often stump-like superior appendages of many *Agrionidae* are a more advanced type, in which the principal part in maintaining the grip becomes shifted on to the inferior appendages, aided by special developments of the pronotum of the female.

Perhaps the most remarkable appendages occurring in the Zygoptera are those of *Hemiphlebia mirabilis* (fig. 14 L). Both pairs in the male are rather long, white, and very conspicuous. The inferior appendages are like a pair of white ribbons. The male uses them to attract the attention of the female by waving them about while at rest on a reed-stem. The female replies to his signals by shewing the whitened underside of the tip of her abdomen [173]. Another Australian Zygopterid, *Synlestes albicauda*, hides on the dark trunks of trees in dense jungle, and can only be seen by the whiteness of its appendages, which it probably uses somewhat after the manner of *Hemiphlebia*.

CHAPTER III

THE WINGS

All Dragonflies possess two pairs of well-developed wings, of either equal or slightly unequal size. They are elongated membranous outgrowths from between the pleura and terga of the synthorax. The fore-wings are attached to the mesothorax, the hind-wings to the metathorax. Each wing is composed of a chitinous non-cellular membrane, usually quite transparent (*hyaline*) and formed of two delicate layers fused together. This membrane is strengthened by numerous thickened rods or bars of hard, dark chitin, known as the *veins* or *nervures*, forming a complicated supporting network known collectively as the *wing-venation*. As in other insects, the basis or foundation of the plan of the wing-venation is formed by a number of main longitudinal veins, some of which are simple and some branched. These separate the wing into areas known as the *wing-spaces*. Between the main veins and their branches, a large number of *cross-veins* are arranged, in such a manner that the wing is finally divided up into a very large number of small quadrilateral or polygonal spaces, known as *cells* or *cellules*.

The main veins in the Dragonfly wing, with few exceptions, are formed along the principal tracheae of the developing larval wing. The cross-veins, on the other hand,—again with one or two notable exceptions,—are not formed about tracheae at all, but appear quite independently of the latter during the last larval instar. The earlier students of Dragonfly venation, amongst whom de Selys was pre-eminent, studied the wing-veins without reference to the larval tracheation. Hence there arose a purely arbitrary terminology, in which homologies were either obscure or ignored. The study of the relationships between the larval tracheation and the wing-venation was begun by Comstock and Needham [48] and continued by Needham [102], whose work stimulated further work along similar lines [176, 180]. De Selys contented himself with giving names to the various veins. Comstock and Needham not only adopted a new terminology, in which the names of corresponding veins and tracheae were the same, but they also introduced the system of *notation*, whereby each main vein, each branch, and all the more important cross-veins and wing-spaces, were also designated by an abbreviation of one or more letters. This system forms the basis of the notation adopted in this book. Only in a few important particulars, where recent investigation has shewn the earlier work to have been either incomplete or at fault, has it been necessary to make alterations. The table on pp. 40–43 gives the names of the principal parts of the wing, together with the Selysian names, the Comstock-Needham notation, and the revised notation. Veins, points, and angles are denoted by capitals; spaces and areas by small letters. Foreign abbreviations, such as *q* for cross-vein (German, *querader*), have been rejected.

In addition to the above, secondary branchings of the veins may be denoted by the addition of the suffix "*a*" for the upper or distal branch, "*b*" for the lower or proximal branch, of the vein affected; e.g. in *Aeschna*, *Rs* bifurcates into Rs_a above and Rs_b below, at the fork. Also, intercalated supplements running to the posterior border, between branches of main veins, may be designated by adding the suffixes A, B, C ... in order, from the apex backwards, to the vein *above* the supplements; e.g. in *Agrioninae* the "postnodal sector" is M_{1A} (not M_{1a}).

In Plate II we have figured either the whole or the more important parts of wings selected as typical of the five families of Odonata. The Revised Notation is printed in red on the Plate. A study of this plate, in conjunction with fig. 17 (which shews the tracheation in the corresponding larval wings), will, we trust, not only explain the venational scheme satisfactorily, but also justify some important alterations from the familiar Comstock-Needham notation. In order to study the various subfamilies in detail, a series of forty-seven illustrations of typical wings is also provided, to be used in conjunction with Plate II and with the venational definitions given in chap. XIV.

Table of the Principal Parts of the Wing (Plate II and fig. 17).

Name	Course, extent, or position	Selysian Name	Comstock-Needham Notation	Revised Notation
Main Veins at Base of Wing:				
(1) Costa	From base to tip, anteriorly	Costal nervure	C	C
(2) Subcosta	From base to nodus, below C	Subcostal nervure ...	Sc	Sc
(3) Fused Radius + Media ...	From base to arculus	Median nervure	$R+M$	$R+M$
(4) Fused Cubitus + Analis ...	From base to anal crossing	Submedian nervure ...	Cu	$Cu+A$
(5) Secondary or Recurrent Analis	From anal crossing back to base, below $Cu+A$	Postcostal nervure ...	A	A'
Parts of the Radius:				
(*a*) Main Stem	From arculus to nodus, below Sc, thence to tip, below C	Median nervure	R	R
(*b*) Radial Sector (in Anisoptera only)	Leaves R at nodus, crosses M_1, runs a little way along M_2, leaves it at O, then below M_2 to a point somewhat below the wing-tip	Subnodal sector	Rs	Rs
Parts of the Media:				
(*a*) Main Stem				
M_{1-4} (Upper part of Arculus)	From $R+M$ downwards to the point where M_{1-3} comes off	... ...	M_{1-4}	M_{1-4}
M_{1-3}	From arculus to origin of M_3	Upper sector of arculus ...	M_{1-3}	M_{1-3}
M_{1-2}	From origin of M_3 to origin of M_2 ...	... ...	M_{1-2}	M_{1-2}
M_1	From origin of M_2 to tip of wing, below R	Principal sector	M_1	M_1
(*b*) Most distal branch ...	From origin to a point somewhat below the wing-tip (above Rs in Anisoptera)	Nodal sector	M_2	M_2
(*c*) Middle branch	From origin to posterior border of wing, beyond the middle	Median sector	M_3	M_3
(*d*) Most proximal branch ...	From arculus to distal angle of triangle; thence to posterior border a little proximal to M_3	Lower sector of arculus, short sector	M_4	M_4
(*e*) Zygopterid Sector (extra branch in the Zygoptera only)	Lies between M_2 and M_3 (comparable with Rs in Anisoptera)	Subnodal sector	Rs	Ms

Parts of the Cubitus:				
(*a*) Main Stem	From anal crossing to discoidal cell, thence along lower border of same to its lower distal angle	Submedian nervure ...	Cu	Cu
(*b*) First (upper) branch ...	From discoidal cell to posterior border of wing, below M_4	Superior sector of triangle	Cu_1	Cu_1
(*c*) Second (lower) branch ...	From discoidal cell to posterior border of wing, below Cu_1	Inferior sector of triangle	Cu_2	Cu_2
Parts of the Analis:				
(*a*) Main Stem = Anal Crossing	From $Cu+A$ to A'	(Submedian nervule) ...	...	Ac
(*b*) Branches (numbered from most distal to basal)	Very variable courses in the anal field	... ...	$\{A_1, A_2, A_3, A_4$	A_1, A_2, A_3, A_4
Supplements:				
(1) Bridge (in Anisoptera) ...	Completes Rs backwards from O to M_{1-2}	Subnodal sector (proximal portion)	B (*br*)	B
(2) Long Bridge (in *Lestidae*)	Completes Ms backwards from O to M_{1-2} or to M_3	Subnodal sector (proximal portion)	B	B
(3) Radial Supplement ...	Lies below Rs	... ...	*Rspl*	*Rspl*
(4) Median Supplement ...	Lies similarly below M_4	... ...	*Mspl*	*Mspl*
(5) Trigonal Supplement ...	Rarely developed, in post-trigonal space close to triangle	... ...	...	*Ts*
(6) Cubital Supplement = Mid-rib of Anal Loop	Divides anal loop (in *Libellulidae* only) from base to toe	... ...	$[A_2]$	*Cuspl*
(7) Anal Bridge (in Zygoptera)	From Ac to Cu_2	... ...	$[A^*]$	*Ab*
(8) Postnodal Sector (in Zygoptera)	Between M_1 and M_2	Postnodal sector ...	...	M_{1a}
Cross-Veins:				
(1) Nodus	Thickened cross-vein uniting distal end of Sc with C, and forming a joint	Nodus	N	N
(2) Subnodus	Thickened oblique vein continuing the nodus downwards between R and M_{1-2} (in Anisoptera, subnodus = basal portion of Rs)	Subnodus	*sn*	*Sn*
(3) Antenodals	Between base and nodus	Antecubital nervules ...	$[Anq]$	Ax
(*a*) First series ...	Between C and Sc	... ...	...	Ax }
(*b*) Second series ...	Between Sc and R	... ...	...	Ax' }

Table of the Principal Parts of the Wing (Plate II and fig. 17) (*cont.*).

Name	Course, extent, or position	Selysian Name	Comstock-Needham Notation	Revised Notation
(4) Postnodals	Between *C* and *R*, from nodus to pterostigma	Postcubital nervules ...	...	*Px*
(5) Arculus	From $R+M$ to *Cu*, near base (only its lower portion is a true cross-vein)	Arculus	*Arc*	*Arc*
(6) Oblique Vein	Represents a small portion of *Rs* (in Anisoptera) or *Ms* (in *Lestidae*); runs from M_2 to origin of *B*	... ...	*o*	*O*
(7) Brace Vein...	Supports pterostigma proximally ...	... ...	...	*Rc*
(8) Cubito-anals	Crossing cubito-anal space (*Ac* appears to be one of them)	Submedian nervules ...	[*Cuq*]	*Cux*
(9) Bridge Cross-veins ...	Above bridge, in bridge-space	... ...	[*Bq*]	*Bx*
Special Areas:				
At base of Wing:				
(1) Costal space	Between *C* and *Sc*, up to nodus ...	Costal space	*c*	*c*
(2) Subcostal space	Between *Sc* and *R*, up to nodus ...	Subcostal space	*sc*	*sc*
(3) Median space	Between $R+M$ and $Cu+A$, up to arculus	Median or basilar space ...	*m*	*m*
(4) Cubito-anal space ...	Lies below *m*, and extends up to discoidal cell	Submedian space ...	*cu*	*cu*
(5) Anal Field	Space below *sm*, from base up to discoidal cell and part of Cu_2 (lower distal limit is Cu_{2b})	... ...	...	*a*
(6) Anal Loop (in Anisoptera) = part of anal field	Enclosed between A_2 proximally and A_1 or Cu_2 distally; not extending beyond Cu_{2b}	Anal loop	*al*	*al*
(7) Anal Triangle (in males of Anisoptera) = part of anal field	Area below *cu*, bounded distally by A_3	Anal triangle	...	*at*

(8) Membranule (Anisoptera)	Opaque membrane bordering *a* at base of wing	Membranule	*mb*	*mb*
(9) Pterostigma	Thickened opaque area between *C* and *R*, towards apex of wing	Pterostigma	*st*	*pt*
(10) Discoidal Cell:		Discoidal cell		
A. In Zygoptera =Quadrilateral	Area enclosed by M_4 above, *Cu* below, lower part of *arc* basally and a strong cross-vein distally	Quadrilateral or quadrangle	*q*	*q*
B. In Anisoptera				
= { Supra-triangle	Space above triangle, from arculus to distal angle of triangle	Supra-triangle, or hyper-trigonal space	*s* or *ht*	*s* (or *ht*)
+ Triangle	Distinct triangular space formed by *Cu* basally, and two strengthened cross-veins costally and distally	Triangle or discoidal triangle	*t*	*t*
(11) Sub-triangle (Anisoptera)	Attached to basal side of *t*, proximally...	Sub-triangle or internal triangle	*st*. *t′* or *ti*	*st*. *t′* or *ti*
(12) Discoidal Field	Area between M_4 and Cu_1, distal from triangle or quadrilateral	Discoidal field	...	*d*
(13) Sub-quadrangle (Zygoptera)	Below quadrilateral from *Ac* to Cu_2 ...	Sub-quadrangle	...	*sq*
(14) Post-trigonal Cells ...	Cells in *d*	Post-trigonal cells ...	...	...
(15) Bridge space (in Anisoptera or *Lestidae*)	Enclosed by *B*, *O* and M_{1-2}	Bridge space	...	*b*
Contour of Wing:				
(1) Anterior Border = Costa ...	From base to apex	Costa	*C*	*C*
(2) Apex or tip	... ...	Apex	...	*Ap*
(3) Anal Border (Anisoptera)	From base of *A′* to anal angle (not a true vein)	Anal border	...	...
(4) Anal Angle	Separates anal from posterior border. Often rounded	Anal angle	...	*Aa*
(5) Posterior Border	From anal angle (Anisoptera) or end of petiole (Zygoptera) to apex. Not a true vein	Postcostal border ...	...	...
(6) Petiole (Zygoptera) ...	Narrowed basal stalk of wing	Petiole	...	...

We shall devote the remainder of this chapter to (1) a short account of the development and histology of the wing in the larva, and (2) phylogenetic studies of the principal features of the imaginal wing.

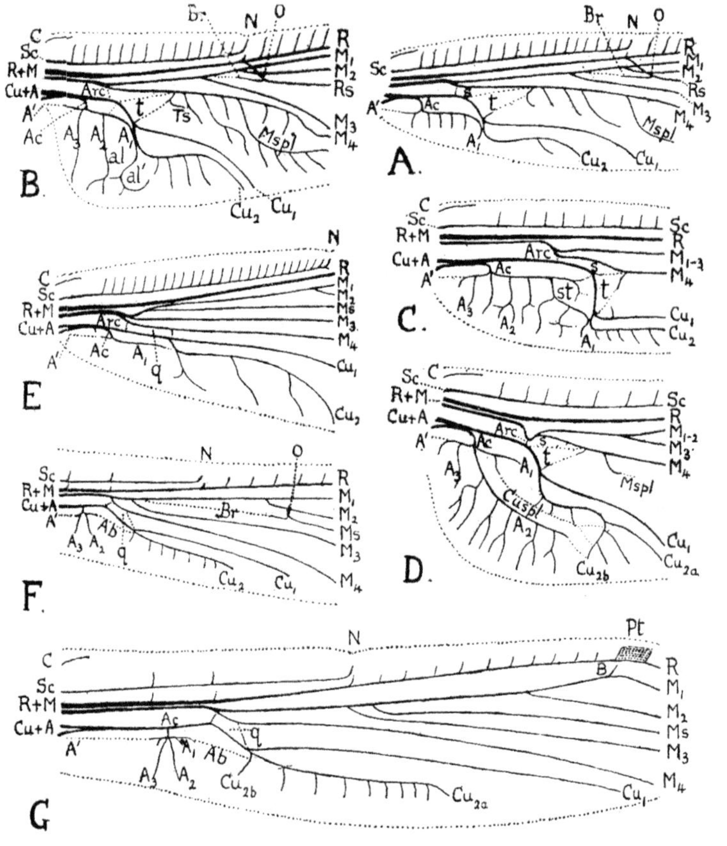

Fig. 17. Tracheation of the larval wing (to be used in conjunction with Plate II). A. *Aeschna brevistyla* Ramb., fore-wing. B. The same, hind-wing. C. *Hemicordulia tau* Selys, fore-wing. D. The same, hind-wing. E. *Calopteryx splendens* Harris. F. *Austrolestes cingulatus* Burm. G. *Caliagrion billinghursti* Martin. The lettering of the tracheae corresponds with that of the veins which replace them in the imaginal wing. Original except E (drawn from photograph sent by Dr F. Ris).

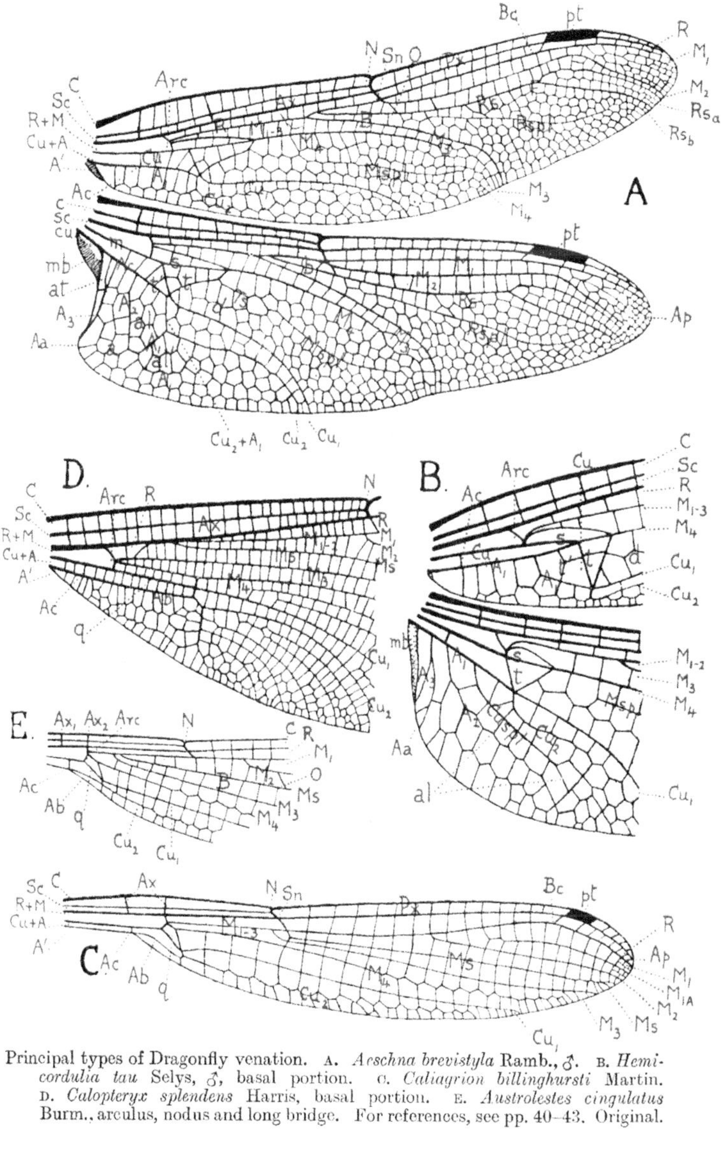

Principal types of Dragonfly venation. A. *Aeschna brevistyla* Ramb., ♂. B. *Hemicordulia tau* Selys, ♂, basal portion. C. *Caliagrion billinghursti* Martin. D. *Calopteryx splendens* Harris, basal portion. E. *Austrolestes cingulatus* Burm., arculus, nodus and long bridge. For references, see pp. 40–43. Original.

Development of the Wing.

At the third or fourth larval ecdysis, the edges of the pleura of the synthorax, bordering the terga, are seen to be slightly raised up, forming two nearly parallel ridges, the *pleural ridges*. At the next ecdysis, four minute swellings appear on these ridges, somewhat on the inner side. One pair arises from the ridges of the mesopleura, the other from those of the metapleura. After the following ecdysis, the forward extension of the mesopleura causes the pleural ridges to become more definitely inclined towards one another anteriorly. The swellings are now like tiny buds, and project upwards and a little inwards. From this point on, at each successive instar, the pleural ridges become more and more oblique, while the wing-buds cease to turn upwards, and come to project slantingly inwards and backwards. They soon take the form of definite triangular flaps, fore and hind-buds lying quite close to one another on either side. Also, the first pair have come to lie quite close to one another along the middle line. The second pair remain far apart at their bases, owing to the divergence of the pleural ridges, but tend to touch at their tips, which project inwards. These buds or flaps may from now on be spoken of as the *wing-sheaths* ("wing-cases," "wing-pads," or "wing-folds"). As growth proceeds, the hind wing-sheaths gradually come to overlap the fore, until only a narrow strip along the costal border of each of the latter is visible. Concurrently, the direction of the sheaths becomes definitely backwards, so that they lie straight along over the abdomen, reaching to the end of the fourth segment or beyond,—when fully developed,—and having their costal borders all parallel to the long axis of the body. The hind wing-sheaths become more or less deeply pigmented. On the fore wing-sheaths,—owing to their being covered by the hind,—pigmentation is very weak or absent, except along the free costal border.

The *alar trunk*, or main trachea supplying the wing, arises at a very early stage as two small tracheae from the dorsal trunk. The more anterior trachea passes to the costal end of the base of the wing-bud, the more posterior to its anal end. The latter trachea is the larger. These two tracheae soon become fused into a single loop, which we call the *alar trunk*. From this loop, six main tracheae pass into the developing wing-sheath. These are the *costal*, *subcostal*, *radial*, *median*, *cubital*, and *anal* tracheae respectively, counting from in front backwards. In the later stages of growth, they not only supply oxygen to the cells of the growing wing, but gradually assume positions and branchings which foreshadow the main venational plan of the imago. Thus we may say that this plan is the combined result of three interacting forces: (1) the necessity of leading up to the imaginal design, (2) the necessity of supplying every portion with sufficient oxygen for development, and (3) the necessity of relying almost wholly on an oxygen supply coming from the *anal* end of the alar trunk (owing to the larval gills being situated at the anal end of the body). The influence of (3) is seen in (*a*) the almost total loss of the costal trachea, (*b*) a great shortening of the subcostal, (*c*) a gain by the median over the radial trachea, (*d*) strong development of the cubital and also of the anal trachea (in non-reduced forms). How far the imaginal design has reacted on the tracheal supply, and vice versa, it is not easy to say. A study of fig. 17, C and D, shews that this interaction has been very considerable.

During the last larval instar, the imaginal venation begins to shew up very distinctly. A definite cuticularization of the wing-sheath occurs a short time before metamorphosis, so that the details of venation are permanently recorded on the sheath, and are left behind on it at metamorphosis. This fact is of great value when we want to establish the identity of any un-named exuviae. Beneath this pattern, the imaginal venation proper begins to form, by the deposition of pigment-bands along the courses of the future veins. These bands are not in general deposited on both upper and under surfaces of the wing-rudiment, but on one or the other alternately, according to whether the corresponding trachea runs closer to one surface or to the other. The margin of the wing, however, including the costa (which is formed quite independently of the rudimentary costal trachea) is represented by pigment bands on both surfaces. Just before metamorphosis the vein-rudiments proper are formed by a weak deposition of chitin along the course of the pigmented bands, as well as along all the future cross-veins. Thus it comes about that the veins of the imaginal wing become alternately convex and concave, according as they are laid down upon one or other surface of the wing-rudiment.

At metamorphosis the tracheae are withdrawn from the wing, and shrivel up. During the rapid growth of the wing, chitin is deposited with amazing rapidity. As each vein becomes stretched to its full extent, the hypoderm cells along its course completely exhaust themselves with chitin formation, so that the vein becomes built up to the necessary strength in the course of a few minutes. When the construction of both veins and membrane is completed, the hypoderm cells shrivel and die; the blood by whose agency the wing was expanded gradually evaporates, as the wing is exposed to the air and sunlight. Finally, the two surfaces of the wing come into contact and fuse together, except along the main veins, where small blood-canals are retained.

Histology of the Developing Wing. The wing-bud is simply an ectodermal evagination, in the form of a small bag lined internally with hypoderm cells, and externally with the cuticle. Even before any sign of an external bud can be seen, the hypoderm cells increase in size, becoming distinctly columnar, and forming a small fold or bag beneath the cuticle. At the next ecdysis, the cuticle becomes folded also, so that the evagination is then visible externally. At first, the layers of hypoderm bordering the inner and outer surfaces of the wing-bud do not meet. Between them runs a narrow prolongation of the haemocoele, filled with blood. Into this space tracheae soon penetrate, taking the courses already described. Fine nerves accompany the tracheae[1].

By the time that the wing-bud has become a definite flap or lamina, a change has occurred in the form of the hypoderm cells. These have become spindle-shaped, with the nucleus lying in the middle. Owing to this elongation, the two layers of hypoderm come into contact along their inner ends. Here they unite, forming a basement membrane, except in the places in which the tracheae and nerves run, small blood-channels are left

[1] The presence or absence, in the wings of Odonata, of the peculiar structure called *Semper's rod*, which accompanies the tracheae in the wings of certain Lepidoptera, does not appear to have been determined with certainty.

surrounding these organs. The hypoderm cells remain spindle-shaped until metamorphosis. Their action at that period has already been described. The hollowness of the main veins is, of course, due to the fact that they are formed along the courses of the tracheae, i.e. either above or beneath a blood-channel. The principal cuticularization is in the form of a circular arc (in transverse section) somewhat greater than a half-circle. A weak flattened cuticularization closes the vein from the other side.

Convex and Concave Veins.

If we cut across the wing of a Dragonfly with a pair of scissors, we shall see that the surface is not a plane, but consists of alternate ridges and hollows, with the veins alternately on the tops of the former and at the bottoms of the latter. A vein placed on the top of a ridge is called a *convex vein*; one at the bottom of a hollow, a *concave vein*. Now, in the larva, the wings lie flat along the abdomen dorsally. Consequently, it is clear that, at metamorphosis, when the wings are spread out to their full extent, it is the *under* surface of the larval wing which becomes the *upper* surface of the imaginal wing, and vice versa. Hence *convex* veins are developed on the *under* surface of the wing-rudiment, *concave* veins on the *upper* [31].

In order to determine the nature of the veins, let us make cuts across the wing of *Aeschna brevistyla* (Plate II, fig. *A*) at various levels. We obtain the following results:

1. *Between base and arculus.*
2. *Half-way between arculus and nodus.*
3. *Between nodus and pterostigma, beyond bridge.*

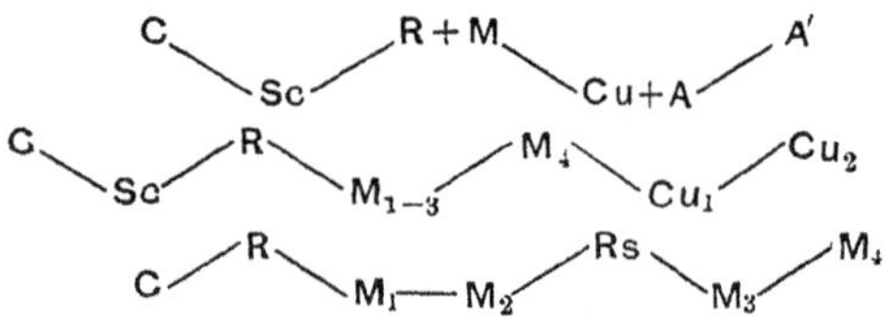

We see that C descends nearly to the level of Sc at the nodus, and from thence onwards lies well below the level of the high ridge of R, until the pterostigma is reached, when, of course, C and R come to lie nearly at the same level. The fact that M_1 and M_2 are consecutive veins, both lying at the bottom of a hollow, suggests that they were originally separated by a convex supplement. Such a supplement occurs in the archaic *Petalura*, starting about eight cells distad from the sub-nodus (fig. 20). In the Zygoptera, the corresponding "post-nodal sector," M_{1A}, is convex.

Phylogenetic Studies.

Comparison between the Wings of Anisoptera and Zygoptera.

The original ancestor of the Winged Insects had, no doubt, the meso- and metathorax separate and of equal size. Consequently, the fore and hind-wings may be conceived of as having been *originally equal*. They were developed as broad lateral expansions or "planing areas," used at first simply on the parachute method. Later on, they took on the active function of flight, and became true wings. Owing to the limited extent of each thoracic segment, the fore-wings in the oldest insects invariably partly overlapped the hind-wings. In insects of strong flight, this is clearly a disadvantage. Hence it came about that the fore-wing, very early in the course of its phylogeny, tended to become somewhat narrower than the hind, i.e. an *anisopterous* condition was set up. The earliest ancestors of the Dragonflies known to us—the Protodonata of the Upper Carboniferous—exhibit such a moderate anisopterous condition (fig. 156). The essential difference between our two living suborders, the Anisoptera and the Zygoptera, is that, in the former, the tendency has been *to increase the anisopterous condition* (for the culmination of the evolutionary effort, see the wings of *Tramea*, fig. 136); while, in the latter, owing to the reduction of all the wings alike into mere "sculling organs," with no power of soaring or "planing," the original anisoptery has tended to disappear, both fore and hind-wings becoming almost equally narrow, as in the *Agrionidae*. In nearly all cases, however, the evidence of original anisoptery is preserved in the fact that the hind-wing is *a little* shorter, *a little* broader, and very slightly *less petioled* than the fore-wing, as careful measurements of almost any species will shew.

Another fundamental difference between the two suborders is found in the condition of the two largest wing-veins *R* and *M*. If we turn again to the Protodonata (p. 304), we find that the then dominant family *Meganeuridae* occurred both in the Upper Carboniferous and in the Permian. In the Carboniferous genus *Meganeura* (fig. 156), *R* in the fore-wing had two very distinct branches situated close together; but in the hind-wing the same vein was quite unbranched. In the closely allied genus *Meganeurula*, *R* was unbranched in both wings. But in the Permian genus *Typus* (= *Tupus*), *R* was found to be branched in both wings, in the same manner as in present-day Anisoptera, the presence of both bridge and oblique vein having been satisfactorily demonstrated [150]. In other words, in the family *Meganeuridae*, the condition of *R* was variable and only of generic value. Unfortunately there is a large gap between the Permian Protodonata and the earliest true Dragonflies, which appear in the Lias. Hence the further course of this development is not capable of definite palaeontological proof. It is sufficient, however, to remark that, in the Lias, no forms with the intermediate condition of *R* (as seen in *Meganeura*) have been found. Every Liassic genus either has *R* once-branched in both wings, or unbranched in both. In other words, in the Lias, the dichotomy between Anisoptera and Zygoptera was complete, through the dying-out of intermediate forms.

As regards the vein *M*, we find that, in Anisoptera from the Lias onwards, it was constant, and gave off only three branches. In Zygoptera however, during the same period, *M* has always had four branches, one of which (*Ms*) played the part of the missing branch of *R*. In the *Heterophlebiinae* (fig. 159) the two most distal branches of *M* (M_{1-2}) came off together by a common stalk. Owing to the big gap thus left between M_{1-2} and M_3, we find here the development of a "long bridge" backwards from M_2 towards the main stem of *M*. This formation parallels that of the shorter bridge of Anisoptera. Our recent *Lestidae* are the direct descendants of forms such as the *Heterophlebiinae*.

A third principal difference between the two suborders lies in the evolution of the discoidal cell. It has, I think, been too strongly insisted on that this cell is highly differentiated only in the Anisoptera, its condition in the Zygoptera being considered much more primitive. The structure of this cell is of such interest that we have dealt with it in a separate section (p. 56).

To sum up, then, we may indicate the basic differences between the two suborders as follows:

	Anisoptera	Zygoptera
1.	Tendency to accentuate the original anisoptery	Tendency to obliterate the same
2.	Basic condition of *R once-branched*, of *M three-branched*	Basic condition of *R unbranched*, of *M four-branched*
3.	Discoidal cell divided into two	Discoidal cell a single complete whole

We may add that, on the totality of evidence, there is no justification for assuming that one suborder is more archaic than the other. Recent writers have invariably assumed that the Zygoptera are more archaic than the Anisoptera. The truth seems rather to be that the Anisoptera have preserved the ancient strongly-built Protodonate type more completely, but have added to it the larger number of caenogenetic developments. On the other hand, the Zygoptera have undoubtedly departed far more from the original "natural conception" of a Dragonfly; but, chiefly owing to the fact that their line of evolution has been a gradual reduction of the basic structural plan, they have not added definite new structures to the same extent as the Anisoptera have. Take away its triangles, and *Petalura* undoubtedly represents the most archaic of all living Odonata, and the nearest approach to *Meganeura* in size, strength, density of venation, and general body-design. Ignore its obvious reduction to a weak type, and *Pseudophaea* may lay claim to be equally as archaic as *Petalura*. Again, the wonderful advance made by the *Eucorduliini* or *Trameini* over the *Petalura*-type stamps these forms as the most highly organized of all Dragonflies. But the advance made by the *Agrionidae* over the *Pseudophaea*-type is every bit as great, though in another direction; and the structure of their wings probably represents, on the whole, an even wider departure from the original type (taking function as well as structure into account) than in the case of the *Trameini*.

The Pterostigma (fig. 18).

A *pterostigma* occurs in other Orders of Insects besides Odonata, but in none is it so constant or so conspicuous. From an evolutionary point of view, a strengthening of the wing on the costal border, close to the tip, at the point of greatest impact with the air during flight, was clearly a necessity. The close approximation of *R* to *C* just at this point in the Protodonata provided the required means, viz. a thickening of the membrane in the narrow space between them. Thus arose the most primitive type of

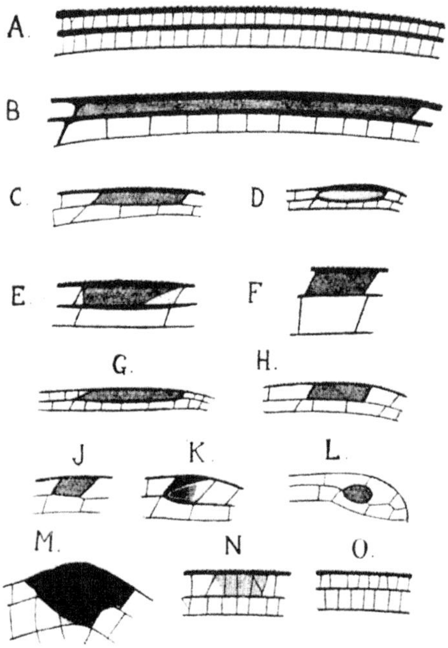

Fig. 18. Various forms of pterostigma. A. Ancestral (Protodonate) condition. B. *Petalura ingentissima* Tillyard. C. *Aeschna brevistyla* Ramb. D. *Hemigomphus heteroclitus* Selys. E. *Nannophya dalei* Tillyard. F. *Macromia terpsichore* Förster. G. *Diphlebia lestoides* Selys. H. *Austrolestes analis* Ramb. J. *Ischnura heterosticta* Burm., ♂, hind-wing. K. The same, fore-wing. L. *Anomalagrion hastatum* Say. (after Needham). M. *Mecistogaster lucretia* Drury, ♂, fore-wing. N. *Calopteryx splendens* Harris, ♀. O. The same, ♂. (B–D, G, H × 3; E, F, J, K, M × 6; N, O × 8.) Original, except L.

Odonate pterostigma, the exceedingly long, narrow, and strongly-braced form still preserved to us in the *Petalurinae* (B). Such a form is, indeed, very little removed from a simple fusion of *C* and *R*, such as is seen in some Protodonata.

A decrease in length and an increase in thickness brought about the next stage, such as we find in the *Aeschninae* (C). In many cases a brace was not developed. In many *Gomphinae* (D) further strengthening was gained by convexity of the posterior border. In the *Libellulidae*, the gradual adoption of a soaring or "skimming" habit of flight, instead of the original method of "darting," lessened the strain on the pterostigma, which finally tended to become both smaller and unbraced in the highest forms (E, F).

In the Zygoptera, a still earlier and more complete departure from the original method of flight quickly relieved the stress on this part of the wing. Hence it is only in the oldest *Calopterygidae* (*Epallaginae*, *Thorinae*, G) and in the *Lestidae* (H) that the archaic regular elongated pterostigma is found. In the *Calopteryginae* (N, O) this organ is either completely absent, or represented by a "false" pterostigma of a whitish texture, traversed by fine veins, and not enclosed on either side by strong cross-veins. In the *Pseudostigmatinae* also, the pterostigma is either absent or false. It exhibits a peculiar abnormality in *Mecistogaster* (M), where it forms a kind of costal hump. In the *Agrionidae* in general there is a decrease in length in the pterostigma from the older to the more advanced forms. The highly developed *Agrioninae* have it lozenge or trapeze-shaped. In a few genera, such as *Anomalagrion* (L), it becomes abnormal, and may sink below the level of the costa. In these advanced forms, the general weakness of the "sculling" flight renders the pterostigma almost unnecessary, as regards its original purpose of strengthening the wing. In *Ischnura* (J, K) a secondary sexual difference is developed, the pterostigma of the fore-wing in the male becoming particoloured, while that of the female remains unicolourous. A similar device occurs in the *Libelluline* genus *Hemistigma*, but in both sexes.

The Nodus (fig. 19).

The phylogeny of the formation of the nodus in the Dragonfly wing has been until quite recently purely a matter for speculation, the palaeontological evidence available being capable of more than one interpretation. On the one hand, it seemed clear that the shortening of the subcosta, to end up at a point somewhere near the middle of the costal border, must have been the essential condition precedent to nodus-formation. But, on the other hand, forms like *Aeschnidium* (fig. 163) and the recent *Telephlebia* (fig. 19 F) seemed to shew the subcosta running through and beyond the nodus.

The recent discovery of the fossil *Mesophlebia* (fig. 160) is a great step in advance. Though it may not be true that all recent Odonate nodi have been actually evolved by passing through a stage resembling that of *Mesophlebia*, yet it seems very clear that this stage gives us a very simple solution of the problem, and one which seems very likely to have been the actual evolutionary solution in the case of the Anisoptera, and probably of most Zygoptera as well.

In order to follow the steps of the evolutionary process, it is essential to bear in mind that the so-called costa of the Dragonfly wing is not really a main vein formed along a preceding trachea, but merely the strengthened anterior border of the wing. Thus its structure is quite different from that of a main vein. When carefully examined it is seen to be regularly ribbed by a series of spines or teeth along its outer border. It seems to be clear that this formation is due to the action of separate hypoderm cells, and that the portion of the costa carrying one such tooth is the product of a single pair of cells. Thus there is no inherent difficulty in understanding how a *joint*, such as the nodus, could be evolved on the costa, though this difficulty would be hard to overcome if we were dealing with a main vein of uniform structure. Granted then that the nature of the costa is such that the joint can be easily formed when required, *Mesophlebia* shews us definitely how it was formed. It is clear from this fossil that the nodus consists of two distinct parts only, viz. the part proximal to the actual joint, and the part distal to it. The former is derived from a strengthening of the original fusion between the costa and the end of the subcosta. The latter is formed simply by specialization of the cross-vein lying distally next to the ending of the subcosta, strengthened by a thickening of the costa itself just distally to the joint. I shall call this cross-vein the *nodal cross-vein* (*nc*). The nodus of recent Dragonflies differs from that of *Mesophlebia* only in two points, (1) by a closer union of parts around the joint, (2) by the alteration in direction of the lower half of the nodal cross-vein. This latter change has been brought about by the need of strengthening the wing below the nodus, either by the recession of M_2 towards the nodus, and the conversion of an intermediate cross-vein into the *subnodus* (*sn*); or else by a similar arrangement on the part of the radial sector, the basal piece of which forms, in

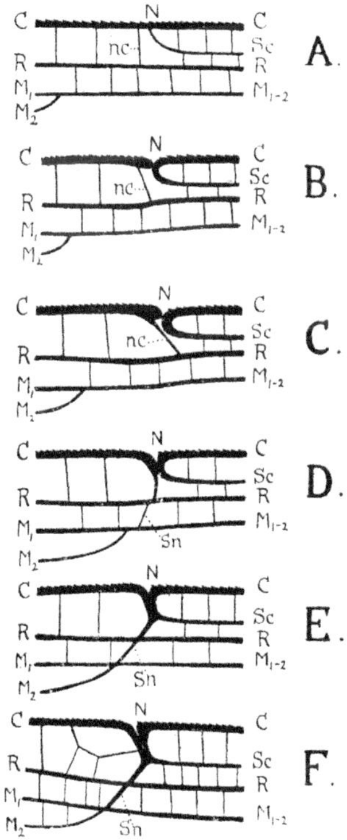

Fig. 19. Phylogeny of the nodus. A, archaic condition with no nodal joint; B–C, development of *Mesophlebia*-stage (subnodus absent); D–E, addition of subnodus and completion of normal nodus formation as seen in recent Dragonflies; F, nodal region of *Telephlebia godeffroyi* Selys, shewing apparent prolongation of *Sc*. *N* nodus; *nc* nodal cross-vein; *Sn* subnodus. Original.

Anisoptera, actually the subnodus itself. In fig. 19 A–E, I have attempted to shew sufficient successive stages in the evolution of the nodus to make the phylogeny clear.

There only remains for discussion the case of the apparent prolongation of *Sc* through the nodus. Fig. 19 F shews the actual state of affairs in the genus *Telephlebia*. It will be seen that, in this case, *Sc* does not pass through the nodus, but that a slight beginning of the formation of a double instead of a single row of cells beyond the nodus is the real cause of the apparent prolongation. This change is associated with the production of wing-pigment, and is known to occur in many other genera and in other parts of the wing. In the case of the fossil *Aeschnidium* (fig. 163) the apparent prolongation is much more marked, but its nature is entirely clear from the presence of a similar extra longitudinal vein formed actually between *C* and *Sc*, proximal to the nodus. This latter could not, from its position, possibly be mistaken for a main vein, and its formation is clearly due to the extreme density of cell-formation in this wing. The same explanation appears to be the obvious one in the case of the apparent prolongation of *Sc* through the nodus.

The Arculus and Discoidal Cell (fig. 20).

Had there been no basal fusion of *R* with *M*, we may safely say that there would have been no arculus. The fusion of *R* with *M* was probably brought about, like the reduction in *Sc*, by the adoption of the aquatic habit by the larva, and the consequent shifting of the course of the oxygen supply of the developing wing from the costal to the anal end of the alar trunk. Originally *R* was the strongest trachea in the wing. When *M* became placed closer to the oxygen supply than *R* was, competition set in. *M* tended to become the strongest trachea of the wing, and finally moved up close to *R*, so that these two large tracheae between them monopolized the "lion's share" of the available oxygen. The formation of the single imaginal vein *R* + *M*, along the contiguous portion of the two tracheae, was the logical result of this. But, as *M* originally supplied a large area of the wing more distally, below *R*, the concomitant of a basal *approach* of *M* to *R* was evidently a more distal *departure* of *M* from *R*, to its natural lower level. This bend or departure of *M* was the rudiment of the future *arculus*. The strengthening of the bend of *M*, by thickening of a cross-vein just below it, completed this interesting formation. This cross-vein would originally of course be attached to M_4 not far from its point of departure from M_{1-3}, the latter being closer up to *R*. Hence, separation of the sectors of the arculus (i.e. M_{1-3} and M_4) is an archaic character. Union of these sectors occurs in many *Libellulidae*, and is evidence of high specialization in this region. Further, the sectors primitively arose from the *upper* portion of the arculus. Movement of their points of origin downwards towards *Cu*, as in *Hemicordulia* (Plate II, fig. *B*), is a sign of specialization.

The essential conditions (fig. 20 *a*) for the formation of the peculiar *discoidal cell* in the Odonata appear to have been two, (*a*) the presence of the arculus, (*b*) a bifurcation of *Cu* at some point distad from the level of the arculus. The second condition is absent in the Protodonata; though *Cu* is generally very beautifully waved, it is always unbranched. The connection between the unbranched *Cu* of these insects and the once-branched *Cu* of the true Odonata seems to be entirely missing. If we grant, however, that conditions (*a*) and (*b*) must be fulfilled before a true discoidal cell can be formed, it is quite easy to see how it *was* formed under those conditions. The development of the cross-vein above and proximally nearest to the bifurcation of *Cu* into a strong support would at once complete the closing-in of a large quadrilateral area with thickened sides, viz. M_4 above, *Cu* below, the lower part of the arculus proximally, and this new support distally (fig. 20 *b*). This is essentially the original *discoidal cell* of the Odonata. Further, since the main veins *M* and *Cu* both arose fairly close together from the narrow wing-base, and diverged as they traversed the wing distad, we may fairly assume that this quadrilateral was not a rectangular one, but of irregular form, *with the distal side somewhat greater than the basal side*. The evidence of palaeontology supports this view, for neither the regular quadrilateral of modern *Calopterygidae*, nor the extremely sharply distally pointed quadrilateral of the *Lestinae* and *Agrioninae*, is found in the Jurassic or Miocene Zygoptera (see fig. 164). These two forms, indeed, represent two divergent lines of specialization from the older type (fig. 20 *c′* to *e′*, and *c′* to *f″*).

As the cross-veins of all ancient forms of Odonata were abundant, it is clear that originally the discoidal cell enclosed a series of parallel cross-veins between its two ends. This was clearly a condition essential to further specialization by sub-division, as in the Anisoptera. The tendency towards broadening the base of the wing in this suborder not only put a strain on the discoidal cell,

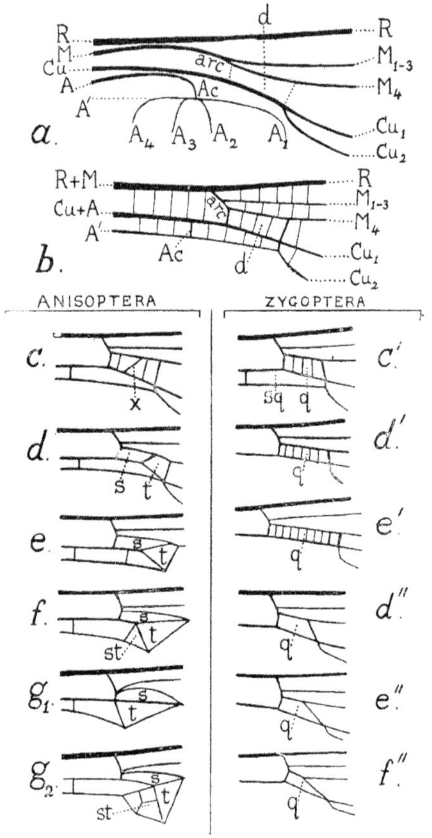

Fig. 20. Phylogenetic diagrams of the discoidal cell. *a.* Essential condition of the wing-tracheae in order that the discoidal cell (*d*) may be formed. *b.* The corresponding archaic wing-venation. Stages in the formation of the triangle (*t*), supra-triangle (*s*), and sub-triangle (*st*) of the Anisoptera are shewn from *c* to g_2; *x* the cross-vein destined to divide the discoidal cell into two; *d* the *Tetrathemis*-stage; *e* incomplete and *f* complete *Gomphine*-stage; g_1 *Libellulid*-stage, hind-wing; g_2 the same, fore-wing. Stages in the formation of the quadrilateral (*q*) of the Zygoptera are shewn from *c′* to *f″*; *c′* archaic type, somewhat irregular quadrilateral: *d′*, *e′* *Calopterygidae*; *d″*–*f″* *Lestidae* and *Agrionidae*; *sq* sub-quadrangle. Original.

resulting in a widening of it, but also tended to pull it obliquely analwards, so that the cross-veins took on a slant. An obvious danger of "buckling" was avoided by the strengthening of one of the cross-veins near the middle of the cell (fig. 20 *c*, *x*). Thus the discoidal cell became divided into two distinct parts, which thence onward underwent different developments. By a slight bending of the lower side at the junction with the thickened cross-vein, and by a slight movement distad along M_4 of the upper end of the same cross-vein, we arrive at what may be termed the *Tetrathemis*-stage[1] of development (fig. 20 *d*; cf. fig. 130). The discoidal cell now consists of (*a*) a more basal and upper portion, the *supra-triangle* (*s*), and (*b*) a more distal and lower portion, the *triangle* (*t*). Both these areas are still quadrilateral in form. From this stage, a further increase in the bend or break of *C*, together with a further elongation of the separating cross-vein—so that its distal end reaches the upper distal angle of the discoidal cell, and its direction becomes nearly longitudinal to the wing-axis—brings us to the *Gomphine*-stage (fig. 20 *e*, *f*; cf. figs. 117, 118). Here the "triangle" becomes for the first time definitely triangular, by the loss of the original costal side, while the supra-triangle assumes the elongated and nearly triangular form seen in most recent Anisoptera.

From this point two main diverging lines of development start off, one leading to the *Aeschninae*, the other to the *Libellulidae*. In the former, fore and hind-wings followed closely similar lines of evolution, the tendency in each case being towards an elongation of the triangle in a direction parallel to the wing-axis (Plate II, fig. *A*). In the *Libellulidae*, however, opposite tendencies set in, in fore and hind-wings. The base of the former became narrower, that of the latter wider. The narrowing of the fore-wing tended to shut the triangle up like a collapsed framework, and at the same time drove it further away from the wing-base (fig. 20 g_2). As the final result of this process, we reach the *Tramea*-stage, where the triangle of the fore-wing becomes excessively narrowed, elongated transversely to the wing-axis, and far removed from the arculus. The

[1] The name does not imply an actual belief that *Tetrathemis* as we know it stopped short at this stage. As the genus is highly specialized in many points, the condition of its discoidal cell may be due to reduction from a more advanced type.

supra-triangle is correspondingly long and narrow (fig. 136). But in the hind-wing, the broadening of the basal portion of the wing had an opposite effect. The triangle was not only stretched longitudinally (somewhat after the manner seen in the *Aeschninae*) but it was drawn towards the wing-base. Finally it took up a stable position with its basal side at the level of the arculus, so that supra-triangle and triangle became coterminous in length (fig. 20 g_1; cf. fig. 136).

Let us now return to the Zygoptera, in which the discoidal cell has remained an undivided quadrilateral. Original differences in the levels of the arculus, and the bifurcation of *Cu*, account for the many variations in the length of the quadrilateral still to be seen. (Contrast *Pseudophaea*, fig. 137, with *Rhinocypha*, fig. 138, or *Calopteryx*, fig. 177.) The cross-veins early disappeared in this suborder, except in a number of *Calopterygidae*, where the original close venation of the wing has persisted up to recent times. Another early tendency was the thickening of a short basal portion of Cu_2 in line with the distal side of the quadrilateral. As the anal bridge (*Ab*) ended up on Cu_2 just at this same point, a special distal portion of the cubito-anal space became cut off between *Ac* and Cu_2, forming a quadrangular area sometimes called the *sub-quadrangle* (*sq* in fig. 20 *c′*).

We can see an archaic type of quadrilateral still extant in the hind-wing of *Epiophlebia* (fig. 141). From this type, a further narrowing of the wing-base leads us to the *Lestine*-type (fig. 20 *f″*; cf. fig. 142). Here the distal side, together with its prolongation along Cu_2, is no longer transverse to the wing-axis, but becomes strongly slanted, while the costal side of the quadrilateral is correspondingly shortened. By a similar method, the sharply-pointed quadrilateral of the *Agrioninae* arose from the older form seen in the fossil *Phenacolestes* (fig. 164) and in many recent *Megapodagrioninae*. The *Lestine* and the *Agrionine*-types are closely parallel; but the former seems to have been in existence for a very long time, whereas the latter is a very recent development.

Quite a different tendency has operated in the case of the *Epallaginae* and *Calopteryginae*, resulting in a gradual loss of the irregularity of the quadrilateral, and the formation of an exact rectangle. Owing to the fact that the sectors of the arculus arise lower down in the latter than in the former subfamily, the quadrilateral of the latter is usually excessively narrow. In the *Thorinae* (fig. 21) the sectors of the arculus are actually attached to *R*. This position has been strengthened, in *Thore*, by a thickening of the antenodal just above it. The effect is to produce a quadrilateral of peculiar and, indeed, unique form, with the basal side more than twice as long as the distal.

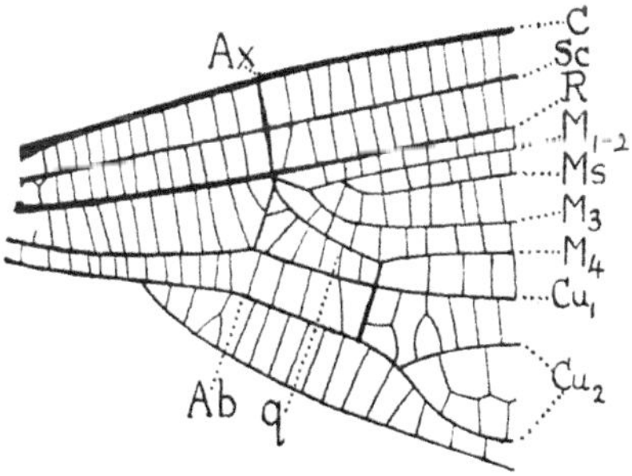

Fig. 21. Base of fore-wing of *Thore gigantea* Selys. *Ax* thickened antenodal; *Ms* Zygopterid sector. Adapted from Needham.

We have seen that, if the level of the arculus be very close to the level of the bifurcation of *Cu*, the quadrilateral will be very short. Such a condition also suggests the possibility of a new arrangement, whereby the arculus, by assuming a sufficient slant, might be continued directly on to the point of division of *Cu*. In this case the lower portion of the arculus would not be needed as a support, and so the discoidal cell would remain open basally. The Jurassic subfamily *Tarsophlebiinae* (fig. 158) adopted this line of development. Quite recently, as the result of extreme reduction,

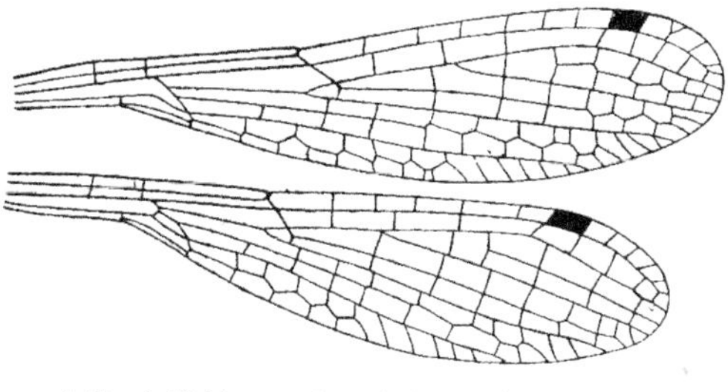

Fig. 22. Wings of *Hemiphlebia mirabilis* Selys, ♂ (hind-wing 11 mm.). Original.

the same tendency has set in in the *Agrionine* genus *Hemiphlebia* (fig. 22). Here of course the arculus has been complete right through the evolution of the genus. But the slanting of the arculus has gone on until it came into line with M_4, and then the distal side of the quadrilateral tended to continue the same line. The final stage of elimination of the basal side has only so far been achieved in the fore-wing of the male, and occasionally in the female also.

The Anal Field (Plate II, figs. *A*, *B*).

(i) *Anisoptera.* We can only briefly trace the many changes in this important area of the wing. Owing to the basal fusion of *A* with *Cu* (due to the same causes as the fusion of *R* with *M*), there was necessarily also a more distal *point of departure* of *A* from *Cu*. This point is the analogue of the arculus, and is represented in the imaginal wing by the *anal-crossing*, *Ac*. This usually only appears as a straight cross-vein, often indistinguishable, except perhaps for a slight slant, from a number of others descending from $Cu + A$, or *Cu*.

Just as *M*, after leaving *R*, divides into a main stem and three branches, so exactly does *A* on leaving *Cu*. But, as the area to be supplied by these branches differs greatly from that supplied by *M*, both in size and shape, we find them correspondingly shortened and directed posteriad rather than distad.

The main stem of *A*, after leaving *Ac*, is represented by A_1. Normally this continues on to meet *Cu* at, or just below, the posterior angle of the triangle. By a slight re-arrangement of one of the cross-veins above A_1, the area known as the *sub-triangle* or *internal triangle* is formed (fig. 20, *st*). In the fore-wing of *Libellulidae* this may become very highly specialized, by means of a sharp bend in A_1. At the same time, the cross-vein forming the original basal side of *st* becomes elongated to continue the line of the proximal part of A_1, thus giving us the complicated structure seen in the fore-wing of *Hemicordulia* (Plate II, fig. *B*), *Tramea* (fig. 136), and many other genera. This is clearly correlated with transverse elongation of the triangle. In the hind-wing, the sub-triangle is usually absent.

In the hind-wings of the older Anisoptera, A_1, after reaching Cu, fuses with Cu_2 for a considerable length (fig. 17 B), and finally bends inwards towards the wing-border. The second branch, A_2, leaves A_1 a little distad from Ac, and runs downwards nearly parallel to $Cu_2 + A_1$, to the wing-border. Thus a distal portion is separated off from the anal field, between A_2 basally and $Cu_2 + A_1$, or A_1, distally. This area remains thus in the *Petalurinae* and most *Gomphinae*. In the *Aeschninae*, however, it becomes definitely closed off below, about half-way between Ac and the wing-border, by development of a cross-vein from the in-turned portion of A_1. Thus is formed the interesting closed area known as the *anal loop* (*al*). In *Aeschna* and a few other genera, this may be supplemented by an additional closed area placed distally to it (*al'*).

The anal loop remains compact and more or less rounded throughout the *Aeschninae*, in the *Macromiini*, and in the *Synthemini*. With the beginnings of a further broadening of the anal field in *Libellulidae*, a new tendency is set up. A_2 shifts close under Ac, and nearly monopolizes the oxygen-supply received from it. A_1 dwindles, and fails to reach Cu. Consequently A_2 grows out strongly into the widened area below it, and tends to assume a slanting course parallel to Cu_2. The latter gives off an inwardly projecting branch Cu_{2b}, tending to run to the wing-border very close to the ending of A_2. Thus, between A_2 and Cu_2, a long loop is formed quite different in shape from the original compact form seen in the earlier *Libellulidae*. This I have recently termed the *cubito-anal loop* [176]. We may, however, retain the term "anal loop," since the formation is clearly homologous with the older type, except in the lack of fusion of A_1 with Cu_2.

For most of its width, this long loop is only two cells wide. In the higher forms, there is developed, along the inner boundaries of these two cell-rows, a strong straight *cubital supplement* (*Cuspl*), whose formation is exactly analogous to that of *Rspl* and *Mspl*. This forms the *mid-rib* of the long loop. Owing to its assuming the form of a complete vein, it was originally mistaken for A_2, while A_2 was regarded as A_3. A glance at the tracheation of this area, however, quickly settles the point.

In the highest forms, the long loop becomes further specialized by the formation of a sharply pointed "toe" (figs. 132–136), ending almost on the wing-border, and a long flat "sole," formed by Cu_{2b}. This form has received the name of the "foot-shaped," "stocking-shaped" or "Italian" loop, owing to its shape.

The third branch of A (A_3) plays an important part in forming the distal side of the anal triangle of the male, in all those Anisoptera in which the anal angle is angulated, and in marking off a weak corresponding basal area next to the membranule in rounded forms. This triangle may be wide or narrow, short or long, and is usually crossed by one or more veins. The shape of the membranule, as well as its size, seems in most cases to be correlated with the shape and size of the area marked off by A_3.

Coming finally to A_4, we find this small branch proceeding *basad* from *Ac*. Along its course, and beyond it to the base, the secondary or recurrent anal vein A' is laid down.

(ii) *Zygoptera*. In this suborder we find trachea A much reduced, so that it rarely reaches *Cu*. The original more distal portion has become quite cut off and fused with *Cu*, at a point where we can still recognize *Ac*, and in a few forms (fig. 17 G) the four branches of A coming off from it. A_1 still runs distad towards *Cu*, and in *Calopteryx* (fig. 17 E) almost reaches it. In most cases, A_1 is very short. The line of A_1 is continued in all imaginal Zygoptera by the formation of a special vein, the *anal bridge* (*Ab*), running from *Ac* to Cu_2, and forming the lower side of the subquadrangle. A' is formed backwards as in Anisoptera. The rest of the anal field is reduced either to (*a*) a triangular area below A' and *Ab* (*Calopteryx* and other non-petiolate forms); (*b*) a single row of cells in the same position (most *Calopterygidae*); (*c*) a narrow uncrossed space below *Ab* (*Lestidae* and most *Agrionidae*); or (*d*) it is completely obliterated (many *Protoneurinae*).

By studying the corresponding tracheation and venation in *Caliagrion* (Plate II, fig. *C* and fig. 17 G) we realize what a large original area of the Odonate wing has been sacrificed in the formation of the *petiole* or stalk. The sub-division of the *Agrioninae* into tribes has been based on the extent of petiolation, with reference to the position of *Ac*.

Sectors, Supplements, Bridges, Transverse Veins.

The term *sector* in Odonata is usually restricted to extra veins developed along definite tracheal branches; while intercalated veins formed in any position without the aid of a definite tracheal branch are called *supplements.* The distinction is somewhat arbitrary, since two or three very distinct tracheae may aid in

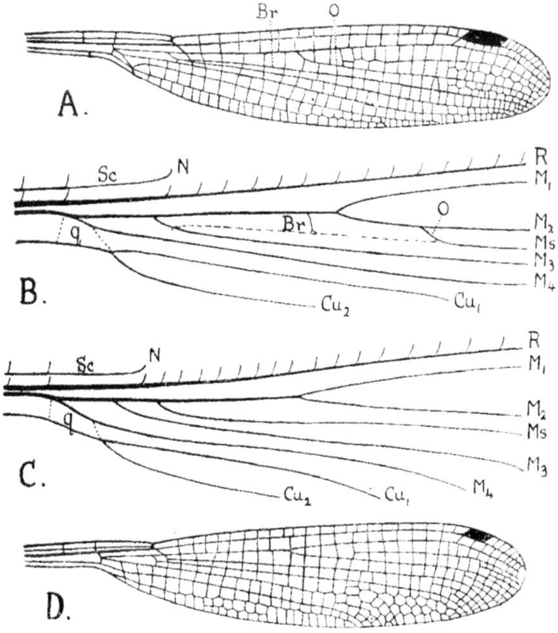

Fig. 23. Comparison of venation of the *Synlestinae* and the *Megapodagrioninae.* A. *Synlestes weyersi* Selys, venation (wing 31 mm.). B. The same, more enlarged tracheation of the larval wing in the region of the long bridge. C. Enlarged tracheation of larval wing of *Argiolestes icteromelas* Selys. D. The same species, imaginal wing-venation (wing 28 mm.); notice absence of bridge. *Br* bridge; *Ms* Zygopterid sector; *O* oblique vein. Original.

the formation of a strong supplement (e.g. *Mspl* in *Aeschna*). A *bridge* is a supplement formed backwards towards the base of the wing in such a way as to continue the line of an important vein, so that this latter may appear much longer in the imaginal venation than it really is when the tracheation is examined.

When a bridge is formed, the original point of departure of the vein so lengthened is marked by a more or less oblique cross-vein, the *oblique vein* (O). The most important bridges in the Odonata are (a) the short bridge of Anisoptera, continuing Rs back to M_{1-2}; (b) the long bridge of the *Lestidae*, continuing Ms similarly backwards, but much further, so that the bridge junctions either with M_{1-2} (*Lestinae*) or M_3 (*Synlestinae*) close to their point of union. A comparison of the venation of *Synlestes* (fig. 23 A) with one of the *Megapodagrioninae* (fig. 23 C) shews how this formation has produced a remarkable resemblance between the two forms. This "hidden convergence" has hitherto led to the inclusion of *Synlestes* in the *Megapodagrioninae*, a subfamily with which it is in no way closely related. [176.]

In the *Agrionidae*, but not so much in the *Lestidae*, most or even all of the cross-veins have become arranged in transverse sets, producing a series of complete transverse veins (Plate II, fig. G). This arrangement, when complete, as in the *Agrioninae*, must be regarded as one of the most beautiful, as well as one of the highest, specializations yet attained by the Odonate wing.

We may conclude this chapter by exhibiting the successive stages of evolution of the Dragonfly wing in consecutive order, as far as the evidence allows us:

1. Original slight inequality of wings (hind-wing broader than fore-wing; Protodonata).
2. Fusion of A with Cu, and formation of A' (completed in Protodonata).
3. Fusion of M with R (nearly completed in Protodonata).
4. Formation of pterostigma.
5. Formation of nodus.
6. Bifurcation of Cu.
7. Completion of arculus and discoidal cell.

	Anisoptera	Zygoptera
8.	Presence of Rs a constant character.	Absence of Rs a constant character.
9.	Sub-division of discoidal cell into two parts.	Strengthening of discoidal cell into a single strong quadrilateral.
10.	Development of the two parts into a well-formed supra-triangle + triangle, either	Development of quadrilateral, either
	(a) of about the same shape in both wings (*Aeschnidae*),	(a) towards a regular rectangular shape (*Calopterygidae*),
	or (b) of widely different shapes in fore and hind wings (*Libellulidae*).	or (b) towards further irregularity by formation of a sharply acute distal angle (*Lestidae* and *Agrionidae*).

CHAPTER IV

THE LARVA OR NYMPH

The term *nymph* is applied to the aquatic larvae of those insects which have an incomplete metamorphosis (i.e. have no true pupa or resting stage), such as Dragonflies, May-flies, etc. The term, however, subserves no useful purpose, since the nymph is, biologically, simply a larval form, and nothing more nor less. It is, also, somewhat inappropriate to the appearance of these creatures. We shall therefore use the term *larva* throughout this book. To those entomologists who hold the belief that a true *larva* must be followed by a *pupa*-stage, we would point out that such is not the case. Larval forms occur throughout the Animal Kingdom (even in some Protozoa). The essential condition of a larval form is that it should differ sufficiently from the perfectly-formed animal to require a metamorphosis in order to reach that final stage. If a new term is wanted anywhere, it is rather to distinguish the larvae of insects with a complete metamorphosis, since these depart most widely from the "general run" of animal larvae.

Hatching of the Larva (fig. 24).

As a typical example, we shall select *Anax papuensis* Burm. [177]. The eggs are placed in the soft tissues of reed-stems, just beneath the water. Before hatching, the embryo (fig. 24 A) lies with its head fitting closely under the pedicel or cap of the egg. The head is well-formed and rounded above; the compound eyes are distinct, black in the centre, surrounded by a reddish zone. All the mouth-parts can be clearly seen. The labium consists of a pair of three-jointed appendages, quite separate except at their extreme bases. The thoracic and abdominal segments are all

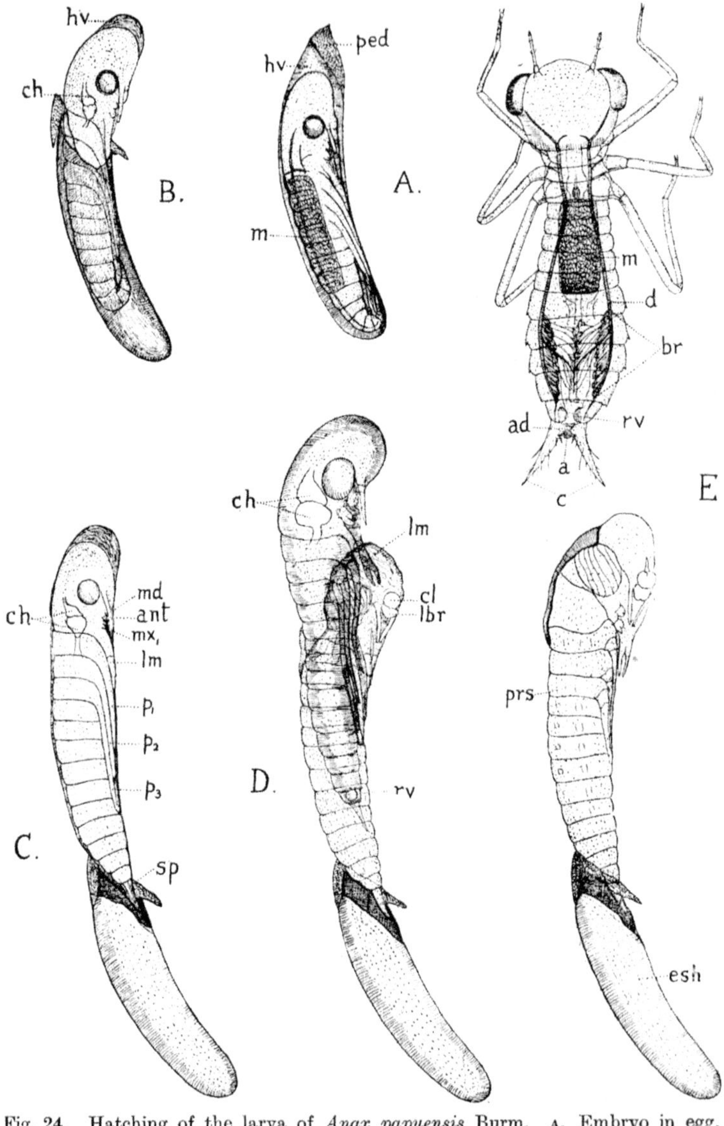

Fig. 24. Hatching of the larva of *Anax papuensis* Burm. A. Embryo in egg, before hatching. B, C. Emergence of pronymph. D, E. Emergence of larva from pronymphal sheath. *a* anus; *ad* appendix dorsalis; *ant* antenna; *br* branchial basket; *c* cerci; *ch* cephalic heart; *cl* clypeus; *d* dorsal tracheal trunk; *esh* egg-shell; *hv* head-vesicle; *lbr* labrum; *lm* labial mask; *m* mid-gut; *md* mandible; mx_1 first maxilla; p_1–p_3 legs; *ped* pedicel of egg; *prs* pronymphal sheath; *rv* rectal valves. (×30.) Original.

distinct, the legs very long. The hind end of the abdomen, together with the cerci and the metatarsi, is bent round ventrally so as to point forwards. The dorsal tracheal trunks, and strong branches to the eyes, can be clearly seen, but are filled with liquid instead of air. A large cylindrical plug of dark yellowish yolk marks the limits of the mid-gut.

Two or three days before hatching, the heart begins to beat, with an irregular pulse of about 30 to the minute. It gradually becomes more regular, and increases to 80 or more per minute. A large vesicle (A, B, *hv*) forms between the head and the pedicel, and appears to be filled with blood. Just before hatching, a very distinct and peculiar pulsating organ, the *cephalic heart* (B–D, *ch*), appears in the head, between the mouth and the aorta. This organ is two-chambered, and pulsates like a simple two-chambered heart. Its exact nature is doubtful; but it is probably only a temporary formation in the oesophagus, by means of which the size of the head is increased, so that it presses hard against the pedicel of the egg. This pressure is the immediate cause of hatching, since it forces the pedicel up like a lid. Thereupon the larva, or *pronymph* as it is now called (C), flows swiftly and easily out of the egg.

The **Pronymph** is very unlike a free larva. It is closely enveloped by a fine chitinous covering, the *pronymphal sheath* (*prs*). This sheath not only invests the body, but closely follows the form of all the limbs, which are held down by it close alongside the body. The labium is only fused basally, and at present is quite unsuited to perform its functions. The head and eyes are moderately large. The head-vesicle still contains some fluid. The pronymphal sheath ends posteriorly around the cerci in the form of a sharp spine (C, *sp*) which usually catches against the broken egg-shell, and serves as an anchor for the pronymph.

Pierre [118] and Balfour-Browne [5] have spoken of this pronymphal sheath as the *amniotic covering*. This is clearly incorrect, since the sheath is a non-cellular chitinous structure. It seems clear that the pronymph is really the *first larval instar*, and its sheath just an ordinary cuticle, such as is cast at ecdysis.

The pronymphal stage in *Anax* lasts only from three to twenty seconds. The cephalic heart increases greatly in size. It pulsates regularly, at about 30 beats to the minute, quickly exhausting the head-vesicle, and driving blood strongly into the dorsal part of the head and into the eyes. These parts swell up with amazing rapidity to twice their original size. The result is a second split, the pronymphal sheath bursting along the back of the head and thorax, so that the young larva slips easily out (D, E).

The moment it is free, the young larva wriggles violently, drawing its legs and labium into their natural positions, ready for use. The latter is now seen to be in the form of a completely fused organ, the *mask* (D, *lm*), very similar to that of the full-grown larva. The cephalic heart quickly collapses, and can no longer be seen. In point of fact, so quickly does the hatching take place, that this organ is only in action for less than a minute, performing about 25 strong pulsations. In the meantime, however, a strong but smaller pulsating organ has appeared between the rectal valves (D, *rv*), and appears to be pumping water into the rectum. At the same time, gas can be seen travelling gradually and evenly down the dorsal tracheal trunks from the region of the mid-gut backwards[1]. On reaching the rectum, the gas passes quickly, first into the larger rectal efferents, then into their branches, and finally around each separate capillary loop, so that the whole beautiful structure of the branchial basket becomes quickly and evenly exposed to view. Almost immediately, active respiration of the basket begins, and the larva uses the ejection of the water as a means of propulsion. The pulsating organ between the valves gradually subsides, but the valves continue to open and shut regularly.

The newly-hatched larva has the appearance seen in fig. 24 E. The head and compound eyes are very large, but no ocelli are present. The antennae are only three-jointed. The labium resembles that of the grown larva, and at no time bears any setae in this species. The segments of the thorax are equal in size, and little larger than those of the abdomen. The legs are

[1] Quite recently, I have shewn that this gas is CO_2, and that it passes into the tracheae by diffusion from the region of the mid-gut. When rectal respiration is fully established. the CO_2 gradually passes out of the tracheae, and is replaced by air derived from the water in the rectum. Rectal respiration is also practised by Zygopterid larvae at this stage, and for the same purpose, their caudal gills being practically useless at first.

large and well-formed, but the tarsi are unjointed. Each tarsus, however, ends in two strong claws. The abdomen ends in two strongly-pointed hairy *cerci* (E, *c*) above which a rudimentary *appendix dorsalis* (E, *ad*) can be distinguished. The mid-gut extends from the first to the fifth segment of the abdomen. It encloses a large barrel-like plug of yolk, which provides nourishment for the young larva, and is gradually absorbed, the large vitellophags composing it (p. 235) breaking up, and becoming replaced by the cells of the permanent epithelium. The ventral nerve-cord has a pair of large ganglia in each of segments 1–8, connected by exceedingly short cords (fig. 58).

The **newly-hatched larva** is the *second* larval instar, the first being the pronymph. Reasons for the quick casting of the first larval skin are not far to seek. Firstly, as the embryo ruptures its amnion at an early stage (p. 238), the formation of a cuticle provides a protection in place of the amnion. Secondly, the secretion of a second cuticle within the first uses up a large amount of waste-products which could not otherwise be disposed of, and thus aids the process of excretion. Thirdly, at the first instar or pronymph stage, the labium has not become sufficiently developed to act as a weapon of prehension. By abbreviating the first instar, the larva practically emerges from the egg fully equipped for the struggle of life.

Variations in the Process of Hatching.

Most Dragonflies hatch out much in the same way as *Anax*. Balfour-Browne [5] has described the emergence of the larva of *Agrion pulchellum*. This differs from that of *Anax* only in the

Fig. 25. "Dragonfly-galls" in an osier stem (natural size). Original drawing by P. Tillyard from specimen lent by Mr K. J. Morton.

increase of the pulse to 100 per minute, and in the fact that the pronymph stage lasts for two or three minutes. The escape from the egg-shell, and again from the pronymphal sheath, does not seem to be so easily accomplished as in *Anax*.

Pierre [118] has carefully described the remarkable method of hatching in the case of *Lestes viridis* (fig. 26). This Dragonfly lays its eggs in autumn on the underside of sallow or osier stems, forming the so-called "Dragonfly-galls" (fig. 25). In the spring, the pro-

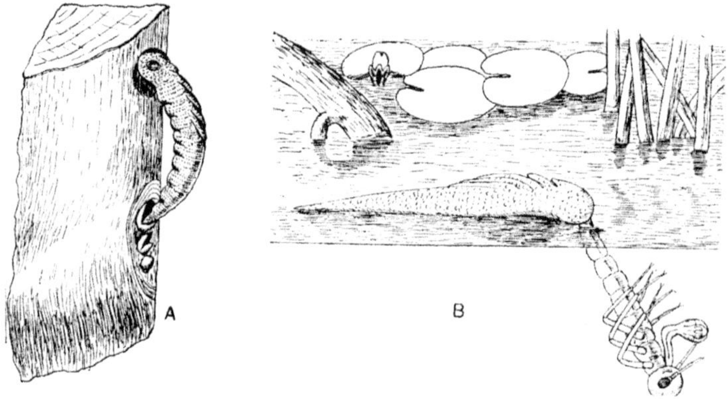

Fig. 26. Hatching of *Lestes viridis* Vand. A. Pronymph emerging from "gall." B. Larva emerging from pronymphal sheath. After Pierre.

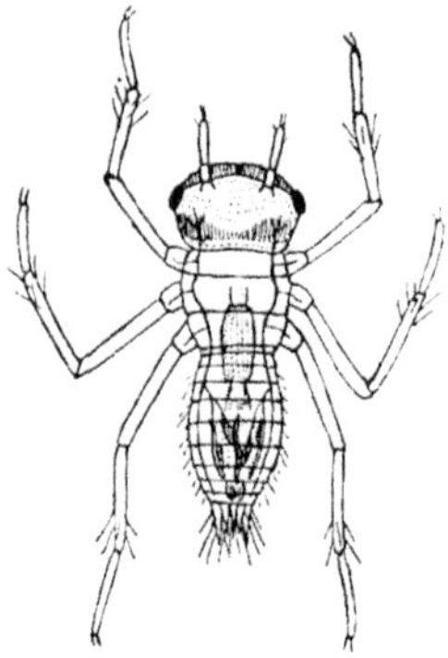

Fig. 27. Newly-hatched larva of *Diplacodes haematodes* Burm. (× 25.) Original.

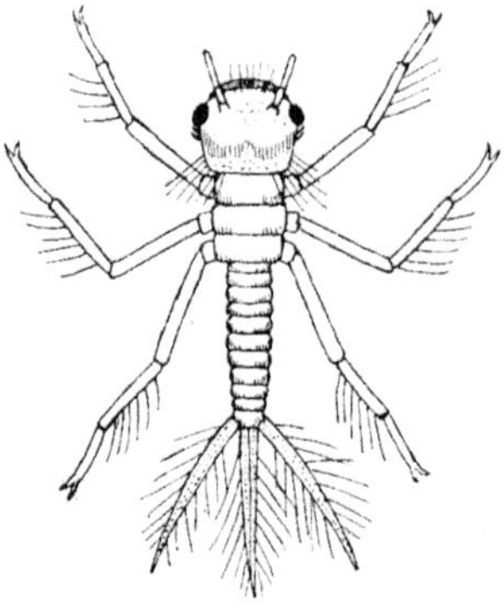

Fig. 28. Newly-hatched larva of *Diphlebia lestoides* Selys. (× 25.) Original.

nymph emerges, enclosed in its sheath, which is armoured all over with small overlapping plates, like a coat of mail. By a strong bending of the body (fig. 26 A), it jumps clear of the stem, and usually falls into the water. If, however, it falls on to the ground, it skips vigorously about, until, aided by the slope of the ground towards the river, it eventually jumps into the water. There it floats with its dorsal surface downwards. The free larva emerges much in the same way as in *Anax* (fig. 26 B).

Other Forms of Young Larva. In the *Libellulidae* (fig. 27) the general appearance differs greatly from that of *Anax*. The eyes are small and black, the head rather squarish; the legs and the short abdomen are more or less hairy. The cerci are short. In the Zygoptera (fig. 28) the larva is much slenderer, the head not unlike that of the *Libellulidae*, the abdomen almost cylindrical, slender and elongated, the legs hairy. The three caudal gills are usually of equal length, and often longer than the abdomen. They are slender pointed processes, almost filiform, and very hairy. They do not appear to be efficient organs of respiration.

Larval Growth: Instars and Ecdyses.

The growth of the larva is marked by a number of stages or *instars*, separated by the act of *ecdysis*, or casting of the cuticle. At each ecdysis, not only the outer cuticle, but also all its internal prolongations, are cast off. Thus the linings of the stomodaeum and proctodaeum (chap. XII)—including the armature of the gizzard—and the intimae of the tracheal trunks, are all cast off at this crisis, and are renewed by the cells underlying them.

The number of instars is not constant, either for different species or for individuals of the same species. It varies between eleven and fifteen, the pronymph being reckoned as the first. The larva may reach maturity within one year of hatching (most Zygoptera, many *Libellulidae*, and *Anax*), or it may take two years, as in *Aeschna*, or even from three to five. External conditions, such as drought, warmth, etc., affect not only the length of larval life, but also the total number of instars. It took me three years to rear the larva of *Austrocordulia*, already more than half-grown when I found it.

The changes which take place during larval growth may be briefly summarized as follows:

1. *Increase in size of the compound eyes.* The number of facets increases at each ecdysis. The eyes never meet on the top of the head (as in many imagines); but, in the case of *Anax*, they

approach very closely to one another. They remain surprisingly small in *Libellulidae*, but become raised up on pyramids in the higher forms, so as to stand clear of the huge mask (fig. 29 F).

2. *Development of the ocelli.* These are absent during most of larval life. During the last few instars, their rudiments appear; and, in the Zygoptera at least, they are well-formed in the full-grown larva. They do not appear to become functional until metamorphosis.

3. *Increase in the number of joints of the antennae.* The antennae at first are three-jointed, there being a short broad base or *scape*, a short but distinct *pedicel*, and a single long *distale*. At the second ecdysis, the distale divides into two. At the next, the more basal of these two again divides. At the fifth or sixth ecdysis, the more basal of these again divides. There are thus, by then, *six* joints, four being distalia. Either at the last or penultimate ecdysis, the *apical* distale divides into two. Thus the full-grown larva usually has 7-jointed antennae, as in the imago. In *Gomphinae*, no sub-division takes place after four joints are formed. Either the fourth or the third joint becomes more or less swollen; in the latter case the fourth may become only a vestigial knob.

4. *Changes in the form of the labial mask.* These differ much in the various groups [19]. The general tendencies are (*a*) more complete fusion of parts, (*b*) repression of galea and lacinia, (*c*) growth and strengthening of the movable hook, (*d*) increase in the number of setae, in setigerous forms. In *Calopteryx* the original cleft in the median lobe becomes intensified instead of closing up. In some species of *Anax*, setae are at first present and afterwards lost.

5. *Changes in the shape of the thorax.* These are the most important of all the changes that take place in larval life, since they are correlated with the development of the wings. The pleural ridges begin to thicken very early (4th to 5th instars). At the next instar, the wing-rudiments appear as tiny buds growing backwards from these ridges. At each ecdysis, these buds increase in size, and soon become triangular flaps. The pleural ridges take up an oblique position. Thus the hind wing-sheath gradually overlaps the fore, until it covers all but the costal

margin. As the wings develop, the synthorax increases in size out of all proportion with the prothorax. The latter, however, is never so small in the larva as in the imago. In the later larval stages, the mesopleurae grow forward to meet each other in front of the wing-sheaths. The latter lie parallel to one another, pressed flatly against the abdomen. Just before metamorphosis, the swelling of the pleurae becomes so pronounced that the wing-sheaths stand up vertically above the abdomen.

6. *Division of the tarsus.* At the third instar a small basal piece separates off from the tarsus. At the fourth or a later instar, a third small joint is added in the same manner. In the *Gomphinae*, the pro- and mesotarsi remain two-jointed.

7. *Growth of the appendix dorsalis.* In those Anisoptera (e.g. *Anax*) where this appendix is at first rudimentary, increase of size takes place only slowly. At the eighth or ninth instar, the appendix dorsalis may be almost as long as the cerci. It generally becomes grooved ventrally, and is often bifid distally. In some forms it never becomes long.

8. *Changes in the rectal gills of Anisoptera.* These are fully dealt with in chap. IX. All Anisopterid larvae start with a *simplex* system of gills of the *undulate* type. In the *Aeschninae* this does not reach the duplex form until the seventh instar. In the *Libellulidae*, the lamellate duplex form is assumed at the third instar.

9. *Changes in the caudal gills of Zygoptera.* The slender processes of the young larva become triquetral at the second or third instar. By flattening (either laterally or dorso-ventrally) subsequent instars form the lamellar gill. Saccoid gills are formed by inflation of the original triquetral form. In *Calopteryginae*, a differentiation in size and shape takes place in the median and the lateral gills.

10. *Changes in the nervous system.* The ventral ganglia do not increase much in size as the larva grows. Thus they become smaller and smaller relative to the segments containing them. The cords connecting consecutive pairs of ganglia rapidly increase in length. A comparison of figs. 54 and 59 shews the extent of this change. The ganglia tend to move forward in each segment. Finally, the ganglion of the second segment moves up into the

first, while that of the first moves up into the metathorax. In the imago, the latter is attached posteriorly to the ganglion of the metathorax.

11. *Increase in the number of Malpighian Tubules.* There are only three in the newly-hatched larva. These increase at each ecdysis, until a total of fifty or more is attained.

The full-grown Larva (fig. 29).

We may now compare the structure of the full-grown larva with that of the imago (chap. II), using one of the large *Aeschninae* for principal reference.

The Head.

The eyes are smaller, and placed more anteriorly in the larva. Consequently the epicranium (*ep*) is much larger, especially the occipital portion, which expands laterally into two very prominent *postocular lobes*. The *frons* (*fr*) is clearly marked off, and shews the frontal ridge in the Anisoptera. The *clypeus* (*cl*) is undivided, the *labrum* (*lbr*) well-formed. The *antennae* (*ant*) are longer and more functionally important than in the imago. Fig. 2 shews some interesting variations in their structure. The *ocelli* may be barely indicated on the flat expanse of the epicranium (*Aeschninae*) or they may be clearly marked (Zygoptera). The mandibles and maxillae closely resemble those of the imago, except in *Pseudophaea* and *Cora*, where the larval mandibles are biramous [31].

Structure of the Labial Mask (figs. 29–32).

The chief difference between the head of the larva and that of the imago is undoubtedly the structure and function of the labium. In the larva, this has become a highly specialized and very efficient organ for the capture of prey, unique in the Animal Kingdom. This organ is known as the *mask* (figs. 29, 30) because it hides the other mouth-parts, and sometimes even covers up the whole face. Its formation is chiefly due to a great lengthening of the mentum and sub-mentum. These come to function much like the fore-arm and arm of the human fore-limb, while the palps are opposed

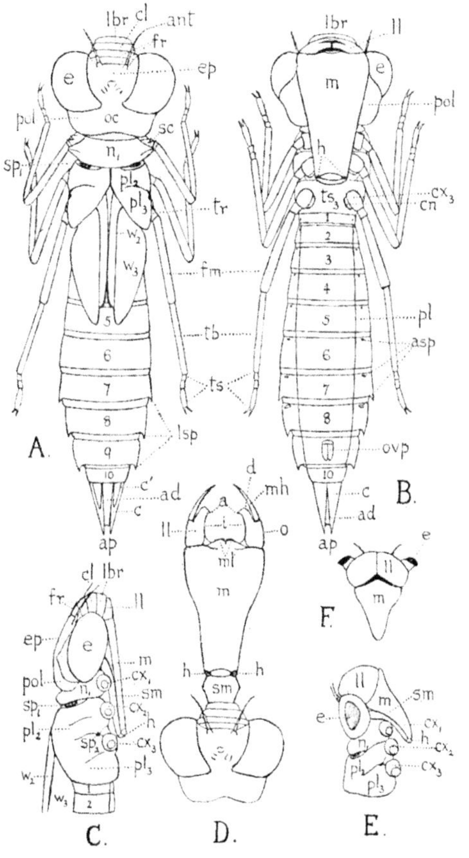

Fig. 29 A–D. Full-grown larva of *Aeschna brevistyla* Ramb. A, dorsal view; B. ventral view; C, lateral view of head and thorax; D, dorsal view of extended mask. E–F. Larva of *Tramea loëwii* Br., to shew mask covering face. E, lateral view of head and thorax; F, head seen from in front. to shew raised eyes projecting above mask. *a* apical hook; *ad* appendix dorsalis; *ant* antenna; *ap* anal pyramid; *asp* abdominal spiracles; *c* cercus; *c'* cercoid; *cl* clypeus; *cn* condyle; cx_1–cx_3 coxae; *d* distal border of lateral lobe; *e* eye; *ep* epicranium; *fm* femur; *fr* frons; *h* hinge; *i* inner border of lateral lobe; *lbr* labrum; *ll* lateral lobe; *lsp* lateral abdominal spines; *m* mentum; *mh* movable hook; *ml* median lobe; n_1 pronotum; *o* outer border of lateral lobe; *oc* occiput; *ovp* ovipositor; *pl* abdominal pleurite; pl_2 mesopleurum; pl_3 metapleurum; *pol* postocular lobe; *sc* supracoxal armature of prothorax; *sm* sub-mentum; sp_1 mesostigma; sp_2 metastigma; *tb* tibia; *tr* trochanter; *ts* tarsus; ts_3 metasternum; w_2 fore wing-sheath; w_3 hind wing-sheath; 1–10 abdominal segments (urotergites in A and C, urosternites in B). ($\times 1\frac{1}{2}$.) Original.

to one another like the thumb and forefinger, and serve to grip the prey. The *sub-mentum* (*sm*), in the position of rest, is directed backward along the ventral surface of the thorax, between the bases of the legs. At its distal end it bears a strong *hinge* (*h*), which functions as an elbow-joint, and serves to articulate the mentum. The *mentum* (*m*) lies ventrally below the sub-mentum, but is directed forward, so that its distal end lies just below the mouth. Lateral squames are absent. Galea and lacinia are absent or vestigial, but the base of an original ligula is represented by the *median lobe* (*ml*), which is fused with the mentum, and forms

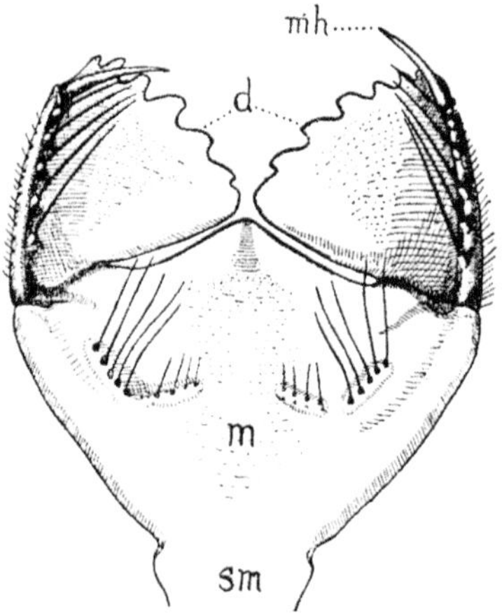

Fig. 30. Concave or spoon-shaped mask of *Synthemis eustalacta* Burm. *d* toothed distal border of lateral lobe; *m* mentum; *mh* movable hook; *sm* sub-mentum. (×8.) Original.

its mid-anterior border. Movably articulated with the mentum, on either side of the reduced median lobe, is a strong *lateral lobe* (*ll*) formed by the palp. These two lateral lobes are very highly specialized, and form the actual organ of prehension. Each carries, at its outer distal angle, a strong *movable hook* (*mh*). The *end-hook* of the imaginal labium, which is fixed, is represented in the labial mask by the *apex* (*a*) furnished with a small hook in some forms, but not in most. The presence of a movable hook and an apex enables us to divide the margin of the lateral lobe into three portions, whose different degrees of development in the various groups lend to them considerable taxonomic importance.

These are (i) the *outer border* (*o*), from the outer edge of the base up to the movable hook; it varies in length, but is always thickened and often armed with spines; (ii) the *distal border* (*d*), or space between the movable hook and apex; originally very short, but probably toothed, it takes on a development of great importance in the concave *Libellulid* mask; (iii) the *inner border* (*i*), running from the apex to the inner edge of the base; it may be short or long, smooth or toothed.

Setae (figs. 30, 32) may be developed both on the mentum (*mental setae*) and on the lateral lobes (*lateral setae, ls*). The mental setae often form two groups (fig. 32 N), an outer group of long *primary setae* (ms_1) and an inner group of short *secondary setae* (ms_2). In *Lestinae* (fig. 32 H) two or three large setae occur on the movable hook. The border of the median lobe carries hairs, small setae, and sometimes tubercles or spines. Small *warts* (fig. 32 N) are very characteristic of the *Libellulid* mask, especially on the surface of the lateral lobes.

Mechanism of the Mask. The larva relies, for the capture of its prey, almost entirely on the mask. Either it remains absolutely motionless until the victim comes within reach, or it stalks it slowly and stealthily until it is close enough to strike. Then the mask is thrown out with lightning rapidity, by means of its strong extensor muscles. At the same time, the extensors of the palps throw the lateral lobes wide apart. Curiously enough, there are no muscles connected with the movable hooks; yet these organs are shot out with great rapidity at the moment of striking. The ingenious mechanism of the hooks has been explained by Amans [1]. A strong ligament on the inner side of the base tends naturally to keep them tightly closed (as in the position of rest). Their sudden extension is due to contraction of the *diaphragm* (p. 211) situated between abdominal segments 4 and 5. The mask being hollow, with a definite prolongation of the haemocoele right up to the bases of the hooks, the forward pressure of the diaphragm forces blood into the cavity, in such a way as to overcome the elasticity of the ligament, and so shoots the movable hooks suddenly outwards. The action is quite momentary, for the pressure is relaxed almost at once, and the hooks spring quickly back, fixing themselves in the body of the hapless victim. When the struggles of the latter become less violent, the flexor muscles come into play, and draw the mask back to its position of rest; so that the mandibles can be brought into action. A more effective or more cruel implement it would be almost impossible to conceive for its purpose.

Homologies of the Labial Mask. Considerable discussion has centred round this point. Our description has followed the views of Butler [19], who supports them with sound ontogenetic evidence. Her views agree with those originally advanced by Rambur and Hagen. The opposing view is that of Gerstaecker [55]. He considers that the lateral lobe represents the fused palpus and galea, while the ligula (median lobe) is formed of the fused laciniae only. His argument is much weakened by the well-known fact that parallel developments of the first and second maxillae occur very frequently. In the Odonata, the inner lobe of the first maxillae is formed by the fused galea and lacinia. We should expect a parallel condition in the labium.

Accepting Butler's views, we may hypothetically construct (fig. 31) a typical archaic form of Odonate labium, from which all present types, both of larval and imaginal forms, can be easily and simply derived. This archaic labium has, of course, the exact structure of the generalized type of labium found in mandibulate insects (e.g. the Cockroach), and corresponds with a fused pair of typical biramous jointed appendages of the Crustacea. It has a pair of fused two-jointed basal protopodites, which carry distally both endopodites (ligula) and exopodites (palpi), the latter being two-jointed also. The following table gives the parts of such a labium, and their respective fates in the larval and imaginal labia of recent Odonata:

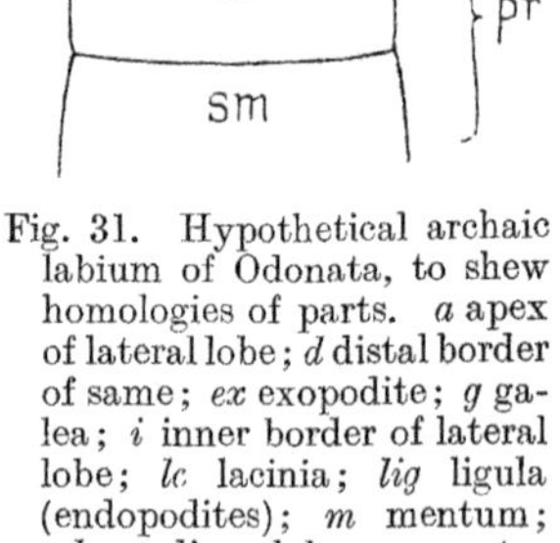

Fig. 31. Hypothetical archaic labium of Odonata, to shew homologies of parts. *a* apex of lateral lobe; *d* distal border of same; *ex* exopodite; *g* galea; *i* inner border of lateral lobe; *lc* lacinia; *lig* ligula (endopodites); *m* mentum; *ml* median lobe; *o* outer border of lateral lobe; p_1, p_2 joints of palpus; *pr* protopodite; *sm* sub-mentum. Original

Table of the Parts of the Labium, or Fused Second Maxillae (R = right, L = left).

Parts	Segments or Joints: Name	Segments or Joints: Position	State	Fate in Larval Mask	Fate in Imaginal Labium
Protopodites (R and L)	*Sub-mentum* (*sm*) ...	Basal ...	Fused R and L pieces	Lengthened ...	Reduced
	Mentum (*m*) ...	Distal ...	Fused R and L pieces	Lengthened ...	Widened, fused with median lobe; side pieces (squames) present
Endopodites (R and L = *ligula*)	*Median Lobe* (*ml*) ...	Basal ...	R and L pieces partly or wholly fused	Reduced, fused with *m*	Not reduced, except in *Libellulidae*; fused with *m*
	Galea (*g*)	Outer distal	Separate, close to middle line	Lost, or vestigial (small spines or tubercles)	Lost
	Lacinia (*l*)	Inner distal	Separate, close to middle line	Lost, or vestigial (small spines or tubercles)	Lost
Exopodites (R and L = *palpi*)	*Lateral Lobe* (*ll*) ...	Basal ...	Separate, wide apart	Strongly developed	Strongly developed, especially in *Libellulidae*
	Movable Hook (*mh*)	Distal ...	Separate, wide apart	Present, usually large	Small, vestigial, or absent

(N.B. Reference should be made to figs. 29–32, especially 31.)

Variations in the Form of the Mask (figs. 29 D, 31, 32). Two main types of mask are easily distinguishable, the *flat* mask and the *concave* or *spoon-shaped* mask. The former is clearly the more archaic form. The following sub-types occur:

(i) *Gomphinae* (fig. 32 D). The mentum is short and square, without setae, the median lobe not usually projecting. The lateral lobes have a rounded apex, a short distal margin, and a more or less denticulated inner margin. The movable hook is strong. Setae are absent.

(ii) *Aeschninae* (figs. 29 D, 32 A to C). The mentum and sub-mentum are greatly elongated, and narrowed basally. The median lobe usually projects slightly. In the older forms (fig. 32 C), the lateral lobes are shaped as in *Gomphinae*; in the rest (A, B), the apex is square or pointed into a distinct hook or tooth. Setae are absent, except in *Gynacantha* (B). In many *Brachytronini* two pairs of tubercles on the median lobe may be remnants of galeae and laciniae.

(iii) *Calopterygidae* (fig. 32 E, R). The median lobe is typically slightly cleft, but secondarily very deeply cleft in *Calopteryginae* (R). The mentum may be long or short, trapezoidal, without setae. The lateral lobes are slender and carry three teeth; one (the largest) at the apex, a second on the distal border (which is much reduced), and a third on the inner border just below the apex. Setae are absent (except a few small ones in *Calopteryginae*). This type passes over without change from the *Epallaginae* to the *Megapodagrioninae*.

(iv) *Agrionidae* (excluding *Megapodagrioninae*) (fig. 32 F). Type (iii) modified. The median lobe becomes prominent and triangular. The mentum remains short. Lateral and often also mental setae are developed. The distal tooth becomes flattened into a serrated ridge; the inner tooth tends to disappear, leaving a convexity only. There are many minor variations on this theme.

(v) *Synlestinae* (fig. 32 G). Intermediate between (iii) and the older *Aeschninae*. Mentum and sub-mentum much elongated, narrow. Median lobe prominently bilobed, with a fairly deep cleft. Lateral lobes as in *Gomphinae*, but the apex projecting in a strong tooth, supported by a second sharp distal tooth. No setae.

(vi) *Lestinae* (fig. 32 H, J). A most remarkable and highly differentiated type. Mentum and sub-mentum of variable length;

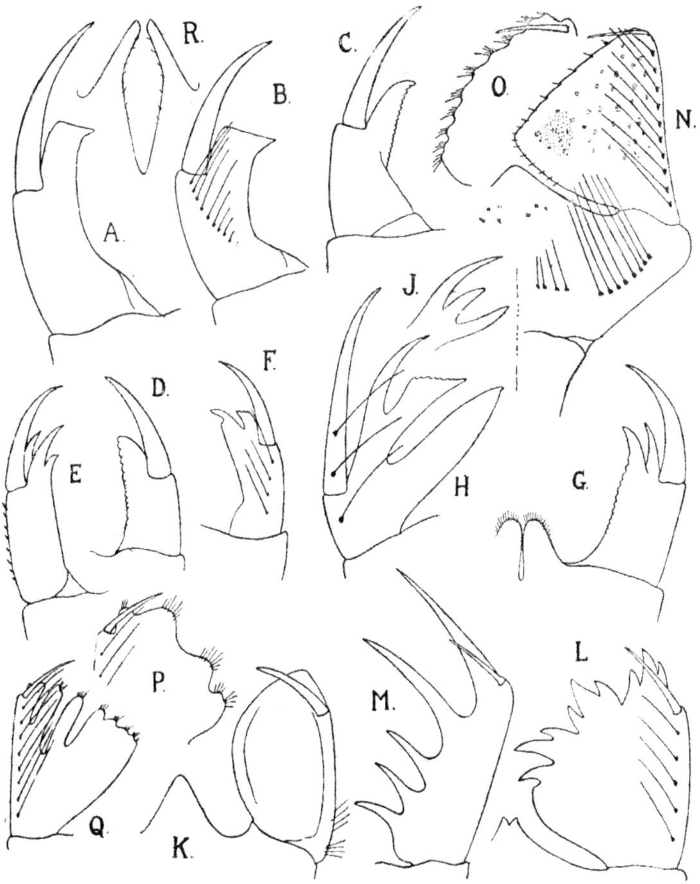

Fig. 32. Variations in the form of the mask, chiefly in the lateral lobes. A. *Aeschna brevistyla* Ramb. B. *Gynacantha rosenberghi* Selys. C. *Austropetalia patricia* Tillyard. D. *Hemigomphus heteroclitus* Selys. E. *Diphlebia lestoides* Selys. F. *Caliagrion billinghursti* Martin. G. *Synlestes weyersi* Selys (with cleft median lobe also shewn). H. *Austrolestes analis* Ramb. J. *Austrolestes psyche* Selys (distal border). K. *Petalura gigantea* Leach (with prominent median lobe also shewn). L. *Cordulegaster annulatus* Latr. M. *Azuma elegans* Br. N. *Diplacodes haematodes* Burm. (with half of mentum also shewn). O. *Hemicordulia tau* Selys (distal border). P. *Austrocordulia refracta* Tillyard. Q. *Cordulephya pygmaea* Selys. R. *Calopteryx splendens* Harris (deeply cleft median lobe). Original.

the median lobe slightly cleft, not prominent. Lateral lobes with the outer and inner borders reduced, the movable hook and distal borders greatly enlarged. Two or three strong setae on the hook; usually one seta on the lateral lobe, and a row on the mentum. A huge apical tooth, supported by a strongly projecting serrated ridge on the distal border. This ridge breaks up secondarily into either (*a*) an outer tooth and an inner ridge (H), or (*b*) three teeth of varying sizes (J).

The beginnings of a concave formation are found in:

(vii) *Petalurinae* (fig. 32 K), a reduced archaic side-branch of very feeble structure. As in all concave labia, the movable hook is reduced in size, and the median lobe is prominent and triangular. It differs however from the typical concave mask in having the inner border elongated, and the distal border short; also, no setae are present. This labium is merely a mud-scoop, and probably helps the larva to shovel out its mud-canals.

Passing on to the true concave forms, we come next to

(viii) *Cordulegastrinae* (fig. 32 L). Here we see the reduced, slender hook and the projecting, triangular, median lobe, which however is bifid at the tip (an archaic character). The inner border is now reduced, and the distal border becomes much elongated and irregularly toothed. By this arrangement, the lateral lobes are enabled to meet together along their distal borders, above the median lobe, in an irregular median line, formed by the teeth, which interlock like cogs. Mental and lateral setae are present.

(ix) *Libellulidae* (fig. 32 N–Q). A single type, with many minor variations, characterizes this huge family. The submentum is narrow, the mentum wide and sub-triangular. The median lobe is projecting and triangular. The lateral lobes are huge, very concave, and roughly triangular in shape, owing to the extension of the distal border to a length about equal to that of the inner border. Mental setae are generally numerous; in the higher forms, they divide up into two sets (primary and secondary). Lateral setae are present along the strongly ridged outer border. The hook is small and slender. The dentition of the distal lobe shews three principal types: (*a*) the *Macromian*, with deep regular incisions (fig. 30); (*b*) the *Eucordulian* (fig. 32 O), with shallower regular crenations, each crest armed with a set of small setae; and (c) the *Libelluline* (fig. 32 N) in which the crenations are smoothed out and the distal border becomes more or less straight, with setae marking the original notches. Three peculiar aberrations from the main developmental line are figured in fig. 32 M, P, Q. *Epophthalmia* and *Azuma* (M) are highly specialized *Macromian* types, in which the distal border has become hypertrophied, forming a set of huge teeth which do not interlock with their fellows on the opposite side. This is a most formidable and peculiar weapon. *Austrocordulia* (P) is the only known *Idocordulian* labium, but is probably not typical. Four or five large irregular rounded lobes, carrying sets of setae, lie along the distal border. *Cordulephya* (Q) is a highly specialized *Eucordulian* type, in which deep narrow indentations are formed along the upper half of the distal border. These do not interlock, but form a kind of sieve for the ejection of pieces during mastication. Warts (fig. 32 N) occur on many masks of type (ix). A reference to fig. 29, E and F, will shew how the concave mask excels the flat form in hiding the whole face. It must, however, be added that the *Aeschnine* labium is far more effective for the capture of large prey, both as to reach and as to the efficacy of its hooks. The concave labium seems more adapted for catching very small animals.

The Thorax (fig. 29 A–C).

The short but well-formed **prothorax** of the larva is always larger than in the imago. It is often armoured with projecting tubercles or spines. In the *Aeschninae* two such spines, placed close together above the procoxa, form the *supracoxal armature*, a structure of considerable systematic importance, especially in the genus *Aeschna* (p. 350 and fig. 180). The plates of the neck or microthorax are well-developed.

The **synthorax** is strongly formed, with the parts closely fused in the well-grown larva. The secondary divisions of the pleurae are only weakly indicated. The *mesostigmata* (sp_1) are placed latero-dorsally between prothorax and mesothorax, on two flattened pieces called the *mesostigmatic laminae*; these fuse together in the imago to form the *spiracular dorsum*. The pleural ridges are placed obliquely, the mesopleurae having grown forward to meet one another in the middle line in front of them. The *wing-sheaths* (w_2, w_3) nearly always lie parallel and flat along the back of the abdomen, and reach as far as the end of the fourth segment or beyond. In the *Cordulegastrinae* and *Synthemini* they are strongly divergent (fig. 33). The *metastigmata* (sp_2) are present, but very small.

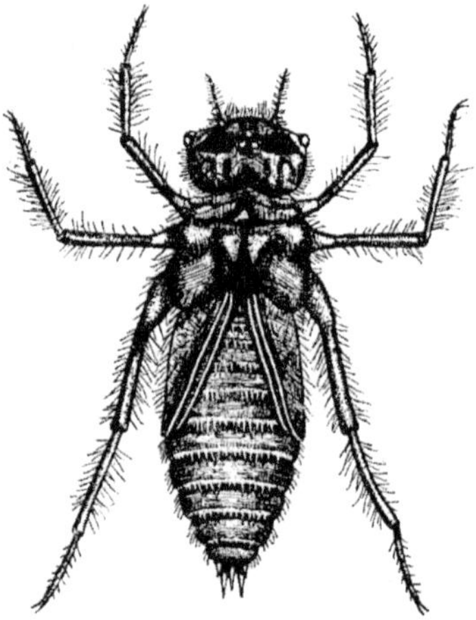

Fig. 33. Larva of *Synthemis eustalacta* Burm., with divergent wing-sheaths. (×2.) Original.

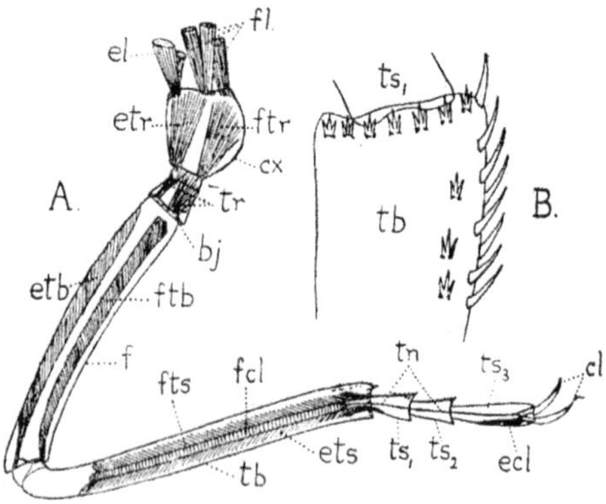

Fig. 34. A. Hind-leg of larva of *Aeschna brevistyla* Ramb. (×6). B. Distal end of tibia, to shew trident-spines (×30). *bj* breaking-joint; *cl* claws; *cx* coxa; *f* femur; *tb* tibia; *tr* trochanter; ts_1–ts_3 joints of tarsus. For explanation of muscles, see p. 208. Original.

The **legs** (fig. 34) are always longer than those of the imago, and their bases placed further apart. The coxae (*cx*) are peculiar in being articulated ventrally with the sternum by means of a special condyle (fig. 29 B, *cn*). The articulation of the femur with the small trochanter is very peculiar, and forms a *breaking-joint* [40]. The trochanteric muscles, instead of being inserted into the base of the femur, are attached to a stretched membrane which keeps the femur in position. By a sudden contraction of these muscles, the membrane can be broken and the femur released, with scarcely any loss of blood. This process of *autotomy* enables the larva to escape whenever a leg is seized by an enemy, or gets hopelessly entangled. The hairs present on the legs of almost all young larvae usually disappear in later instars, so that the legs become quite smooth. Some larvae, however, have very hairy legs (fig. 33). The imaginal armature is absent, or only rudimentary, in the latest instars. Small spines are, however, frequently developed distally on the tibia, and on the tarsal joints. In *Aeschna*, a series of beautiful *trident-spines* (fig. 34 B) is found on the tibia. The tarsal claws are always well-developed, but the tooth found on them in the imago is either absent or rudimentary.

Interesting specializations in the form of the legs are correlated with special habits in some larvae. Thus, rock-dwelling larvae (e.g. *Diphlebia*, fig. 35) have broad, flat, blade-like femora which can be closely applied to the rock surface. The burrowing larvae of *Gomphinae* have strongly ridged surfaces on femur and tibia, while the tarsi of the fore and middle legs are reduced to two joints (fig. 36 A, B).

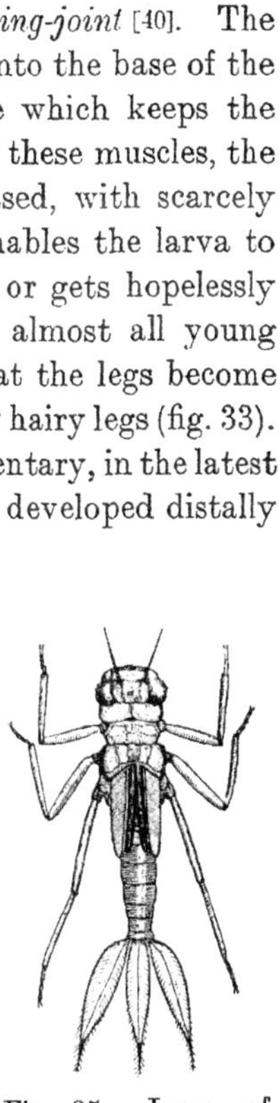

Fig. 35. Larva of *Diphlebia lestoides* Selys. (Natural size.) Original.

The Abdomen (fig. 29 A, B).

In the larva, the abdomen is always shorter and broader than in the imago, the difference being very pronounced in *Libellulidae* and some *Gomphinae*. The segmental sclerites are more easily distinguished. The *tergites* (A, 5–10, C, 1–2) are always more or less convexly arched, though they may become rather flattened in very broad forms, especially at the sides. A definite *trilobation* of the abdomen is produced in the *Petaliini* (fig. 37) by the development of rounded or sub-triangular upwardly-projecting lobes from the sides of the terga. Extreme elongation of the last few segments is a characteristic of some *Gomphine* larvae (fig. 186). The *sternites* (B, 1–10) are rectangular plates, either flat or slightly convex. The *pleura* (B, *pl*) differ from those of the imago in being strongly chitinized. In the Anisoptera they form flat side-pieces to the

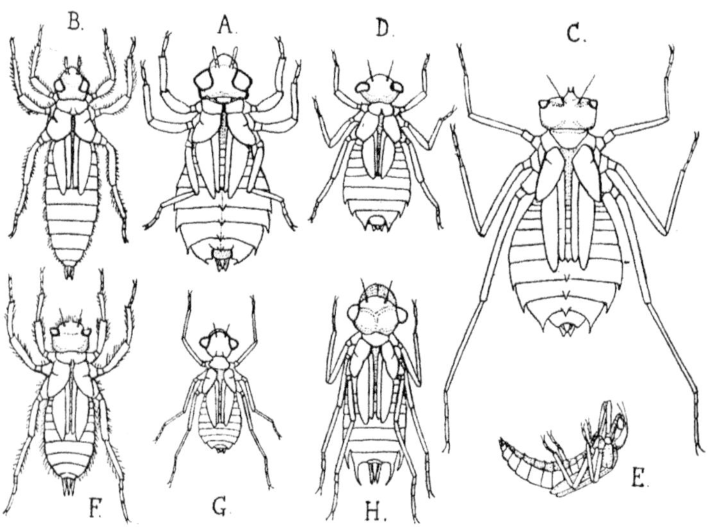

Fig. 36. Types of Anisopterid larvae. A *Ictinus australis* Selys. B. *Gomphus vulgatissimus* Linn. C. *Macromia* sp. (Indiana, U.S.A.). D. *Austrocordulia refracta* Tillyard. E. The same, feigning death. F. *Orthetrum caledonicum* Br. G. *Diplacodes bipunctata* Br. H. *Tramea loëwii* Br. (Natural size.) Original.

sterna; in the Zygoptera they are infolded. Eight pairs of abdominal spiracles (B, *asp*) are present, the eighth being the largest. Their positions, near the anterior margins of the pleura, close to the terga, correspond with those of the imago. They are, however, small and non-functional, the tracheal cords connected with them remaining solid until metamorphosis.

Dorsal spines may occur on the terga of some or all of the segments, especially in running-water forms. *Lateral spines* frequently occur at the postero-lateral margins of the more posterior segments (A, B, *lsp*).

Heymons [72] has investigated the posterior end of the abdomen, with a view to shewing the presence of reduced eleventh and twelfth segments. His conclusions, which will be more fully appreciated by a study of the appendages (see below), determine the eleventh segment as being represented by a single tergite

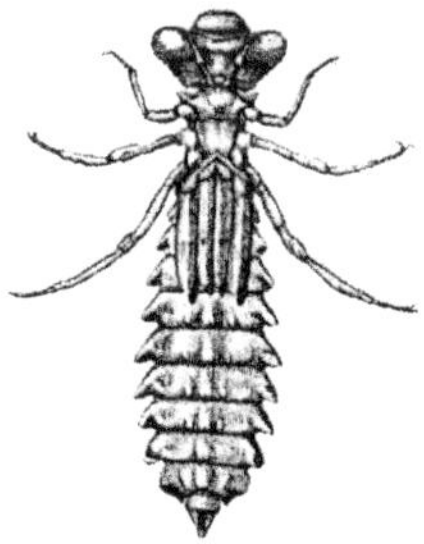

Fig. 37. Larval exuviae of *Austropetalia patricia* Tillyard. (Natural size.) Original.

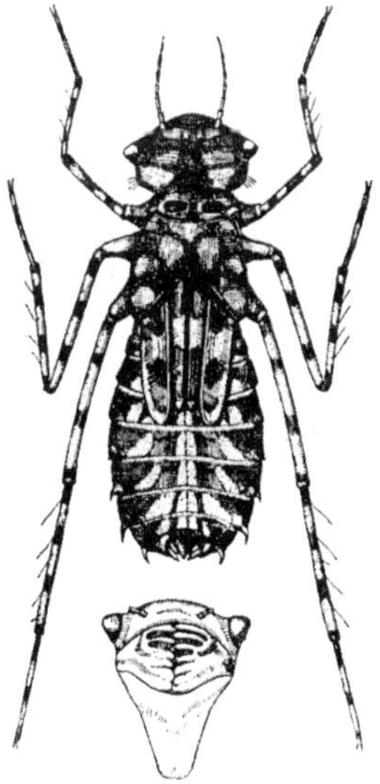

Fig. 38. Larva of *Cordulephya montana* Tillyard. Below, the head seen from in front. (×3.) Original.

(base of the appendix dorsalis) and a bipartite sternite (bases of the cerci). If these appendages, especially in young larvae, be forced apart, three small processes can generally be made out surrounding the anus. Of these, the mid-dorsal one, or *lamina supra-analis* (fig. 39 G, *la*), is the reduced twelfth tergite; while the two lateral ones, or *laminae sub-anales* (*la′*), represent the bipartite sternite. Heymons has also traced these parts in the imago.

The Anal Appendages (fig. 39).

The whole question of the complicated anal appendages of the Dragonfly larva is one of some difficulty. The famous Heymons-Handlirsch controversy has thrown a flood of light upon it. But, while Handlirsch's original determinations were undoubtedly erroneous, and Heymon's corrections of these a very great step

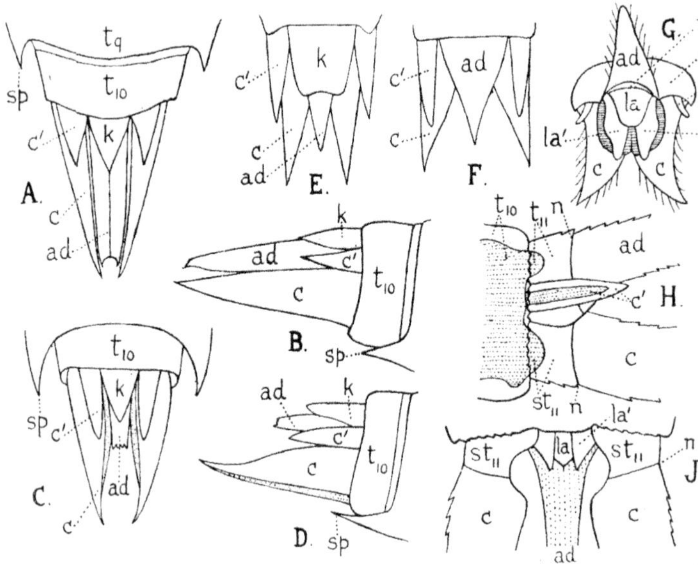

Fig. 39. Anal appendages of Odonate larvae. A. *Aeschna brevistyla* Ramb., ♂, dorsal view. B. The same, lateral view. C. *Dendroaeschna conspersa* Tillyard, ♂, dorsal view. D. The same, lateral view. E. *Epitheca bimaculata* Charp., ♂, dorsal view. F. The same, ♀. G. The same opened out and viewed from behind. H. *Agrion* sp., lateral view, at metamorphosis (imaginal parts shaded). J. *Agrion* sp., ventral view. *ad* appendix dorsalis; *c* cercus; *c'* cercoid; *k* process from which inferior appendage of male imago is developed; *la* lamina supra-analis; *la'* lamina sub-analis; *n* breaking-joint; *sp* lateral spine of segment 9; st_{11} eleventh urosternite; t_9–t_{11} urotergites. (A–D × 6.) A–D original, E–J after Heymons.

in advance, yet Handlirsch's Parthian shot (in which he claims that Heymons' "cercoids" are only "re-developed or imaginal cerci") raises a nice point which is best left to individual opinion for settlement. For the purpose of clearly understanding the anatomy of this region, we shall here select for study well-grown male and female larvae of the subfamily *Aeschninae* (in which the appendages are best developed).

Three large appendages can be easily seen—the same three that we have already noticed in the young larva. These are the *appendix dorsalis* (*ad*), mid-dorsal in position, and the two latero-ventral *cerci* (*c*). When closed, the three together form the *anal pyramid* (fig. 29 A, B, *ap*). These appendages are the same in both male and female larvae, and are the *true larval appendages*. In the Zygoptera they form the three *caudal gills* (fig. 39 H, J). In most Anisoptera they are much shorter than in the *Aeschninae*.

From the fourth or fifth instar onwards, a second set of appendages develops. These are destined, at metamorphosis, to give rise to the imaginal appendages. In both sexes, these consist of a pair of small pointed appendages, the *cercoids* of Heymons, lying one on either side of the appendix dorsalis, above the cerci. They give rise to the "superior appendages" of the male imago, or the "anal appendages" of the female. Thus, in larvae of both sexes, there are five appendages, of which the three largest are truly larval; the two smallest, imaginal. Now if we examine a male *Aeschnine* larva, we shall see, placed directly above the appendix dorsalis, a short projecting piece overlying the base of the latter (A–D, *k*). This is the developing "inferior appendage" of the male imago. In most Anisoptera it is not distinctly separated off from the appendix dorsalis, and simply appears as the swollen base of the latter (E, *k*). As Heymons states that this appendage is really formed from the base of the appendix dorsalis, we forbear to name it here. It is, however, clear that its condition in the more archaic *Aeschninae* (C, D, *k*) raises the question of its having a separate identity, and gives further point to Handlirsch's remarks about the relationships of the cercoids with the cerci.

In the Zygoptera, the position is much simplified. Besides the three caudal gills, which are the true larval appendages, we can distinguish the two cercoids as small processes (H, *c'*), either blunt or pointed, placed as in the Anisoptera. No other process corresponding to the basal swelling *k* in male Anisoptera is to be found.

At metamorphosis, the following changes occur:

(i) In all cases, the *appendix dorsalis* is cast off. But, in the males of Anisoptera, it leaves behind it, basally, the *inferior appendage of the male*, developed from the process *k*. In all other cases, a rudimentary eleventh tergite is all that is left of the base (H, t_{11}).

(ii) In all cases, except in the males of Zygoptera, the *cerci* are cast off, leaving only rudimentary eleventh sternites behind (H, st_{11}). In the males of Zygoptera, they leave behind them the two "inferior appendages" of the imago, developed within their bases. Thus these latter are in no sense homologous with the "inferior appendage" of the Anisopterid male, which lies dorsally above the anus.

We may summarize our results in the following table:

Table of the End-segments and Appendages of the Larval Abdomen.

Segment	Name of Part	Larvae: *Anisoptera*	Larvae: *Zygoptera*	Imagines: *Anisoptera*	Imagines: *Zygoptera*
10	Tergite	*	*	*	*
	Sternite	*	*	*	*
	Cercoids (2) ...	*	*	* ♂, (superior appendages) ♀, (anal appendages)	* ♂, (superior appendages) ♀, (anal appendages)
11	Tergite	×	×	×	×
	Bipartite sternite	×	×	×	×
	Appendix dorsalis	*	* (median gill)	* in ♂ (inferior appendage) — in ♀	—
	Cerci (2)	*	* (lateral gills)	—	* in ♂ (inferior appendages) — or × in ♀
12	Tergite (= lamina supra-analis)	×	×	×	×
	Sternite (= two laminae sub-anales)	×	×	×	×

* present, × rudimentary, — absent.

The chief characters of the larvae in the various groups of Dragonflies are given in chap. XIV. The principal types are shewn in figs. 29, 33, 35–38, 40–42, which should prove more useful than verbal descriptions. The rectal gills of Anisoptera and the caudal gills of Zygoptera are dealt with in chap. IX. Since our knowledge of the internal anatomy of Odonata is chiefly derived from a study of the larva, the internal larval organs will be found dealt with in the chapters devoted to those parts.

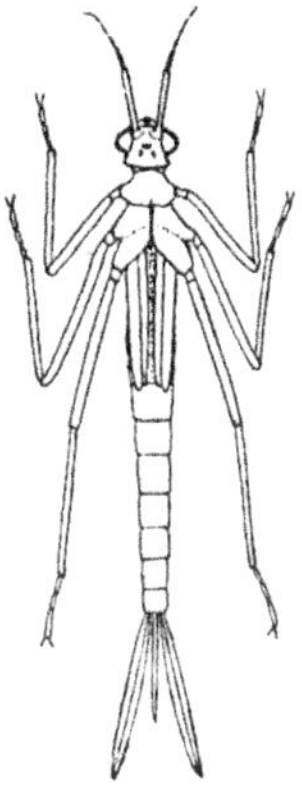

Fig. 40. Larva of *Calopteryx virgo* Linn. (× $1\frac{1}{3}$.) Original.

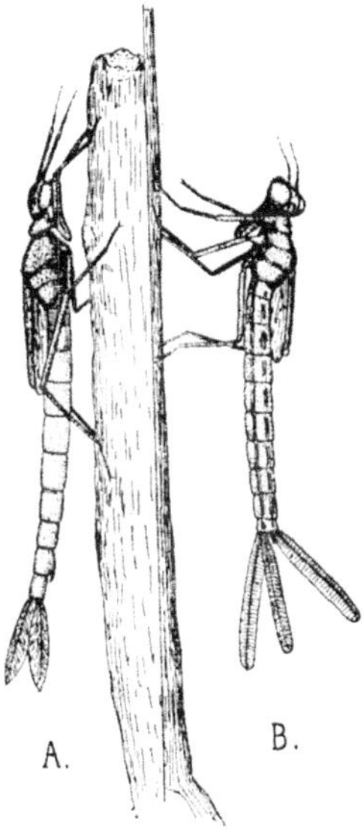

Fig. 41. Larvae of *Lestidae*. A. *Synlestes weyersi* Selys (× $1\frac{1}{4}$). B. *Austrolestes cingulatus* Burm. (× $1\frac{1}{2}$). Original, drawn from photographs of living larvae.

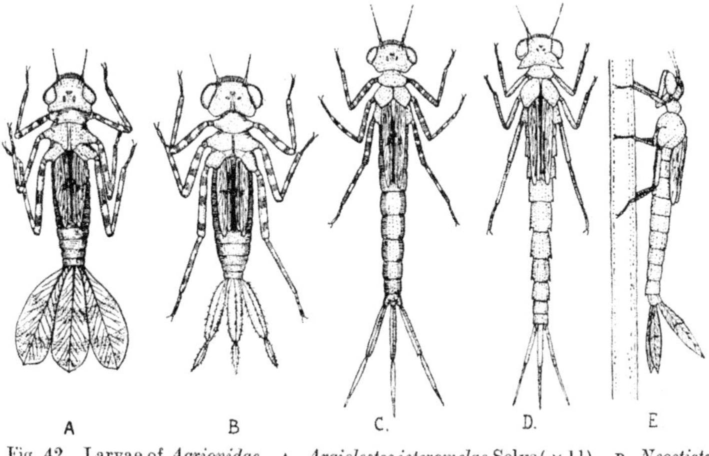

Fig. 42. Larvae of *Agrionidae*. A. *Argiolestes icteromelas* Selys (× $1\frac{1}{3}$). B *Neosticta canescens* Tillyard (× 2). C. *Isosticta simplex* Martin (× 2). D. *Caliagrion billinghursti* Martin (× $1\frac{1}{3}$). E. *Ischnura heterosticta* Burm. (× $1\frac{2}{3}$). Original, drawn from photographs of living larvae, except A, which was chloroformed.

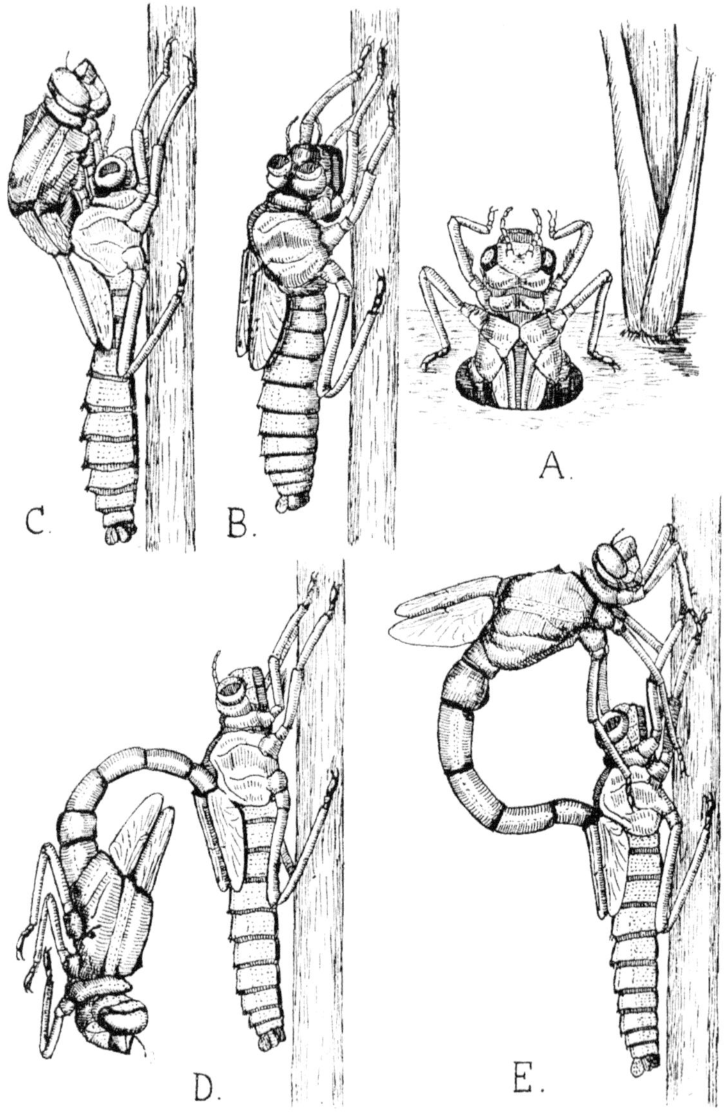
C.
B.
A.
D.
E.

Metamorphosis (fig. 43).

The emergence of the imago from the larval skin or *exuviae*, usually spoken of as the "metamorphosis," is in point of fact only the consummation of an *internal* metamorphosis which begins

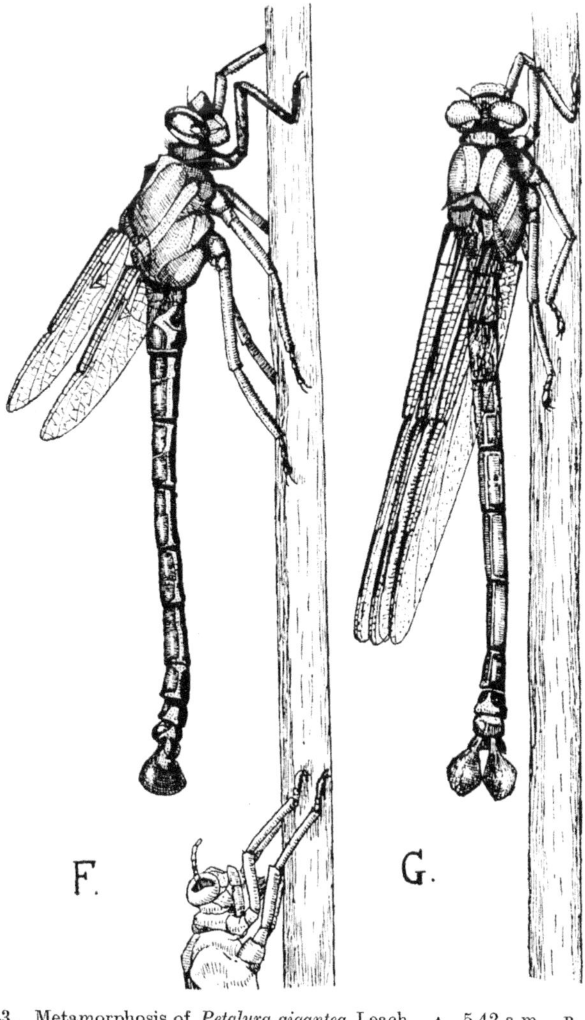

Fig. 43. Metamorphosis of *Petalura gigantea* Leach. A. 5.42 a.m. B. 5.50 a.m. C. 6 a.m. D. 6.12 a.m. E. 6.18 a m. F. 6.30 a.m. G. 6.45. (Natural size.) Original, drawn from sketches and notes in the field.

a considerable time before. The beginning of this change is marked by an alteration in the colour and behaviour of the larva. The colour darkens considerably, greenish larvae becoming a dull opaque brown. The larva becomes listless and refuses to feed. A rapid proliferation of the hypoderm cells, preparatory to the formation of the imaginal exoskeleton, causes the larva to appear tense and swollen. In particular, the thorax swells up so that the wing-sheaths stand vertically off from the abdomen. This is partly due to the growth of the great wing-muscles, partly also to increased growth of the pleurae forcing the meso- and meta-notum backwards into an oblique position. The compound eyes begin to swell, owing to rapid formation of new separate elements. The imaginal hypodermis and cuticle soon begin to shew through the larval skin, and are generally pinkish in colour. Soon the gills cease to be functional. At the same time, the thoracic stigmata, either of one or both sides, become functional, and the larva takes up a position with its head and part of its thorax protruding from the water, so as to breathe air directly.

The emergence of the imago was first described by Réaumur [129], whose beautiful account has never been excelled. Numerous subsequent writers have described and figured the process accurately in the case of an *Aeschnine*, *Libelluline* or *Agrionid* Dragonfly. We have here introduced a touch of novelty by figuring the emergence of *Petalura gigantea* (fig. 43), which differs from the more usual type in several very interesting points. The description, however, will be general, the peculiarities of *Petalura* and other genera being noted by the way.

As soon as the internal changes are practically complete, the larva climbs out of the water, usually up a stick, rock, reed-stem or other suitable object. Some species climb to a considerable height, but most of them only travel from a few inches to a yard. Many larvae (*Gomphinae* and some *Libellulidae*) cannot climb, and just crawl out of the water up a shelving beach. They are sometimes compelled to emerge when only partly out of the water. *Petalura*, which forms burrows or canals in the peat or mud of small mountain-swamps (Plate IV, fig. 1), hauls itself out of its hole (A) by means of its legs, and ascends a convenient reed-stem for about a foot. It is thickly coated with mud.

When the larva has selected its position for transformation, it fixes its tarsal claws so firmly in position that the exuviae remain clinging tightly to their support long after the imago has emerged. The larva then becomes quite motionless, while the internal swelling of parts increases (B). This swelling becomes particularly pronounced in the region of the eyes, and on the back of the thorax. The actual splitting of the skin first appears along the mid-dorsal line of the thorax, and quickly extends forward on to the head. It seems to be caused by the insect arching its back strongly, in an attempt to withdraw its head into the thoracic cavity. The thorax bulges out through the crack, so that the withdrawal of the head is quickly accomplished. As soon as head and thorax are free, the legs and wings are gradually withdrawn from their sheaths (C). By the time this is accomplished, the abdomen is only partly free. The insect next usually hangs for some time with its head downwards, while the action of sun and air hardens its legs (D). As soon as these are strong enough, the insect waves them about in an effort to find something to catch hold of. Nothing being usually available, it attempts to jerk itself upwards, and sooner or later succeeds, by a great effort, in seizing the head of the exuviae, or the reed-stem above it (E). Next, holding firmly on to the stem, the insect withdraws its abdomen from the larval skin, and usually climbs a little way above it, so as to be quite free. The *Gomphinae* crawl straight out of their skins, often with barely any rest for the drying of the legs.

The transformation is completed by the elongation of the abdomen and the expansion of the wings (F, G). The abdomen is usually broad, short, and very soft when first withdrawn. In *Petalura* and in most *Aeschninae*, it is already a considerable length, and reaches its full extent before the wings have completed their growth. In other forms, the abdomen usually completes its lengthening after the wings are fully expanded. The process is, however, always correlated with the growth of the wings. There seems to be no doubt that the mechanism of expansion is a combination of a rapid pulsation of the blood with the swallowing of a large amount of air into the alimentary canal, and, particularly, into the crop. Thus, not only is the abdomen considerably expanded, but blood is displaced by air within it, so that a large supply is available for pumping into the wings. These latter appear at first as small, crumpled, greenish or yellowish bags. While the blood is being forced in, the wings are alternately slightly opened and shut, thus aiding the pulsations of the heart. The expansion of the hollow bag of the wing takes place with remarkable rapidity, and is a wonderful example of quick growth. The basal portions expand first, the wing becoming gradually smoothed out as far as a big transverse crinkle marking the nodus. The apical half expands with great rapidity. When the wing is fully expanded—a result accomplished in many cases in less than ten minutes—the two wing-surfaces lie parallel against one another, only actually fusing along the veins. The space between the two is otherwise filled with blood plasma. The presence of this fluid gives the wings a pale greenish colour, while the additional refraction of the rays of light through the still separated media gives a beautiful iridescence to the surface. This does not disappear until the wings are quite dry and strong,—a consummation which takes from an hour or two to several days, according to the state of the weather. The wings are generally protected from damage during expansion by an outward arching of the thorax, so that they hang clear of the abdomen (G).

The assumption of the imaginal colour-pattern may be either slow or rapid. In *Petalura*, the larval skin is very soft and unpigmented. The imaginal colour-scheme begins to shew up even before emergence. By the time the abdomen has become fully extended, the coloration is very vivid, rich brown and pale yellow. As the insect matures, the colours become duller, the yellow markings finally turning greyish, or merging into the darkened brown. Many *Gomphinae* assume a bright black and yellow colour-pattern with great rapidity, the yellow later on turning to greenish or to yellowish brown. But, in most cases, the colour-pattern is only slowly assumed, the imago appearing at first of a peculiar pinkish fleshy tint. Brilliant heliochromes, such as bright blues and reds, are often not developed for weeks. An imago which has not developed its mature colouring is called a *teneral* form. Teneral males usually resemble their corresponding females in colour.

The times given for the different stages of metamorphosis in fig. 43 are perhaps a little shorter than usual for most Dragonflies. The times vary not only with different species, but also according to the strength of the sun's rays, temperature, humidity, and also for both local and individual causes in each separate case. Most Dragonflies emerge at or near dawn. Dragonfly larvae which have been previously disturbed, and have retreated again to the water, generally emerge about midnight, as do also some of the large *Aeschninae* habitually. Many Dragonflies emerge just before or even during thunderstorms. The most rapid emergence ever witnessed by the author was that of a large number of larvae of *Hemigomphus heteroclitus* just before a storm. The imagines appeared with great rapidity, crawling straight out of their exuviae. Within ten minutes, many were fully coloured and expanded, and made off quickly into the forest before the storm broke.

An examination of a cast larval skin shews clearly how the linings of the great tracheal trunks are withdrawn. The dorsal intimae are withdrawn through the mesostigmata, the viscerals through the metastigmata. They are to be seen projecting from the thoracic split in a twisted and shrivelled manner. The ventral intimae are withdrawn through the eighth pair of abdominal spiracles, and can be seen lying tightly stretched along each side of the abdomen, just above the pleurae. From them, branches proceed to each of the other seven pairs of abdominal spiracles. All the spiracles become fully functional at metamorphosis.

At metamorphosis, as at each ecdysis, the whole of the stomodaeal and proctodaeal linings are cast off, and shrivel up. The dental folds of the gizzard, which are thicker and more resistant, may often be found within the alimentary canal of the freshly-emerged imago. The gizzard itself becomes reduced in size, and often loses all its teeth. Non-functional remnants of the rectal gills of Anisoptera remain over in the imago.

CHAPTER V

THE ALIMENTARY AND EXCRETORY SYSTEMS

As the alimentary and excretory systems of the Dragonfly are closely related, both morphologically and physiologically, we shall deal with them together, in a single chapter.

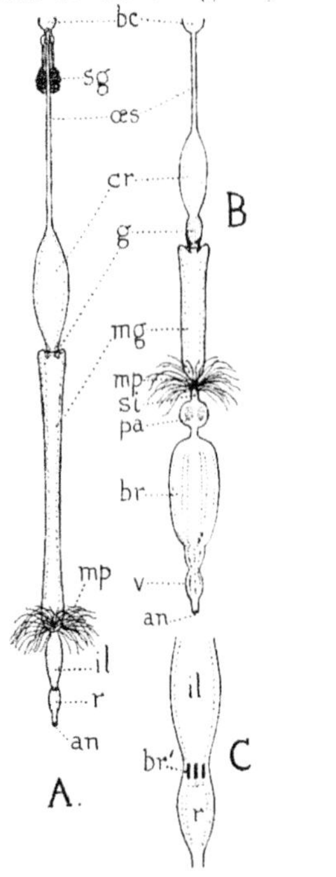

Fig. 44. Alimentary canal of *Aeschna brevistyla* Ramb. A. Imago (natural size). B. Larva (× $1\frac{4}{5}$). C. Hind-gut of imago (× 3), to shew remnants of rectal gills (*br′*). *an* anus; *bc* buccal cavity; *br* branchial basket; *cr* crop; *g* gizzard; *il* ileum; *mg* mid-gut; *mp* Malpighian tubules; *œs* oesophagus; *pa* pre-rectal ampulla; *r* rectum; *sg* salivary glands; *si* short intestine; *v* vestibule. Original.

The Alimentary Canal (fig. 44).

In the Dragonfly, as in all insects, the alimentary canal is formed of three main regions. The most anterior of these (*fore-gut*) is formed as an invagination of the ectoderm in the region of the mouth. Hence its epithelium is continuous with the hypodermis around the mouth, and is covered internally by a fine chitinous lining or *intima*, corresponding and continuous with the cuticle. In like manner, the most posterior region (*hind-gut*) is a similar invagination of the ectoderm, and its epithelium is similarly covered by a fine internal chitinous intima. Separating these two regions lies the *mid-gut* or *mesenteron* ("chylific stomach," "ventriculus") formed from the endoderm. In the Dragonfly it is the mid-gut alone which carries on the processes of digestive secretion and absorption. The fore-gut simply serves to prepare the food for digestion, while the hind-gut is concerned with the removal of the waste left over.

The boundary between fore and mid-gut is marked by the zone of invagination of the gizzard, which projects into the anterior end of the mid-gut (fig. 46). The boundary between the mid and hind-gut is marked by the entry of the Malpighian tubules, which open into the alimentary canal at the extreme anterior end of the hind-gut. The following table shews the principal divisions of the alimentary canal in the larva and imago:

Region	Division	In Larva	In Imago
Fore-gut	Buccal cavity and Pharynx	Undivided, without salivary glands	Undivided, with salivary glands
	Oesophagus		
	Crop		
	Gizzard (or Proventriculus)		
Mid-gut	Annular Zone		
	Digestive Zone		
Hind-gut	Small Intestine or Ileum	Short Intestine; Pre-rectal Ampulla	Undivided
	Rectum; Anus	Branchial Basket[1]; Anal Chamber or Vestibule[1]	Undivided

The relative lengths of these parts in larva and imago differ very greatly, chiefly because of the difference in the lengths of the segments containing them. Thus, for example, the mid-gut, normally contained in abdominal segments 3–6, is the shortest region in Anisopterid larvae, but the longest in all imagines (cf. fig. 44 A and B).

As our present knowledge of the alimentary canal in Dragonflies is principally derived from a study of the larva, and as its structure also presents more points of interest in the larva than in the imago, the following account will be understood to apply chiefly to the larva. The more important differences in the imago are noted at the end of each section.

The Fore-Gut.

1. *Buccal Cavity and Pharynx.*

The mouth-parts, which properly form part of the digestive system, in so far as they prepare the food by chewing, have been already dealt with (chap. II). Within the mouth is a wide undivided cavity, bounded above by the *labrum* and *epipharynx* (p. 14), below by the *labium* and *hypopharynx* (p. 17), and laterally by the jaws. The

[1] Only in Anisopterid larvae. In Zygopterid larvae the rectum is undivided.

hypopharynx stands up in the cavity like a tongue, but appears to be an organ of touch rather than of taste. The epipharynx, on the other hand, is only slightly convex, and does not project appreciably into the cavity. The base of insertion of the hypopharynx lies far back on the labium. Quite close to it, in well-grown larvae and in imagines only, opens the duct of the *salivary glands*.

The Salivary Glands (fig. 45) have been investigated by N. Poletaieff [122, 123] and Bordas [12]. They are two in number, very small, and placed laterally in the prothorax, on either side and close to the oesophagus. Each complete gland consists of a large number of tiny *acini* or grape-shaped glands (fig. 45 D), ranging

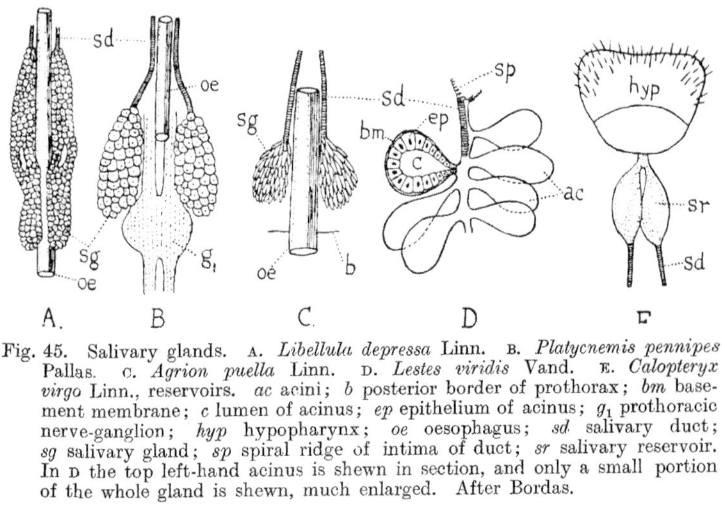

Fig. 45. Salivary glands. A. *Libellula depressa* Linn. B. *Platycnemis pennipes* Pallas. C. *Agrion puella* Linn. D. *Lestes viridis* Vand. E. *Calopteryx virgo* Linn., reservoirs. *ac* acini; *b* posterior border of prothorax; *bm* basement membrane; *c* lumen of acinus; *ep* epithelium of acinus; g_1 prothoracic nerve-ganglion; *hyp* hypopharynx; *oe* oesophagus; *sd* salivary duct; *sg* salivary gland; *sp* spiral ridge of intima of duct; *sr* salivary reservoir. In D the top left-hand acinus is shewn in section, and only a small portion of the whole gland is shewn, much enlarged. After Bordas.

from fifty in the smaller Zygoptera to over one hundred and fifty in the Anisoptera. The acini are usually broadly ovoid or pear-shaped. In *Agrion* (C), however, they are elongated. Each is lined internally by a single layer of secreting cells (*ep*), hexagonal, and resting on an outer basement membrane (*bm*). A delicate chitinous intima (not visible in the figure) lines the lumen of the acinus internally. This is continued into the narrow mouth or duct of the acinus, and beyond into the main *salivary duct* (*sd*), where it becomes raised up into a *spiral ridge* (*sp*) resembling the spiral thread of tracheae. The *salivary ducts* are narrow, and pass forwards, one on either side of the oesophagus, towards the base of the hypopharynx. Just before reaching this, each duct dilates into a *salivary reservoir* (E, *sr*). Finally, the two ducts unite immediately in front of the reservoirs, and enter the pharynx by a single median opening.

The salivary glands of the Dragonfly are peculiar (*a*) for their small size, (*b*) in having the reservoirs situated along the course of the ducts. Where salivation is of more importance, and the glands consequently more highly developed (as in the Cockroach), it is usual for the reservoirs to be developed apart from the ducts of the glands, with special ducts of their own.

The salivary glands are usually confined to the prothorax. In *Lestes*, however, they pass backwards into the mesothorax, and become united across the oesophagus into a single mass.

2. *The Oesophagus* (fig. 45).

The pharynx passes posteriorly into a very narrow tube, the oesophagus. This tube passes upwards and backwards between the brain and the suboesophageal ganglion, as explained on p. 15, and leaves the head via the upper division of the posterior foramen. It then passes back through the microthorax and the whole length of the thorax to the base of the abdomen, which it enters for a short distance before expanding into the crop.

The oesophagus is wrinkled or ridged internally by a number of longitudinal folds. Histologically it consists of the following layers: (*a*) a fairly strong resistant cuticle or *intima*, lining the lumen internally, (*b*) a layer of cubical or slightly flattened *epithelial cells*, which secrete the intima, (*c*) an incomplete internal muscular tunic formed of longitudinal fibres, collected into cords running along the courses of the longitudinal folds, (*d*) a complete and fairly strong external muscular tunic, formed of transverse circular fibres. At certain points along the course of the oesophagus, the circular fibres are strengthened and arranged in a double layer, so as to form a series of weak *sphincters* for contracting the lumen [1].

3. *The Crop.*

This is a large dilatation of the oesophagus, occupying part of the first and second abdominal segments in the larva, but usually the whole of segments 1–2 and part of 3 in the imago. Histologically it is an undifferentiated portion of the oesophagus, with the muscular tunic slightly stronger, and allowing of greater dilatation. Posteriorly it is closed off from the gizzard by a strong sphincter. At metamorphosis, it becomes greatly dilated with air, as described on p. 97. In those Anisoptera in which the abdomen is swollen basally, the crop remains distended with air for a long time, thus allowing the swollen contour of the abdomen to become fixed and hardened.

4. *The Gizzard* (figs. 46, 47).

In the larva, the gizzard lies in the second abdominal segment. It is a fairly wide bottle-shaped chamber, very highly specialized. If the gizzard be slit open and cleaned, there will be seen, arranged towards the posterior end of the organ, a number of raised yellowish longitudinal areas, carrying teeth. These are the *dental folds* or *fields*. In Dragonflies they are nearly always either four, eight or sixteen in number. The number of the folds and the structure of the teeth are points of considerable systematic importance, which have been carefully studied by Ris [131] and Higgins [74].

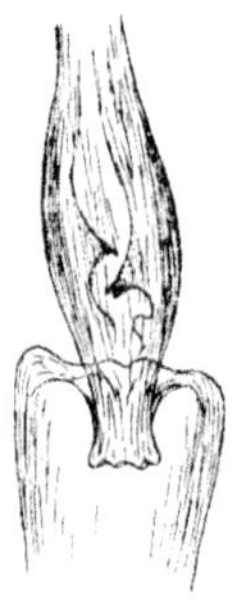

Fig. 46. Invagination of gizzard into mid-gut, in larva of *Cordulephya pygmaea* Selys (× 9). Original, cedar-oil preparation.

The structure of the gizzard may be comprehended by a glance at a transverse section through the region of the teeth in any highly developed type, such as *Cordulephya* (fig. 47 B). We see that the epithelium and cuticle remain weak and undifferentiated between the dental folds. Beneath the teeth, however, the epithelium becomes raised up into a strong ridge of tall columnar cells (*ep*). The cuticle is immensely thickened, and attached to the epithelium in many places by numerous *fibrillae* (*fi*). Just above the columnar epithelium the cuticle is thick, soft and transparent (being apparently newly-formed). Further away, it forms a distinct darkened layer, which passes finally into the strong dark brown external layer of the teeth (*t*, *t′*). The whole outer layer of cuticle, including the teeth, is cast off at each ecdysis, and a fresh layer (represented by the soft inner cuticle in fig. 47 B) takes its place, becoming quickly hardened [145].

The *muscular tunic* of the gizzard is excessively strong, owing to the enormous development of the circular muscle-fibres (*cm*). The longitudinal fibres remain weak, forming only eight or sixteen separate strands (*lm*) placed *externally* to the circular tunic. The latter consists of several layers, greatly increased and strengthened below the dental folds, where the fibres run directly inwards towards the cuticle. By insinuating themselves between the columnar cells,

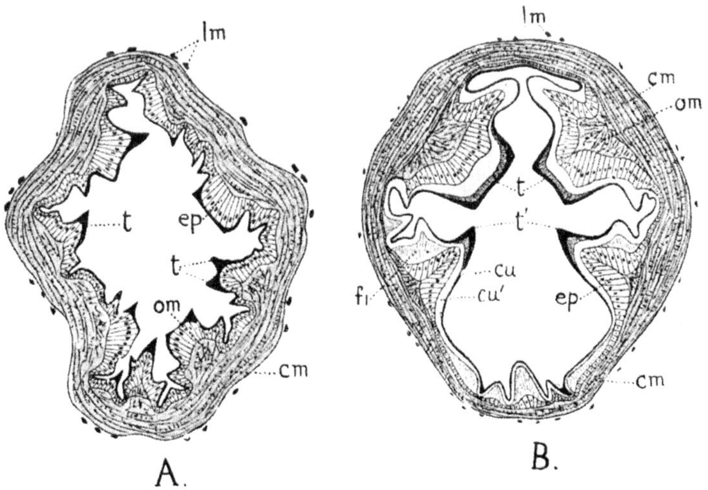

Fig. 47. T.S. through larval gizzard. A. *Synlestes weyersi* Selys (×80). B. *Cordulephya pygmaea* Selys (×48). *cm* circular muscle; *cu* cuticle; *cu'* soft cuticle; *ep* raised epithelium; *fi* fibrillae; *lm* longitudinal muscle; *om* oblique muscle; *t*, *t'* teeth. Original.

these fibres become attached directly to the soft inner portion of the cuticle. Such an arrangement is very exceptional, and points to the powerful nature of the operations of this organ.

For purposes of comparison, a transverse section through the less highly specialized gizzard of *Synlestes* is shewn in fig. 47 A.

The Dentition of the Larval Gizzard (fig. 48).

In the oldest type of gizzard, the dental folds are narrow and elongated, bearing numerous small teeth, all of equal size, and pointing backwards towards the posterior end. The folds are arranged into equal numbers of larger and smaller kinds, placed alternately. These are known as *major* and *minor* folds respectively. Thus the gizzard has a distinct radial symmetry. Sixteen appears to have been the original number, eight being major folds, and eight minor. A reduction to eight, and then to four major folds took place in some forms. In the Anisoptera, four is the regular number, except in the *Petalurinae*, where there are eight. In the Zygoptera, sixteen is the usual number, but there are many cases of reduction to eight.

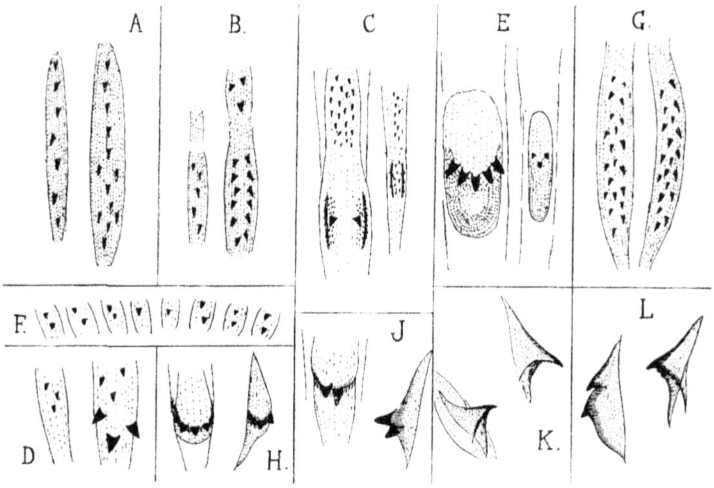

Fig. 48. Dental folds of larval gizzards in Odonata. A. One major and one minor fold from gizzard of *Diphlebia lestoïdes* Selys. B. The same from *Caliagrion billinghursti* Martin. C. The same from *Ischnura heterosticta* Burm. D. The same from *Synlestes weyersi* Selys. E. The same from *Austrolestes analis* Ramb. F. Complete dentition of *Petalura gigantea* Leach. G. Two folds from *Hemigomphus heteroclitus* Selys. H. The same from *Austroaeschna multipunctata* Martin. J. The same from *Anax papuensis* Burm. K. The same from *Cordulegaster annulatus* Latr. L. The same from *Cordulephya pygmaea* Selys. K. after Ris; the rest original, chitin preparations.

Many beautiful specializations of the teeth may be briefly noted. In the Zygoptera, the archaic sixteen-fold gizzard is found (with occasional reductions to fourteen or less, and in a few cases to eight) throughout the *Calopterygidae* and all the *Agrionidae*, except the *Platycneminae* and *Agrioninae*, where the minor folds become excessively reduced, while the major folds develop the specialized dentition shewn in fig. 48 C. In the *Lestinae* there are only eight folds, four major and four minor, the former having a specialized dentition of three or five very large teeth, with very numerous smaller ones (E).

Amongst the Anisoptera, the *Petalurinae* (F) have a very reduced archaic dentition of eight folds, each carrying only from one to six small teeth. The *Gomphinae* (G) have four large (major) folds, of the original elongated form, bearing numerous undifferentiated teeth. In the *Aeschninae* (H, J) there are four large folds, each of which has become shortened, and carries a few strong teeth close together. The tendency in this subfamily is towards the accumulation of a few teeth on a raised central area, forming a kind of molar. There is also a *slight* difference of level between the two folds on one side of the gizzard and those on the other, the one pair lying more anteriorly.

In the *Cordulegastrinae* (K) there are four folds, two on one side lying considerably more anteriorly than the other two. A single large tooth, in the form of a pyramid on a rather flattened triangular base, stands up from each fold. The apex of the tooth is very sharp, and somewhat curved. The two posterior edges are strongly denticulated. This form of gizzard is, of course, no longer radially symmetrical, but exhibits a distinct *bilateral symmetry*, due to the difference of level in the two pairs of folds. The gizzard of the *Libellulidae* (L) is essentially of the same type, but the more posterior pair of folds has become somewhat altered. The denticulation of the posterior ridges is lost, the apical tooth is strong and blunt, and a second sharper tooth has appeared on the anterior edge, not far from the summit.

In the imago, the gizzard is a comparatively small, weak structure, apparently merely a survival of the larval gizzard in most forms, without any real functional use. The dentition is either completely lost, the folds being simply areas of slightly thickened chitin roughened with numerous fine points, or else it is reduced to a simpler form. The reduction is least in *Calopterygidae* and *Agrionidae*. In the *Lestidae* and the Anisoptera, the dentition is quite lost in the imago.

Using Higgins' notation [74], a major fold is denoted by F, a minor fold by f, followed by the number of teeth on the fold. Where the teeth are of two kinds, a $'$ denotes larger, a $''$ smaller teeth. Very numerous teeth are denoted by n. If each fold is divided into two distinct tooth-bearing regions, a fractional device is used, the numerator denoting the more anterior region, the denominator the more posterior. The number of folds is denoted by a multiplying factor. We can now exhibit the characteristics of the gizzards in the various groups, by means of the following table, placing the Zygoptera[1] first:

Group	Symmetry	No. of folds	Shape of folds	Dental Formula
ZYGOPTERA				
1. *Calopterygidae* + *Agrionidae* (except 2)	Radial	16 (8 *F*, 8 *f*)	Elongated	8 (*F* 7—20, *f* 3—10)
2. *Platycneminae* + *Agrioninae*	Radial	16 (8 *F*, 8 *f*)	Elongated	$8\left(F\,\frac{10\text{—}20}{2'+n''},\ f\,\frac{2\text{—}10}{n''}\right)$
3. *Lestinae*	Radial	8 (4 *F*, 4 *f*)	Dental area reduced	4 (*F* 3′—5′ + *n*″, *f* 0′—1′ + *n*″)
4. *Synlestinae*	Radial	8 (4 *F*, 4 *f*)	Much reduced	4 (*F* 3′—8′ + 0″—1″, *f* 2″—4″)
ANISOPTERA				
5. *Petalurinae*	Radial	8	Much reduced	8 (*f* 1—6)
6. *Gomphinae*	Radial	4 (*F*)	Elongated	4 (*Fn*)
7. *Aeschninae*	Slightly bilateral	4 (*F*)	Of molar form	(2 + 2) (*F* 2—8)
8. *Cordulegastrinae*	Bilateral	4 (*F*)	Large pyramids	(2 + 2) (*F* 1′ + *n*″)
9. *Libellulidae*	Bilateral	4 (*F*)	Large pyramids	2 (*F* 1′ + *n*″) + 2 (*F* 2′)

N.B. In 7–9, bilateral symmetry is denoted by separation of 4 into 2 + 2. In 8–9 *n*″ indicates denticulation of the posterior edges of the pyramid.

Function of the Gizzard. In the less specialized forms, the gizzard appears to act merely as a strainer, and not as a triturator of the food. The great development of the teeth and muscles in the higher forms points to the definite assumption of a triturating function, though this has been denied by some authors. The adoption of bilateral symmetry, whereby one pair of folds could work against one another at a different level, and without interference from the second pair, is strong evidence of a triturating function in these forms. The "molars" of *Aeschninae* in particular seem to be definitely formed for trituration, but of very little use for straining. The presence of a mass of elastic connective tissue beneath these "molars," as also beneath the pyramidal teeth of *Libellulidae*, is strong evidence in the same direction. We might add that the

[1] There are a number of exceptions to the general type in this suborder. Thus, *Hetaerina* has only 8 folds; *Caliagrion* has 8 folds with a number of strong teeth only; while a number of cases shew reduction from 16 to 14 folds or less. *Pseudophaea* has 18 folds.

gizzard is probably also of use in compressing the triturated particles into a form suitable for passing into the mid-gut.

Posteriorly, the gizzard narrows to a short portion which, histologically, closely resembles a part of the oesophagus. This is deeply sunk into the anterior end of the mid-gut, the arrangement suggesting the top of a "safety" ink-pot. It is known as the *oesophageal valve.*

The Mid-gut (fig. 49).

The *mid-gut* of the Dragonfly, both in the larva and imago, is exceptional in possessing no caeca. It is very clearly marked off from the gizzard above, and from the small intestine below, the change in the character of the epithelial cells taking place quite suddenly along a definite transverse plane. The following description applies to the structure and functions of the mid-gut in the larva [98, 187].

The mid-gut is generally of a distinct yellowish colour, owing to the presence of yellow pigment granules in its epithelial cells. Anteriorly there is a short *annular zone*, which takes no part in the process of digestion. The cells of this region remain normal (i.e. they resemble the cells of the other part of the mid-gut during the period of rest). Their function is simply to elaborate the anterior end of the peritrophic membrane (see below).

The rest of the mid-gut is made up of the following layers:

(*a*) A layer of mostly large, columnar *epithelial cells*, bounding the internal cavity. These cells have large spheroidal nuclei (A, *nu*). The yellow pigment granules (*yp*), already mentioned, occur towards their inner or free border. This free border ends in a peculiar delicate lamina (*ib*), destined to form part of the peritrophic membrane. There is no chitinous intima. Externally, the epithelial cells rest on a fine, delicate, but distinct *basement membrane* (*bm*).

(*b*) A weak tunic of circular muscle-fibres (*cm*), external to the basement membrane.

(*c*) A still weaker incomplete tunic of longitudinal muscle-fibres (*lm*), forming a loose investment external to (*b*).

In order to understand the very remarkable physiology of digestion in the Dragonfly larva, and the extraordinary changes undergone by the cells of the mid-gut during the process, it must be borne in mind that the normal state (A) represents a *state of fasting*. This is broken at irregular intervals by large and rapid meals, when the larva is fortunate enough to secure a victim. The larva can fast for long intervals without any danger of starvation[1]. In fact, as will presently be seen, the longer the larva fasts, the bigger becomes its capacity for dealing with the next meal when it does eventuate.

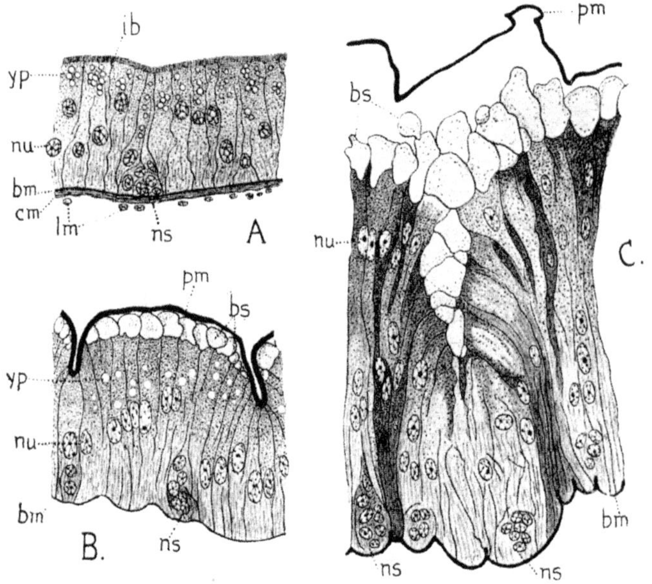

Fig. 49. Part of epithelium of mid-gut in larva of *Aeschna* sp. (× 330). A. During period of rest. B. During period of preparation. C. During period of activity. *bm* basement membrane; *bs* balls of secretion; *cm* circular muscle; *ib* internal lamina of epithelium; *lm* longitudinal muscle; *ns* nests of developing cells; *nu* nucleus of epithelial cell; *pm* peritrophic membrane; *yp* yellow pigment granules. After Voinov.

We may divide the process of digestion into three periods, (i) the normal period, or period of rest, (ii) the period of preparation, and (iii) the period of activity.

(i) *The period of rest.* Fig. 49 A shews the appearance of the cells of the mid-gut, after the completion of digestion of a heavy meal. We have already described this state. We shall only add that, here and there, just within the basement membrane, there

[1] I have kept larvae of *Synthemis* alive for three months without food.

can be seen small nests (*ns*) of developing cells, destined to take the places of those destroyed during the period of activity. These nests were for a long time regarded as being of a glandular nature, and were termed "crypts."

(ii) *The period of preparation* (B). After a sufficient period of rest, the larva begins to prepare for another meal. This is marked by the beginnings of a distinct swelling-up of the epithelial cells, owing to the elaboration and storage of digestive ferments within them. The cells become distinctly more elongated, their nuclei sharing in this process. The digestive products collect at the internal or free ends of the cells, in the form of "balls of secretion" (*bs*). These push the internal border away from the rest of the cell. This border is then seen to consist of a fine continuous membrane, perforated by numerous very minute pores. As the balls of secretion increase, this membrane is pushed quite clear of the cells. It is then called the *peritrophic membrane* (*pm*). Its function is to envelope the food masses received from the gizzard, so as to prevent injury to the delicate epithelium of the mid-gut. If the arrival of the meal is delayed, elaboration of the digestive ferments continues unabated, so that, in the case of a larva that has been fasting for two or three weeks, the mid-gut may be swollen to three or four times its original size, as shewn in C.

(iii) *The period of activity* (C). When the food reaches the mid-gut, the epithelial cells are discharged. Some are thrown off whole, with nucleus and contents complete. Others burst and cast out huge balls of secretion, some of which contain only liquid ferments, while others contain the yellow globules originally present in the cells. Thus the yellow colour of the mid-gut is lost after a heavy meal. The cells which are discharged whole mix with the food and are lost altogether. Those which only burst retain their nuclei, and are capable of returning to the quiescent stage, and of reforming their contents again and again, until finally they too become thrown off complete. The digestive ferments may be forced up to meet the food while it is still in the gizzard.

Regeneration of the Epithelium. The loss of so many cells during the period of activity is quickly repaired on subsidence by the activity of the nests. Their nuclei and cytoplasm divide rapidly, forming numerous cells which move outwards from a common centre, taking the place of the lost cells. These new cells quickly develop a new internal lamina to take the place of the lost peritrophic membrane. The latter surrounds the food-mass, and remains around the faecal pellets after all the nourishment has been extracted.

The Process of Absorption. Voinov [187] shewed conclusively, not only that it is the cells of the mid-gut alone which produce the digestive ferments, but that these same cells can and do carry on the process of absorption of the resulting products at one and the same time. The cells of the mid-gut mature in longitudinal bands separated by nests of younger cells. By feeding the larva with methylen blue, the mid-gut was made to shew blue longitudinal stripes along the lines of the mature cells, with clear areas in between. By use of the same reagent, Voinov also shewed that, if the blood becomes overloaded with nourishment, the cells of the mid-gut *can absorb nourishment from without*, and pass it back into the lumen for temporary storage. It may be inferred that they can deal similarly with impurities in the blood, if the latter becomes surcharged with them. Voinov also shewed, by a clever use of various colouring matters, that the cells of the mid-gut can exercise the power of choice, absorbing certain kinds and rejecting others.

The Peritrophic Membrane. This peculiar formation has given rise to much discussion. According to Voinov, the cytoplasm of each separate cell elaborates an independent internal filament. When secretion begins, these filaments fuse into a single complete lamina, which is cast off, as already described, in the form of the *peritrophic membrane.* The cells then proceed to form a new set of filaments. Thus we see that the food pellets of the larva are only at first slightly shorter than the length of the mid-gut, each being enclosed in a complete peritrophic membrane, as the result of a single act of digestion. After the extraction of nourishment, a considerable contraction, of course, sets in, so that the ejected faecal pellets are considerably shorter.

It should be added that other observers deny the formation of the peritrophic membrane in the manner described by Voinov, and hold that it is formed entirely from the anterior annular zone, in the form of a gradually descending sleeve.

Mid-gut of the Imago. This differs from the mid-gut of the larva in being exceedingly long (fig. 44) and in varying, during the process of digestion, in length rather than in width. Thus it may extend sometimes only from the fourth segment to the sixth, while at other times it is somewhat wider, and reaches from the beginning of the second to the end of the seventh segment. It is always widest at its anterior end, where it receives the invaginated gizzard. For the rest of its length it is practically cylindrical. The physiology of digestion has not been closely studied, as in the larva. We know, however, that the imago feeds more often and in smaller quantities than the larva does. Hence we may conclude that the processes of digestion are not marked by such violent changes as in the larva, though they are of the same type. The peritrophic membrane is evidently not thrown off along the whole length of the mid-gut after each meal, since the faecal pellets are comparatively short and numerous.

The Hind-gut.

1. *Small Intestine or Ileum.*

The mid-gut contracts sharply at its posterior end, to open by a narrow orifice into the anterior portion of the hind-gut, known as the *small intestine.* The extreme posterior end is closed by a strong sphincter muscle, which effectively prevents the passage of any impurities from the hind-gut back into the mid-gut. Immediately below this muscle lies the *zone of entry of the Malpighian tubules* (p. 117). These tubes do not enter separately, but unite into bundles of five or six, forming a common conduit. These conduits become excessively narrowed, and open into the intestine by a ring of minute pores. Below this zone, another sphincter muscle is strongly developed.

In the larva, the anterior portion of the small intestine is narrow, and may be called the *short intestine.* The posterior part is dilated into a spherical bulb, called the *pre-rectal ampulla* [145]. The differences between these two portions appear to be only those of degree, and not of kind. In each case we note the return, for the most part, to the familiar formation already described for the fore-gut, and consisting of the usual four layers of cuticle, flat epithelium, circular muscle tunic and weaker longitudinal muscle tunic, from within outwards. The latter tunic is arranged in six bands. In the short intestine, however, nearly the whole of the ventral half of the epithelium is swollen up into an immense pad. In the pre-rectal ampulla, the same pad is present, but is divided into two parts projecting latero-ventrally into the cavity of the ampulla, one on either side (fig. 44, *pa*) and separated ventrally by a wide zone of ordinary flattened cells. These pads are easily seen from without in the intestine of the dissected larva. They are formed of high columnar cells, remarkable in having their nuclei all placed close under the cuticle, at the inner ends of the cells (fig. 50). Strong tracheal branches penetrate into these pads on either side.

The function of the pads is not known for certain. In the pre-rectal ampulla they are extremely mobile, and can be protruded into the cavity, or retracted from it, by strong muscular action.

In the imago, the small intestine (fig. 44 A, *il*) forms a single complete whole, intermediate in form between the short intestine

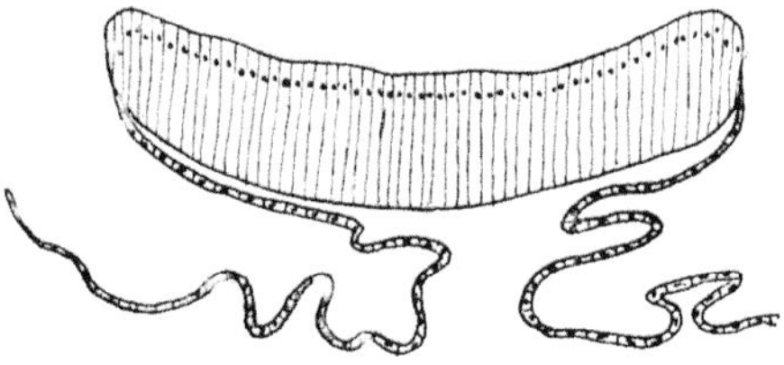

Fig. 50. Lateral sagittal section through epithelium of one of the pads of the pre-rectal ampulla, from larva of *Libellula* (× 50). After Sadones.

and the pre-rectal ampulla of the larva. It is of moderate length, and shews the six longitudinal muscle-bands very clearly developed. The ventral pad appears to be divided into two portions, not so far apart as in the pre-rectal ampulla.

The small intestine ends posteriorly with a strong sphincter, which is capable of closing it off entirely from the rectum.

2. *The Rectum.*

This portion of the hind-gut is of very small extent in the imago, occupying less than the last three segments of the abdomen, and is practically of the same structure throughout. In the larva it is considerably longer, extending from the seventh segment backwards.

It is in the larvae of Anisoptera that the rectum becomes highly specialized, and it is in these also that our knowledge of this part of the hind-gut is most complete. In this suborder the larvae breathe by means of internal tracheal gills, developed in the anterior portion of the rectum, which is greatly dilated to form a *branchial basket* (fig. 44 B) of great complexity. As this organ belongs physiologically to the respiratory system, we have dealt with it fully in chap. IX. We shall only remark here that the gills are projections of the wall of the rectum, involving the cuticle or intima, which covers them completely, and the epithelial cells, which become specialized for their respiratory function by fusion into a complete *syncytium*. An enormous number of tracheal capillaries ramify in these gills. The circular muscular tunic remains complete, except for penetration of it by tracheae. The external longitudinal muscles are arranged in six bands, as in all other portions of the hind-gut. Reduced remnants of the gills persist in the imago (fig. 44, *br*), but the basal pads (p. 188) are cast off [109].

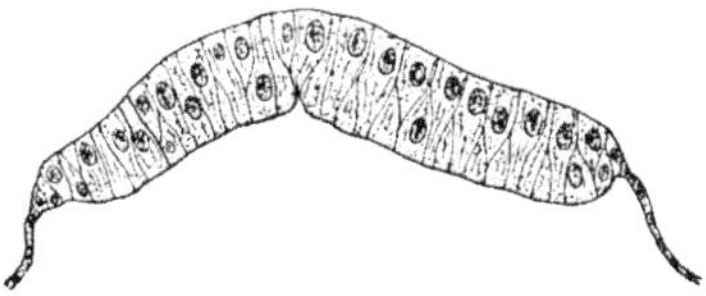

Fig. 51. T.S. through one of the three rectal pads, from larva of *Aeschna brevistyla* Ramb. (×210). Original.

The hinder portion of the rectum, consisting of about one-third or less of its entire length, remains undifferentiated, and exactly resembles the same portion in the Zygopterid larvae. This part, in the Anisopterid larvae, may be termed the *vestibule*, or *stercoral chamber* (fig. 44 B). In the *Aeschnidae* the branchial basket narrows posteriorly, and passes insensibly into the vestibule. In the *Libellulidae*, a strong sphincter muscle definitely separates the two. A posterior sphincter, in both cases, closes the vestibule just in front of the anus.

The epithelium of the vestibule carries anteriorly three longitudinal pads, formed usually of high columnar cells (fig. 51). These pads have for long been known as the *rectal glands*. As they are obviously not of a glandular nature, we follow Sadones [145] in rejecting this term, and shall adopt the term *rectal pads*. We know nothing certain of their functions, any more than we do concerning the pads in the small intestine.

In the larvae of Zygoptera, the rectum carries, throughout its entire length, three longitudinal pads of high columnar cells, closely resembling those found in the vestibule of *Aeschna* (fig. 51). One of these lies mid-dorsally, the other two latero-ventrally. Anteriorly, they are separated by short portions of flattened pigmented epithelium. Posteriorly, these pigmented portions become larger, and project as folds into the rectal cavity, being supported beneath by a mass of cells derived from the fat-body, and carrying tracheae. Folds closely similar to these separate the rectal pads in the vestibule of *Aeschna*.

In the larva of *Calopteryx*, Dufour and Hagen both recognized by dissection, long ago, the presence of three longitudinal folds, which they considered to be gills. In sections of the rectum of this larva, these folds are seen to be the three pads described above, but somewhat specialized in form. The epithelium is only about half as high as that shewn in fig. 51, though otherwise closely similar to it. The pads are thrown out into the rectal cavity in the form of three long double folds or loops, which overlap one another at their free ends. Calvert has recently discovered very similar folds in the rectum of *Thaumatoneura*.

It would appear that Zygopterid larvae do not rely wholly upon their caudal gills for respiration, but use the rectum as an auxiliary respiratory chamber. The rectal pads or folds are not, however, true *tracheal* gills, and must be carefully distinguished from the highly specialized structures found in the rectum of Anisopterid larvae, since in no case are they known to possess any *specially developed* tracheal supply. Their action must be rather that of *blood-gills*, liberating carbonic acid gas from the blood into the rectal water supply, by diffusion, and receiving air in exchange.

In the imagines of all Dragonflies which I have so far examined, the rectum contains *six* large longitudinal pads, one being mid-dorsal, one mid-ventral, two latero-dorsal, and two latero-ventral. These are homologous with the so-called "rectal glands" of other insects. Their epithelium closely resembles that seen in fig. 51. Curiously enough, the three pads which are not developed in the larva (viz. the mid-ventral and the two latero-dorsals) tend in the imago to be more strongly developed than the other three. There is also a tendency for each pad to become constricted into two parts, mid-longitudinally, so that the rectum may shew, apparently, in transverse section, *twelve* pads closely pressed together.

The Anal Region. The *anus* is closed exteriorly by three small valves, representing the reduced parts of a twelfth segment, and known as the *laminae anales* (p. 32). Internally, the anal end of the hind-gut is furnished with six radiating *dilatator muscles*, inserted between the posterior ends of the rectal pads. Besides this, when a separate vestibule is present, the anus may be closed off by strengthening of the circular muscle tunic into a sphincter.

The Excretory System.

1. *The Malpighian tubules* (fig. 52).

These are the principal organs of excretion in all insects. In the imago of the Dragonfly, as well as in all except the first few larval instars, they are very numerous, ranging from fifty to seventy in number. The newly hatched larva has only three; but, at each instar, an increase in the number takes place, so that they soon become numerous.

The structure of a Malpighian tubule is very simple. Each is a slender, very elongated tube, of about 0·05 mm. diameter. It lies freely in the haemocoele, and is thus bathed by the blood on all sides. The free end is quite closed. A small trachea accompanies each tube, and supplies its cells with oxygen, by ramifications over its outer surface. A delicate *peritoneal investment* (p) forms the outer sheath of the tube. Next to this fine cellular layer, there is an exceedingly delicate *basement membrane* (bm), very difficult to detect. Within this are arranged the excretory cells (ex), of somewhat irregular polygonal form, bounding a narrow central lumen or canal. The cells are large, with conspicuous nuclei (nu_1). In transverse

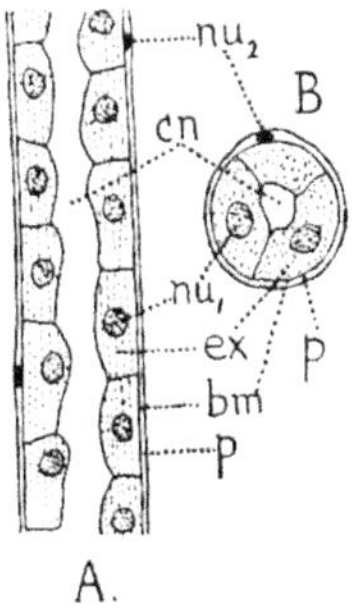

Fig. 52. Malpighian tubule from *Ischnura aurora* Br., imago (× 167). A. Longitudinal section. B. Transverse section. bm basement membrane; cn canal; ex excretory cells; nu_1 nucleus of same; p peritoneal investment; nu_2 nucleus of same. Original.

sections, usually only from three to five cells are met with in the case of the Dragonfly, though six or more appears to be the usual number in most insects. The fine chitinous intima, which bounds the lumen in some insects, has so far not been seen to exist in the Dragonfly.

2. *Physiology of Excretion.*

It is the function of the Malpighian tubules to extract nitrogenous and other waste products from the blood, in the form of uric acid and other compounds. Griffiths [59] has conclusively shewn this in the case of the Dragonfly, both by obtaining the crystals direct from the tubules, and also by the more delicate murexide test. It is supposed that the crystals are got rid of simply by being passed down the tubules into the hind-gut. The passage even of very minute crystals down the irregular canals, in the case of the Dragonfly, would be a matter of considerable difficulty.

The fat-body plays a secondary part in the physiology of excretion, by reason of its general function as a store-house for excess products. When the blood is so overloaded with impurities that the Malpighian tubules cannot adequately deal with them, the fat-body comes to the rescue by storing up the excess in certain of its cells. These impurities are later on given out again, when the tubules are able to deal with them.

Voinov [187] has shewn that the cell-wall of the heart can remove foreign matter from the blood, and pass it to the haemocoele. Such an action does not really rid the blood of impurities, but only serves to clear the passage of the heart.

The Fat-body.

The fat-body is more conspicuous in the larva than in the imago. In neither, however, does it form so large a mass as in the majority of insects. It is a lobed mass of a whitish or yellowish colour, forming a kind of sheath or envelope for the various internal organs. It is especially noticeable in the region of the heart, the alimentary canal, the large tracheal trunks, and more particularly the gonads.

The fat-body of the Dragonfly appears to consist of only two kinds of cells, known as *adipose cells* and *pericardial cells.* The uric-acid cells and bactericidal cells found in Orthoptera are absent, or at any rate have not yet been noticed.

The *adipose cells* are very numerous, forming almost the whole bulk of the fat-body. They are of the usual type, closely packed together, so as to assume an irregular polygonal form, but appearing round and distended wherever their surfaces are free. The nucleus is large and rounded, and the cell itself filled to overflowing with a store of fatty matter, held between the meshes of a fine network of cytoplasm. In all sections previously treated with soap-alcohol, this fat is completely extracted by the reagent, leaving a very characteristic structure, the "skeleton" of the fat-cells (fig. 53, *fb*).

The *pericardial cells* (fig. 53, *pc*) are large oval or rounded cells of a brownish or yellowish colour, and frequently having two large nuclei. They stain darkly, and are not affected by soap-alcohol. They are collected in two small bands along each abdominal segment, close together and just below the heart. Each band contains only a few cells, and not more than three or four are cut in any given section. They appear to be supported by two loops of connective tissue, which partly close them off into two parallel channels beneath the heart. They frequently contain brown or yellow excrement received from the blood, which collects in small crystals, usually towards the ends of the cells.

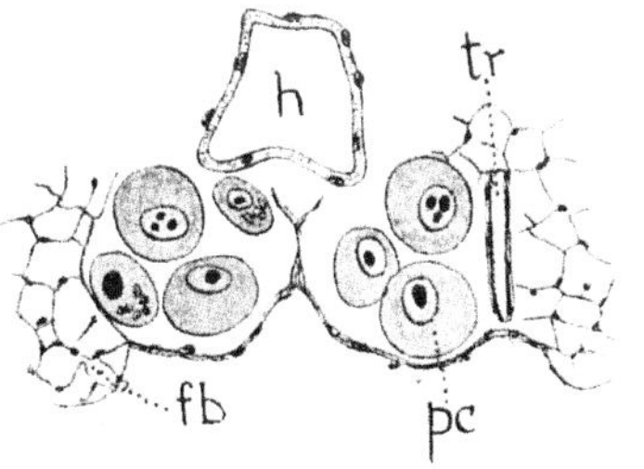

Fig. 53. T.S. through region of heart in fifth abdominal segment of half-grown larva of *Synlestes weyersi* Selys (× 160). *fb* fat-body; *h* heart; *pc* pericardial cells; *tr* trachea. Original.

The fat-body has been shewn to arise from the mesoderm, by differentiation of cells in the walls of the coelomic sacs. When the haemocoele develops, the fat-body forms a lining to it. By proliferation of the cells in certain regions, this lining may come to be several layers of cells thick. Thus the fat-body is everywhere in direct communication with the blood stream, from which it receives the nutritive products of digestion for storage, as well as any excess of impurities which the excretory system may not be able to deal with for the time being. It is therefore an exceedingly important adjunct, both to the alimentary and to the excretory system.

CHAPTER VI

THE NERVOUS SYSTEM

The Nervous System of the Dragonfly (fig. 54) consists of the following principal parts:

(1) The *Brain* or *Supraoesophageal Ganglion*, situated in the head, above the oesophagus.

(2) The *Suboesophageal Ganglion*, also situated in the head, but below the oesophagus, and connected with the brain by a pair of *circumoesophageal commissures*.

(3) The *Ventral Nerve-Cord*, running ventrally through the thorax and abdomen, and swollen into paired ganglia at intervals.

(4) The *Sympathetic Nervous System*, innervating a portion of

---

TABLE OF PARTS OF THE CENTRAL NERVOUS SYSTEM.

Segment	Ganglion	Parts Innervated	Remarks
HEAD			
1st or Ophthalmic	Protocerebrum	Compound Eyes and Lateral Ocelli	= Brain
2nd or Antennary	Deutocerebrum	Antennae and Median Ocellus	
3rd or Intercalary	Tritocerebrum	[gives off Sympathetic System]	
4th or Mandibular	Sub-oesophageal	Mandibles	
5th or Maxillary		1st Maxillae	
6th or Labial		Labium	
THORAX			
1st or Prothoracic	Prothoracic	Fore legs	
2nd or Mesothoracic	Mesothoracic	Middle legs, fore-wings	
3rd or Metathoracic	Metathoracic	Hind legs, hind-wings	Fused in imago
ABDOMEN			
1st	1st abdominal	Segmental muscles, dorsal vessel, reproductive organs (from 8th), etc.	
2nd	2nd ,,		Moves up into seg. 1 in imago
3rd	3rd ,,		
4th	4th ,,		
5th	5th ,,		
6th	6th ,,		
7th	7th ,,		
8th to 10th	8th ,,		Innervates segs. 8–10 and reproductive organs

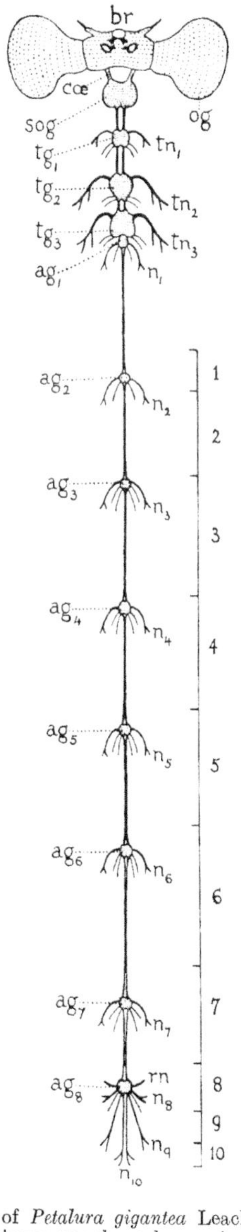

Fig. 54. Nervous system of *Petalura gigantea* Leach ($\times 2$). ag_1–ag_8 abdominal ganglia; *br* brain; *cœ* circumoesophageal commissure; n_1–n_{10} segmental nerves of abdomen; *og* optic ganglion; *rn* nerve to reproductive organs; *sog* suboesophageal ganglion; tg_1–tg_3 thoracic ganglia; tn_1–tn_3 segmental nerves of thorax; *1–10* limits of abdominal segments. Original, nitric acid preparation.

the mouth and oesophagus, and connected with the brain at two points.

The first three of these are often classed together as the *central nervous system.*

In the imaginal Dragonfly, the ventral nerve cord apparently contains only ten pairs of ganglia, though in the larva we can count eleven. This is due to the fact that the first abdominal ganglion moves up into the thorax, and fuses with the metathoracic ganglion posteriorly. The ninth and tenth abdominal segments contain no ganglia; but the ganglion of the eighth segment is large, and innervates the last three segments.

The table on p. 120 shews the details of the Central Nervous System in a convenient form.

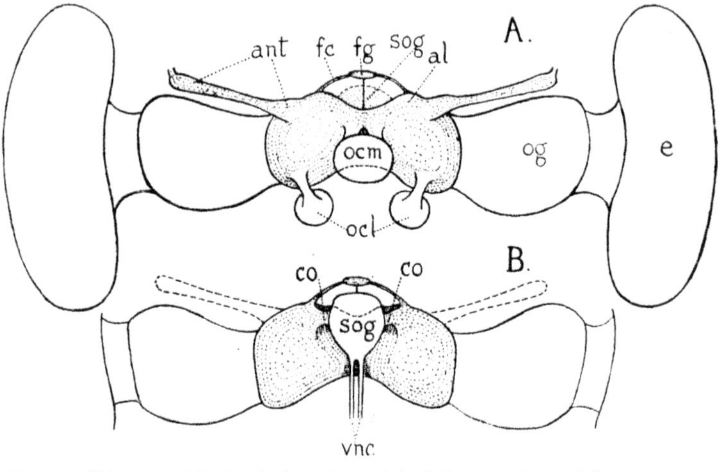

Fig. 55. Diagram of brain of *Austrolestes leda* Selys, constructed from transverse sections (×20). A. Dorsal view. B. Ventral view. *al* antennary lobe; *ant* antennary nerve; *co* circumoesophageal commissure; *e* eye; *fc* frontal commissure; *fg* frontal ganglion; *ocl* lateral ocelli; *ocm* median ocellus; *og* optic ganglion; *sog* suboesophageal ganglion; *vnc* ventral nerve-cord. Original.

The Brain (figs. 55, 56, 57).

The brain is a large mass of nervous matter situated in the head, above the oesophagus, and extending transversely across the whole space between the compound eyes. Its exact position with regard to the head-skeleton has already been explained

(p. 15). The three parts composing it are (1) the *protocerebrum*, or ganglion of the first head segment, which forms the greater portion of the whole brain, (2) the *deutocerebrum*, or ganglion of the second head segment, which gives off nerves to the antennae, (3) the *tritocerebrum*, or ganglion of the third or intercalary segment, from which the sympathetic nervous system takes its rise. These three parts are closely fused together, so that no definite boundaries can be made out.

As the *optic ganglia*, which are connected with the protocerebrum by means of the optic nerves, are exceedingly highly developed in the Dragonfly (exceeding both in bulk and complexity the whole of the true brain or *cerebrum*), we shall deal with them in a separate section.

In order to grasp the relative positions of the three principal parts of the brain, we must recall what we have already said about the changes in the position of the parts of the head. When the mouth became anterior in position (p. 10) the originally ventral portions of the second and third segments (i.e. the clypeus and labrum) became pushed up dorsally above it. In like manner, the deutocerebrum and tritocerebrum have become pushed up above the oesophagus, and have come to lie *in front of* the protocerebrum. Also, just as the clypeus lies above the labrum, so the deutocerebrum lies above the tritocerebrum. In fact, the latter does not really lie *above* the oesophagus in the Dragonfly, but partly on either side of it and partly beneath it (fig. 56 B).

Fig. 55 shews a diagram of the brain of *Austrolestes leda*, constructed from transverse sections through the head of that species. These were 200 in number, and 10 μ in thickness. The numbers given for the seven separate sections shewn in fig. 56 refer to these sections, so that the reader will be able to place them, mentally, at their correct respective levels.

The Protocerebrum. This consists of two large lateral lobes, the *procerebral lobes* (*pl*), formed of an immense number of ganglion cells (the *cortex*) enclosing a mass of nerve-fibres (the *medulla*). The cortical cells are most abundant dorsally, and are of the usual unipolar form, with large nuclei. The two lobes are connected across the middle line by two commissures, (i) the *central body* (*cb*) and (ii) the *pons* or *bridge of the procerebral lobes* (*br*). The central body in the Dragonfly's brain is small, and forms merely a thick mass of transverse nerve-fibres enclosing a few small groups of

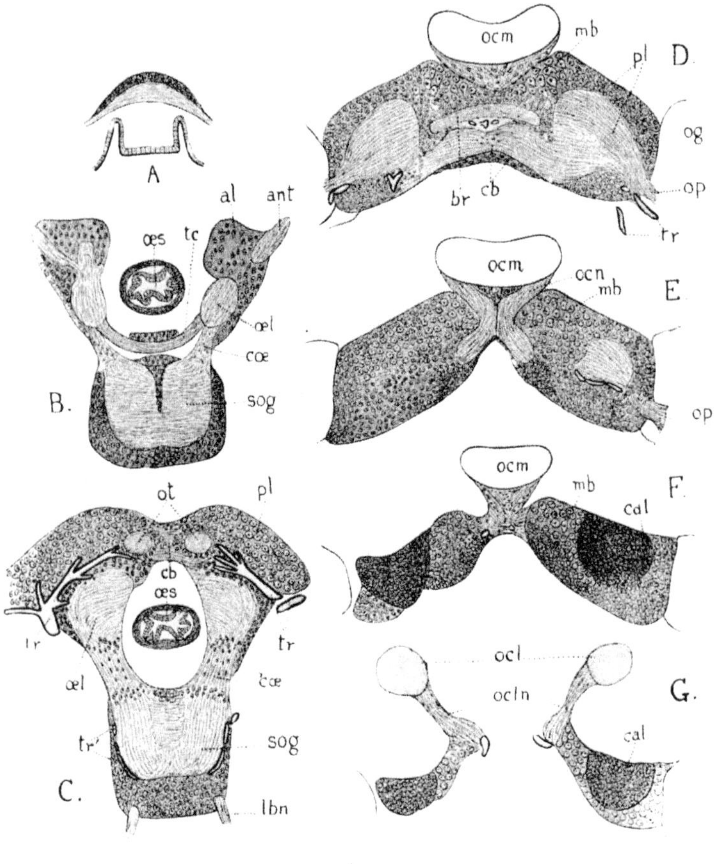

Fig. 56. T.S. through region of brain of *Austrolestes leda* Selys (× 41). The numbers give the order, from before backwards, out of a total of 200 sections of the whole head. A (No. 88). Frontal ganglion. B (No. 103). Deuto- and trito-cerebrum. C (No. 108). Protocerebrum (anterior part) and suboesophageal ganglion. D (No. 127). Protocerebrum, median ocellus and optic nerve. E (No. 130). The same, further back. F (No. 135). The same, with calyx (*cal*). G (No. 138). The same, with nerves to lateral ocelli. *br* bridge of procerebral lobes; *cal* calyx; *cb* central body; *lbn* labial nerve; *mb* giant cells of mushroom body; *ocl* lateral ocellus; *ocln* nerves to same; *ocm* median ocellus; *ocn* nerves to same; *oel* oesophageal lobe; *oes* oesophagus; *op* optic nerve; *ot* optic tubercle; *pl* procerebral lobe; *tc* commissure of the sub-oesophageal ring; *tr*, *tr'* tracheae. Other references as in fig. 55. Original.

nerve-cells. The pons is a more slender commissure passing in the form of a flattened arch above the central body, and separated from it by an area carrying numerous tracheae.

The large and deeply pigmented *cerebral trachea* (fig. 56 C, *tr*) ramifies in a fan-like manner upon and into the cortex of the brain, penetrating chiefly along the plane which appears to mark the original boundary between proto- and tritocerebrum.

The most conspicuous formation in the procerebral lobes of the Dragonfly is the much discussed pair of bodies known as the *stalked bodies*, *mushroom bodies*, or *corpora fungiformia*. Although these do not shew in the Dragonfly the very peculiar and specialized development which they attain in some other insects (e.g. the social Hymenoptera), yet they are very strongly developed,

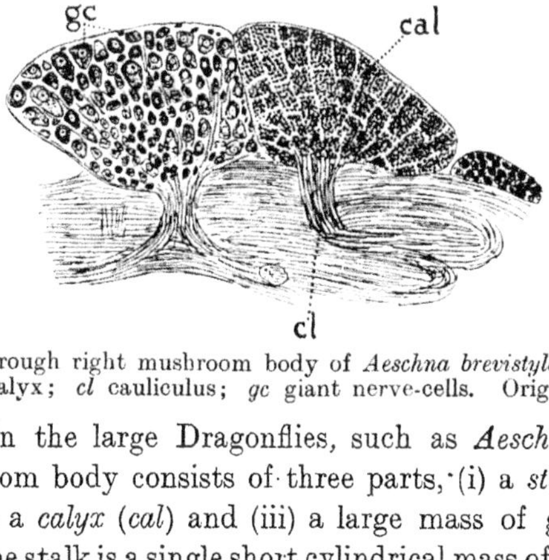

Fig. 57. T.S. through right mushroom body of *Aeschna brevistyla* Ramb. (×60) *cal* calyx; *cl* cauliculus; *gc* giant nerve-cells. Original.

particularly in the large Dragonflies, such as *Aeschna* (fig. 57). Each mushroom body consists of three parts, (i) a *stalk* or *cauliculus* (*cl*), (ii) a *calyx* (*cal*) and (iii) a large mass of giant nerve-cells (*gc*). The stalk is a single short cylindrical mass of nerve-fibres arising vertically upwards from the medulla of the procerebral lobe, and terminating in the calyx. The calyx is a large undivided mass of small ganglion cells taking a very deep stain throughout. The cells of the calyx are arranged in more or less radiating lines, all closely packed together. The calyx forms a kind of external support or "raised back" for the giant cells, and is not folded into the "cup" shape more usually seen in the higher insects. The giant cells of the mushroom body are collected into a large mass placed internally to the calyx. These cells are flask-shaped, with large clear nuclei, each carrying a large darkly-staining nucleolus. Many of them are of very great size. They have been compared with the cells of Purkinje in the cerebellar cortex of Mammals.

Various authors have stated (i) that the mushroom bodies of the Dragonfly's brain are rudimentary, (ii) that this indicates, or corresponds with, the possession of a very low order of intelligence. Neither of these statements is justified by the facts. The mushroom bodies of the Dragonfly's brain are certainly not rudimentary. On the contrary, they are very well developed, though of a generalized type which makes them *appear* poorly developed when compared with the highly-specialized structures to be seen, for instance, in the brain of the wasp. Nor is the intelligence of the Dragonfly by any means of a low order, in comparison with that of most insects. "Intelligence" may be developed through the use of the sense of sight as well as through other senses, but such intelligence is not to be placed on a lower order than that developed, for instance, through the senses located in the antennae. Judged broadly on its activities and powers of perception, the Dragonfly must certainly be classed as a more "intelligent" insect than many of those which possess more complicated mushroom bodies, if indeed the word "intelligent" can be properly applied to these creatures. The implied assumption that the mushroom bodies are the sole seat of "intelligence" seems to be quite unwarranted.

The protocerebrum gives off the following nerves:

(i) The *optic nerves*, described in connection with the optic ganglion (p. 130). The *optic tubercle* (*ot*) is the internal termination, in the procerebral medulla, of the superior portion of the optic nerve.

(ii) The *nerves to the lateral ocelli* (*ocln*). These are rather elongated, and about as thick as the antennary nerves. They arise from two small cell-masses placed in the upper posterior portion of the procerebral cortex.

The paired nerves to the median ocellus are said by the latest observers to arise from the deutocerebrum. They are best dealt with here, however, since, in the Dragonfly at any rate, their whole course is to be followed within the protocerebrum. Arising far apart and posterior to the level of the central body, they run obliquely upwards and inwards to meet in the middle line, well behind the pons. They then diverge again to enter the ocellus on either side. The diverging portions enclose between them a fair number of nerve-cells.

The Deutocerebrum. This consists of two rather small lobes, the *antennary lobes* (*al*), forming slight rounded bosses anteriorly on either side of the middle line (fig. 55 A). These are connected by a transverse commissure lying above and a little in front of the anterior limit of the central body, and representing the so-called *dorsal lobe* of the insect brain. The *antennary nerve* (*ant*) arises from the antennary lobe, and passes transversely, and a little obliquely, outwards and forwards to the base of the antenna, where it divides into two. The nerves to the median ocellus, which are also supposed to arise from the deutocerebrum, are dealt with on p. 126.

The Tritocerebrum. The two lobes of the tritocerebrum are known as the *oesophageal lobes* (*oel*). Each lies below the corresponding antennary lobe, with which it is closely fused. The two oesophageal lobes are connected by a slender transverse commissure, in the form of an inverted arch, called the *commissure of the oesophageal ring* (*tc*). This passes between the brain and the suboesophageal ganglion, beneath the oesophagus. Hence it cannot be seen in fig. 55 B.

The oesophageal lobes give off (i) the arched nerves to the frontal ganglion, known as the *frontal commissures* (fig. 55, *fc*), (ii) slender oblique commissures to the anterior ganglia of the paired sympathetic system, and (iii) the very thick and short *circumoesophageal commissures* (*co* fig. 55, *coe* fig. 56) which connect the brain with the suboesophageal ganglion.

The Optic Ganglion (fig. 58).

As we have already stated, the *optic ganglion* or *optic tract* is the largest and most complicated portion of the cephalic nerve-mass. This, of course, is only to be expected, when we consider the marvellous development of the sense of sight in the Dragonfly, since that sense supplies the stimulus to almost every act of the insect's life.

We shall follow Viallanes [186] in dividing the optic ganglion into six main parts, which will be dealt with, in order, from the compound eye inwards to the procerebral lobe.

1. *The layer of post-retinal fibres*, or *terminal anastomosis* (*pr*). Each ommatidium (p. 141) rests upon the fenestrate membrane (*f*), and sends through it delicate nerve-fibres, which at once collect together in bundles. These bundles quickly pass into a narrow layer of nerve-cells, which form with them a plexus. From this

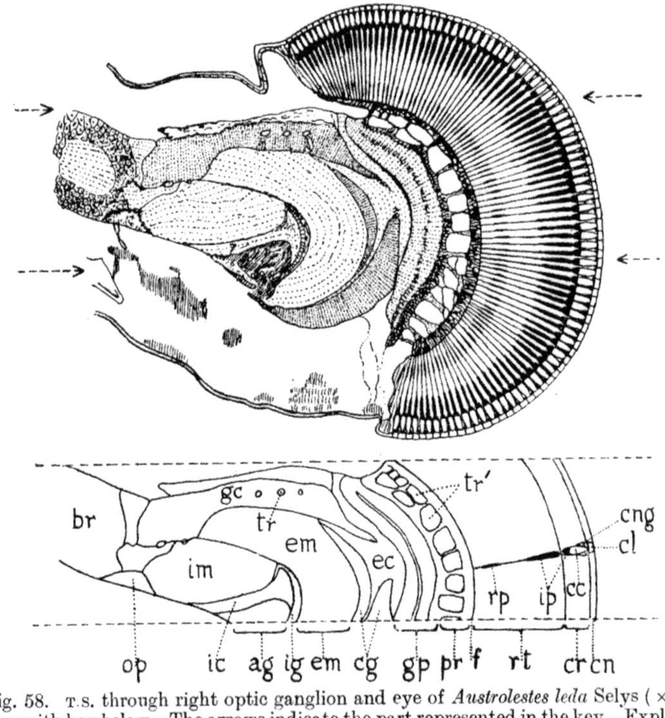

Fig. 58. T.S. through right optic ganglion and eye of *Austrolestes leda* Selys (× 50), with key below. The arrows indicate the part represented in the key. Explanation in text. Original.

plexus the bundles emerge irregularly, being separated from one another by large and thin-walled tracheae (*tr′*). They next enter another small layer of nerve-cells, forming a second plexus, lying just outside the ganglionic plate. All this complex structure is called the *terminal anastomosis.* The portion lying nearest to the fenestrate membrane is heavily charged with black pigment, which exceeds in intensity the layer of retinal pigment in the eye itself.

2. *The ganglionic plate, ganglionic layer*, or *perioptricon* (*gp*). This is a kind of screen placed across the path of the post-retinal fibres. It is chiefly formed of an outer layer of cylindrical columns of nervous matter without nuclei (except for a nerve-cell here and there), and lying more or less in line with the ommatidial axes. The number of columns, however, bears no definite relation to the number of ommatidia. Somewhat separated from this is an inner and much narrower layer of granular nervous matter enclosing numerous nuclei. The post-retinal fibres pierce both layers, and fill up the space between them. Thus this second division of the optic ganglion is, like the first, somewhat complex.

3. *The external chiasma* (*ec*). After leaving the ganglionic plate, the nerve-fibres from the eye cross one another very completely, forming a distinct chiasma.

4. *The external medullary mass*, or *epiopticon* (*em*). This forms a large and very conspicuous U-shaped gutter, with its concavity towards the brain. It consists chiefly of a mass of granular nerve-matter, which receives the nerve-fibres from the external chiasma, and passes them on again to the internal chiasma. Its structure is, however, complicated by the attachment of four ganglionic masses, known as the *ganglionic crown* (*gc*), the *corner ganglion* (*cg*), the *anterior ganglionic mass* (*ag*) and the *internal ganglionic mass* (*ig*). Each of these is formed of small unipolar nerve-cells, with prolongations passing into the main medullary mass. Their names sufficiently indicate their relative positions.

5. *The internal chiasma* (*ic*). The nerve-fibres from the external medullary mass, on passing out, cross again completely, forming a second chiasma.

6. *The internal medullary mass*, or *opticon* (*im*). This is a large semi-ovoid mass, whose convex surface receives the fibres of the internal chiasma. It consists of three separate capsules of granular nervous matter. The most external and posterior of these is directly united with the external medullary mass by fibres quite independent of the internal chiasma. The whole of the opticon is covered with unipolar nerve-cells, and is formed from

their prolongations. It gives rise internally and posteriorly to the optic nerve.

The Optic Nerve. This is a short but very distinct nerve, formed of two separate bundles of nerve-fibres. Of these, the *superior optic bundle* arises from the posterior face of the external capsule of the opticon, and enters the brain by its upper anterior region. It completely penetrates the cortex of the procerebral lobe, and enters the medulla of the protocerebrum, where it terminates in the optic tubercle (fig. 56, *ot*).

The *inferior optic bundle* is much larger, and arises from the whole of the internal capsule of the opticon. It enters the brain by its lower lateral region (fig. 56 E, *op*).

The Suboesophageal Ganglion (figs. 55, 56, *sog*).

This is a large and well-formed ganglion situated below the anterior portion of the brain, and connected with the tritocerebrum by the short stout circumoesophageal commissures (*co*). It is very compact, but still shews, in sections, evidence of its paired origin. As in the thoracic and abdominal ganglia, the nerve-cells are chiefly collected ventrally below the nerve-fibres. This ganglion gives off three pairs of nerves, (i) the *mandibular nerves*, from its extreme upper anterior corners; (ii) the *maxillary nerves*, from its lower anterior corners, at a level slightly behind (i); (iii) the *labial nerves* (fig. 56 C, *lbn*), which arise latero-ventrally from the broadest part of the ganglion, well behind (ii).

This ganglion narrows posteriorly, and finally gives origin to the two exceedingly slender ventral nerve-cords, running very close together backwards through the lower half of the posterior foramen, into the neck and prothorax.

The Ventral Nerve-Cord (fig. 54).

The two fine nerve-cords forming the double ventral cord may be either distinct enough to be seen separately under a low power, or they may be so closely united (in the abdomen) as to appear single, though of course their double nature is easily seen in transverse sections. These conditions appear to be correlated, not with the degree of specialization of other parts, so much as with the tenuity and length of the abdomen. Thus the very archaic *Petalura* (fig. 54) has the two cords so closely fused that it is quite impossible to separate them, whereas in the highly-specialized but short-bodied *Orthetrum* they lie distinctly apart.

The Thoracic Ganglia. Soon after entering the prothorax, the ventral nerve-cord passes into the fused pair of ganglia called the *prothoracic ganglion* (tg_1). This ganglion gives off nerves to the fore-legs (*first crural nerves*, tn_1) and to the prothoracic muscles. From this ganglion the two nerve-cords run backwards into the mesothorax, generally diverging slightly from one another, to enter the large *mesothoracic ganglion* (tg_2), a fused structure similar in nature to the prothoracic ganglion. It lies just above the mesosternum, and gives off nerves to the middle legs (*second crural nerves*, tn_2), to the muscles of the mesothorax, and to the fore-wings (*first alar nerves*). The last are best developed in the well-grown larva. In most Dragonflies, this ganglion is considerably larger than the prothoracic; in the *Libellulidae*, the two are approximately equal in size. The *metathoracic ganglion* lies close behind the mesothoracic (with which it is sometimes fused), just over the metasternum. It gives off nerves corresponding with those of the mesothoracic ganglion.

We have already mentioned that, in the *Calopterygidae*, the division between meso- and metapleura (first lateral suture, p. 23) remains distinct and complete, whereas in all other Odonata it becomes more or less obliterated. The condition of the meso- and metathoracic ganglia appears to be correlated partly with this fact, and partly with the degree of forward movement of the meso- and metasternum. We may exhibit in a short table three recognizable stages in the evolution of the thorax and its nervous system, arranged in phylogenetic order.

Stage	First Lateral Suture	Synthoracic Ganglia	Groups
1	Complete	Distinct	*Calopterygidae*
2	Partly obliterated	Distinct	*Petalurinae, Aeschninae*
3	Partly obliterated	Fused	*Gomphinae, Libellulidae* *Lestidae, Agrionidae*

The Abdominal Ganglia. In the Dragonfly imago, the abdominal ganglia are arranged as shewn in fig. 54, i.e. one in the posterior half of seg. 1, none in seg. 2, and one in the anterior half of each of segments 3–8,—the position being close to the base of seg. 3, and gradually moving a little backwards in each succeeding segment, until in seg. 8 the large ganglion lies close to half-way along that segment.

As, in the larva, there seem to be always eight abdominal ganglia, one situated in each of segments 1–8, it has long been a puzzle to entomologists as to what happened to the missing abdominal ganglion at metamorphosis. The problem was at once solved when I dissected the archaic *Petalura* (fig. 54). For there, very distinct, but attached posteriorly to the metathoracic ganglion in the thorax, was the missing abdominal ganglion (ag_1). It seems quite clear, though not yet definitely observed, that this fusion must take place either before or during metamorphosis, and is correlated with (*a*) the decrease in the size of the first abdominal segment to a narrow ring, and (*b*) the great elongation of the abdomen of the imago as a whole. This stretching out of the abdomen is also probably the cause of the other ganglia lying so far forward in each segment; and, in particular, the cause of the second ganglion moving up into the first segment. Stated in terms of the numbered segments, the positions of the ganglia appear highly irregular. But if we look at the figure of an actual dissection (fig. 54) we shall see that it is the lengths of the segments which are highly irregular, and that the tremendous stretching undergone by the nerve-cords at metamorphosis has been so fairly distributed, as to give approximately equal intervals between the ganglia. It seems clear, too, that the first abdominal ganglion moved forward in correlation with the whole forward movement of the synthoracic sterna, when the thoracic obliquity of the imago became so great that the nerve-cord could not otherwise have stood the strain.

The first seven abdominal ganglia in *Petalura* are all closely similar in size and structure. Each gives off a pair of strong branching nerves to the segmental muscles of its own segment. Branches from these pass to the dorsal vessel. Smaller nerves are given off to the muscles of the alimentary canal. The eighth abdominal ganglion (ag_8) is much larger than any of the others, and clearly represents the three completely fused ganglia of segments 8–10. It gives off three strong pairs of nerves (n_8–n_{10})

to the segmental muscles of these three segments respectively. A strong branch from each of the first pair (eighth segment nerves, n_8) runs forward along the heart. A branch from each of the last pair (tenth segment nerves, n_{10}) passes to the appendages. A fourth pair of strong nerves, placed anteriorly to the above-mentioned three pairs, supplies the reproductive organs. These are called the *genital nerves* (*rn*).

The thoracic and abdominal ganglia are closely similar in structure. Each is seen, in transverse section, to be clearly compounded of two fused ganglionic masses, placed side by side, and closely fused along the median vertical plane. The nerve-cells in all cases lie chiefly ventrally, the nerve-fibres chiefly dorsally. The nerve-cells are unipolar, and the cords are formed from their united prolongations or fibres.

The ventral nerve-cord of the newly-hatched larva (fig. 59) is very different from that of the imago. Relatively to the size of the animal, the ganglia are here immense, while the nerve-cords are almost non-existent. In most *Libellulidae*, the ganglia actually touch one another like a row of flat beads. At each instar, the nerve-cords lengthen slightly, while the ganglia do not appear to increase to any appreciable extent. Hence the ventral chain becomes gradually attenuated, and the ganglia smaller and smaller by comparison with the sizes of their segments.

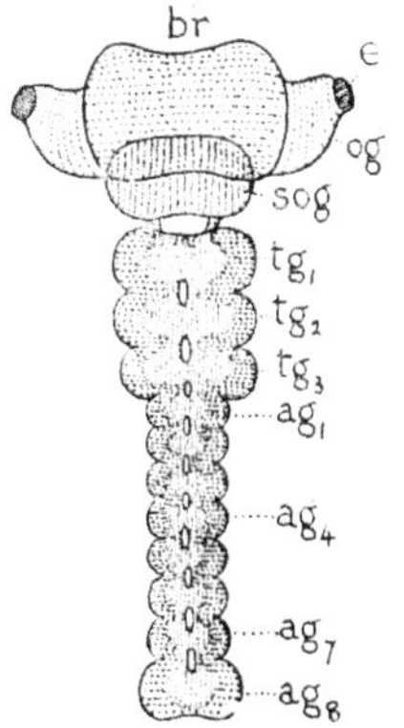

Fig. 59. Nervous system of newly-hatched larva of *Diplacodes haematodes* Burm. (× 53). References as in fig. 54. Original, stained preparation

The Sympathetic Nervous System (fig. 60).

The presence of a complex and well-developed sympathetic nervous system in the Dragonfly was demonstrated by Brandt [15], whose work still remains the chief source of our knowledge of these delicate structures. The system may be divided into two principal parts:

1. *The unpaired Frontal System.* This consists of a small triquetral *frontal ganglion* (fig. 56 A, fig. 55, *fg*, fig. 60, *fr*), situated in the middle line, just in front of the brain. It is connected with the tritocerebrum by a pair of arched *frontal commissures* (*fc*), which enter the brain a little below and internally to the origins of the antennary nerves. The frontal ganglion gives off two fine nerves (*m*) to the mouth parts, and a long and very delicate posterior median nerve-cord (nv_1) which passes beneath the brain, and runs dorsally along the oesophagus, crop and gizzard. The nerve-cells of the frontal ganglion lie dorsally over the nerve-fibres (fig. 56 A).

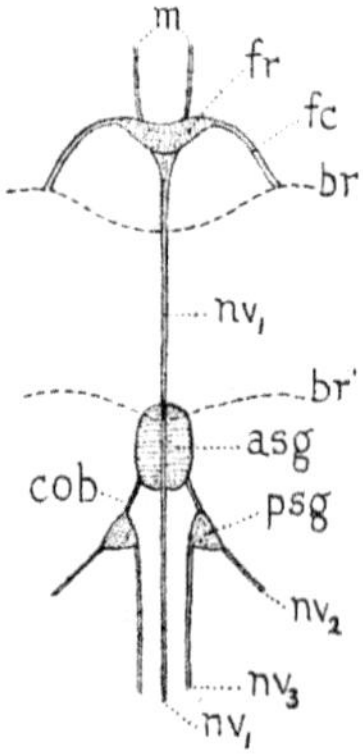

Fig. 60. Sympathetic nervous system of *Libellula*. Explanation in text. *br*, *br'* (dotted lines) boundaries of brain. After Brandt.

2. *The paired Splanchnic* or *Stomatogastric system.* This is formed by two pairs of small ganglia lying upon the oesophagus. The *anterior splanchnic ganglia* (*asg*) are fused together in the middle line. They lie very close behind the brain, and are connected with the tritocerebrum by two very fine commissures. A pair of *oblique commissures* (*cob*) connects them with the *posterior splanchnic ganglia* (*psg*). These are separate, and lie one on either side of the oesophagus, a little behind the anterior pair. They give off (*a*) a pair of nerves (nv_2) obliquely backwards to the muscles of the oesophagus, and (*b*) a second pair (nv_3) running straight back along its lateral walls.

Peripheral Nerves.

In some insects, or at any rate in their larvae, the presence has been demonstrated of a peripheral nervous system quite unconnected with any part of the central nervous system, but forming a kind of general nerve-plexus beneath the hypodermis and the muscle-layer. Zawarsin [202], after examining the various peripheral nerve-endings in the larva of *Aeschna*, found that they were all finally connected with the ventral ganglia. Some of these fine endings innervate small sense organs, such as minute sensillae, cones, or tympanulae, on special parts of the cuticle (p. 156). Others end in special enlarged hypoderm-cells, and serve to produce the effect of general sensation over the body-surface. There is no *separate* peripheral nervous system.

Physiology of the Nervous System.

We can only briefly summarize the results obtained by experiments upon Dragonflies:

1. *The Brain and Suboesophageal Ganglion* control the power of coordination of movement, and almost the whole of the sensory functions of the nervous system. According to Faivre, the brain is the chief seat of sensation, the suboesophageal ganglion alone controlling the coordination of movement. A decapitated Dragonfly does not, however, at once lose the power of movement. Large *Aeschninae*, whose heads have been accidentally struck off with the net, frequently live for as long as two days, and can vigorously flutter their wings, flex their abdomen, or even climb up curtains by means of their legs. If, however, the head be *crushed*, the shock causes partial paralysis, from which the insect recovers after the lapse of a time, to lead a feeble existence for a few hours, and then to die.

2. *The Ventral Ganglia*. Each of these acts as a motor centre for its own segment, controlling the muscles of that segment. Each is also a very feeble sensory centre, in connection with the peripheral endings of fine sensory nerves mentioned above. The movements of the separate segments are clearly coordinated together. The fact that the effect of a sudden shock is diffuse [6] seems to prove that the power of coordination does not rest entirely with the suboesophageal ganglion. The behaviour of decapitated Dragonflies points in the same direction.

3. The *mechanism of feeding* is almost certainly centred entirely in the suboesophageal ganglion. If this ganglion be injured, without damaging the brain, the insect is quite unable to feed. Again, it has been shewn, by at least two competent

observers, that the terminal segments of the abdomen of a large *Aeschnine* Dragonfly may be snipped off with scissors and offered to the head, which will proceed to devour them with apparent relish. This points in the same direction, while it also emphasizes the independence of the ventral ganglia.

4. The effect of mutilation, shock, decapitation, etc., on respiratory movements has been studied in detail by Barlow [6]. The results shew that numerous external influences affect the rate and power of the respiratory movements; that the respiratory movements of each abdominal segment are to *some extent* independent of the rest, provided the ganglia are not damaged; that decapitation does not stop these movements, but diminishes their frequency and force; and, finally, that even severe shock, though it may stop them for a time, cannot prevent a feeble revival of them later on.

It seems quite clear that much more remains to be done in the way of devising interesting experiments on this subject, and that there is still a great deal that we do not understand in the working of the very peculiar nervous system of the Dragonfly.

CHAPTER VII

THE SENSE ORGANS

In the Dragonfly the sense of sight is extraordinarily well-developed, and is probably keener than in any other insect. This sense alone suffices for the chase and capture of prey. It is, however, supplemented by a fairly well-developed sense of taste, located in the labrum and epipharynx. The sense of touch is well-developed in most parts of the body, but particularly in the mouth-parts, legs and abdominal appendages.

Correlated with the great power of sight is the almost total absence of the sense of smell. The antennae, which were originally the seat of this sense (if indeed it was ever well-developed in these insects), are much reduced, and incapable of anything beyond feeble sensation. The sense of hearing, as we understand it, is also practically absent; though there are certain minute organs in the legs which are capable of receiving vibrations, and may be looked upon as feeble auditory organs.

The *eyes* are of two kinds, simple eyes or *ocelli*, and *compound* or *facetted eyes*, made up of numerous separate elements. The ocelli are used for near vision, their range probably lying between a few millimetres and about an inch. The compound eyes, on the other hand, observe objects from an inch or so up to several feet off; while their perception of *movement*, as distinct from *form*, ranges very much further, probably up to fifty feet or more in some cases.

The Ocelli (fig. 61).

These beautiful organs have so far received little attention from biologists, though much has been written upon the compound eyes. They are, nevertheless, of very great interest, and deserve a more detailed study. They are three in number, and are so placed on the epicranium that they isolate a small central triangular area called the *vertex*. This, in many Anisoptera, is raised up into a tubercle or vesicle. The median ocellus lies at the apex of the triangle, which is anterior, while the two lateral ocelli occupy the base angles. The angle of the apex is always obtuse, lying usually between 120° and 170°.

The median ocellus is so placed as to look forward, while the laterals look sideways. This difference is especially marked when the vertex is swollen. The median ocellus then becomes sunk into a pit, with the wall of the vertex rising like a cliff behind it. Thus all lateral and posterior rays of light are cut off. Again, the lateral ocelli become slightly raised up on the sides of the tubercle, so as to face somewhat horizontally outwards. Thus they receive chiefly lateral rays.

The median and lateral ocelli differ considerably in structure. The former is always the larger, and is either oval or reniform in shape, with its major axis transversely placed. Both by its structure and nerve-supply it shews signs of having been originally double. It is therefore to be regarded as the product of the fusion of two originally separate ocelli in the middle line. In structure it is bilaterally symmetrical. The lateral ocelli, on the other hand, are smaller simple oval bodies, not bilaterally symmetrical. Their parts are much contorted and shifted out of place, probably in order to accommodate the organ to the reception of lateral rays of light only.

Structure of the Median Ocellus (fig. 61 A, C, D).

Fig. 61 A gives a section of this organ in *Austrolestes leda*, passing obliquely through the middle of the lens and through the proximal set of rhabdomes, and including several layers of cells. The ocellar nerves are also shewn, though these are actually not quite in the same plane as the other structures. The ocellus consists of the following parts:

(i) A thick convex *corneal lens* (*cnl*). When examined in thin sections, this lens is seen to consist of a very large number of delicate layers of chitin, all of which are transparent, but not all colourless. The outer portion is absolutely clear, and extends to the border of the organ around its whole circumference, where it is separated from the ordinary cuticle of the head by a shallow groove or depression (*gr*) of a deep red colour. The central part of the lens is much swollen and sunk inwards. Its component layers are very closely fused together, so as only to be distinguished with great difficulty. They shade from a transparent pink colour above to a deep clear ruby red below.

(ii) A layer of very elongated *corneagen cells* (*cng*). These are continuous with the hypoderm cells of the head, but differ from them in being quite colourless and transparent. They are the

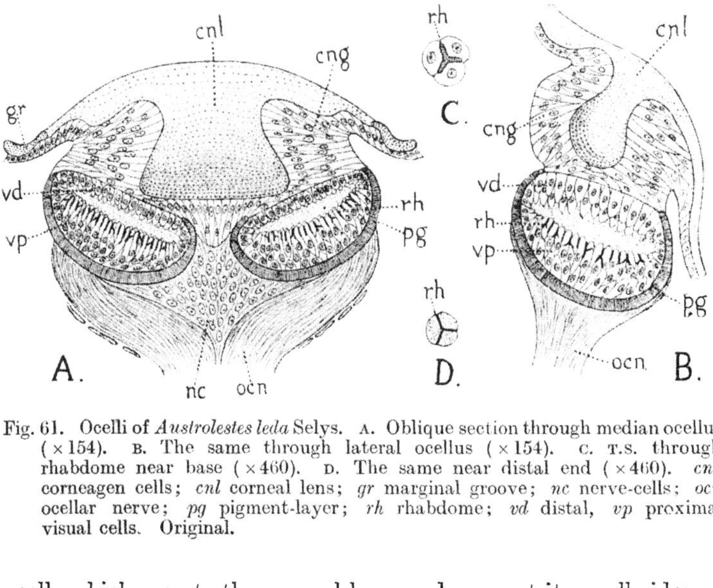

Fig. 61. Ocelli of *Austrolestes leda* Selys. A. Oblique section through median ocellus (×154). B. The same through lateral ocellus (×154). C. T.S. through rhabdome near base (×460). D. The same near distal end (×460). *cng* corneagen cells; *cnl* corneal lens; *gr* marginal groove; *nc* nerve-cells; *ocn* ocellar nerve; *pg* pigment-layer; *rh* rhabdome; *vd* distal, *vp* proximal visual cells. Original.

cells which secrete the corneal lens, and support it on all sides, as shewn in the figure. These cells, together with the lens, form the *dioptric* portion of the ocellus.

(iii) A *retina* formed by two series of *visual cells* or *retinulae*, viz. the distal retinulae (*vd*) and the proximal retinulae (*vp*), and surrounded by a layer of black pigment (*pg*). In the median ocellus the retina forms a thick half-ring, with the two ends directed forwards, so that the anterior portion is seen in section as two symmetrical parts, with two separate and diverging nerve-supplies. This arrangement is strong evidence in favour of the generally accepted view that this organ is formed by the fusion of two originally complete and separate ocelli in the median line.

The visual cells are elongated, with the more basal portion swollen and containing the nucleus. They are arranged in sets of three, closely fused together (fig. 61 C, D). Each set or triad gives rise, along the inner junctions of its component cells, to a very remarkable organ called the *rod* or *rhabdome* (*rh*), which is consequently, in transverse section, seen to be of triradiate form. It is semi-transparent and of a pale yellowish colour. Its function is generally held to be that of a *light-transmuter* (see p. 148). Each rhabdome is of moderate length, extending practically the whole length of the triad of cells forming it. It is somewhat thicker basally than distally, but its three radii are of about the same length throughout. In fig. 61 A, only the upper portions of the proximal rhabdomes are shewn cut [70].

The visual cells of the proximal group appear to be much better developed than those of the distal group. The latter are considerably shorter. In *Austrolestes* I have not succeeded in detecting any rhabdomes connected with these cells, though such are certainly present in the larger ocelli of *Aeschna brevistyla*. This suggests that, in some Dragonflies, the ocelli can only focus objects clearly at one fixed distance, while in others the same organs are effective for two separate distances. In both cases, the form of the lens shews that the objects viewed must be brought very close up to become visible.

The retina is surrounded by a layer of black *iris pigment* (*pg*) developed in cells which are continuous with the corneagen cells, and hence with the hypodermis.

The *ocellar nerve* (*ocn*) is, as already explained, of a double nature, the two diverging portions being separated by a considerable number of nerve-cells (*nc*).

Structure of the Lateral Ocellus (fig. 61 B).

This organ shews exactly the same parts as the median ocellus. It differs from the latter, however, in being unsymmetrical in all sections, and is obviously a single and not a double organ. We may remark that a section through the median ocellus taken along a vertical plane through the letters *cnl* and A in fig. 61 A would very closely resemble fig. 61 B. The lateral ocellus clearly has its dioptric apparatus distorted for the purpose of lateral vision—in fact, it is set at a permanent and very effective "squint." The corneal lens is for the most part quite colourless, but the innermost part of the central "bulb" is coloured ruby-red.

The Compound Eye (figs. 58, 62–64).

We have already discussed (p. 11) the size and shape of these eyes in relation to the head. We have now to consider the structure of the eye itself. Each of the hexagonal facets already noticed on the outer surface of the eye corresponds with a single element of the compound eye, known as an ommatidium. An ommatidium has been compared with an ocellus; but we shall see that there is sufficient difference in structure to make the comparison inadvisable. Moreover, it was until recently quite generally held that the compound eye arose as an aggregation of simple eyes similar to ocelli, and that the hexagonal form of the facets was produced by mutual pressure between numerous originally rounded elements crowding closely together. Though we must admit that crowding would in any case produce hexagonal facets, we must at the same time negative the previous assertion on palaeontological grounds, if on no other. For the compound eye is an exceedingly archaic structure, found (in the Trilobites) well developed right back to the Lower Cambrian. Moreover, in the Dragonflies at any rate, the compound eyes develop in the embryo, whereas the ocelli do not begin to develop until near or at the close of larval life. As these insects are amongst the most ancient of the class, this fact is very significant.

In the compound eye of the Dragonfly, the number of facets, and hence the number of ommatidia, ranges from about 10,000 in the smaller Zygoptera to 28,000 or more in the largest *Aeschninae*.

Structure of an Ommatidium.

Fig. 62 shews a longitudinal section through four ommatidia from the compound eye of *Austrolestes leda*. The hexagonal *facet* (*f*) is the external surface of the **corneal lens** (*cnl*). The latter is deposited by the agency of the *corneagen cells* (*cng*), which are easily made out in the eye of the freshly-emerged imago. The corneal lens is slightly biconvex, and is made up of a thinner outer layer, which stains deeply with eosin, and a thicker inner layer

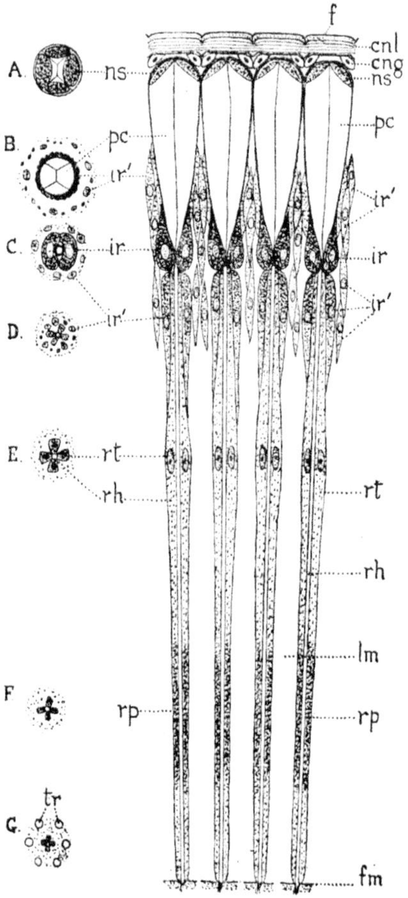

Fig. 62. L.S. through four ommatidia from fig. 58 (×330). At A–G are shewn seven T.S. through one ommatidium at the respective levels at which they are placed. *cng* corneagen cells; *cnl* corneal lens; *f* facet; *fm* fenestrate membrane; *ir* primary, *ir′* secondary iris cells; *lm* lumen; *ns* nucleus of Semper; *pc* pseudocone; *rh* rhabdome; *rp* retinal pigment; *rt* retinal cell; *tr* trachea. Original.

not so easily stained. This inner layer is found to be in reality made up of several fine laminae superimposed one upon another.

The outer layer represents the original cuticle; the inner laminae are formed as a series of definite strata secreted by the corneagen cells. These latter appear to degenerate rapidly after the completion of the lenses, and their nuclei cannot be made out in the eyes of mature imagines.

Beneath the corneal lens lies a very beautiful structure, the **crystalline cone** or *pseudo-cone* (*pc*). This is a highly refractive body of a vitreous appearance, in the form of a cone with its apex inwards, and its broad outer end convexly rounded. It is formed from four elongated transparent cells, two being larger and two smaller (fig. 62 A, B). The nuclei of these cells lie close up to their outer or distal ends, and are known as *Semper's nuclei* (*ns*). These very peculiar nuclei appear to be filled with small clear refractive bodies closely packed between large chromatophil granules. Above each nucleus is a dense plate of deeply-staining material. The inner end or apex of the pseudocone does not actually end in a point, but the ends of the four cells stand slightly apart, forming a slight hollow, into which the distal end of the retinula is inserted.

Embracing between them the inner tapering end of each pseudocone are two large cells shewing a semi-annular form in transverse section. These are the *primary iris cells* (fig. 62 C, *ir*). Each has a large clear nucleus, embedded in a dense mass of black pigment granules which completely fill the rest of the cell. Outside these two cells are arranged numerous slenderer pigment cells of very elongate form. These are the *secondary iris cells* (*ir'*). These cells fall into two groups, according to the approximate positions of their nuclei. One set lies around the pseudocone at a higher level than the primary iris cells; their nuclei are very clearly to be seen lying between the inner halves of the pseudocones. A section through the pseudocone at about the level of fig. 62 B shews twenty or more of these secondary iris cells forming a close ring around the pseudocone, which thus appears to be encased in a ring of black pigment. Outside this ring other secondary iris cells are arranged, while small pigment granules also form a kind of hexagonal pattern around each pseudocone. Lower down, at the level of the primary iris cells, the number of secondary iris cells is much less, usually from six to ten (fig. 62 C). Passing next to a level below the inner ends of the pseudocones, we find a second set of these pigment cells embracing the distal ends of the retinulae. Four of these are of considerable size, and very densely pigmented, especially at their distal ends, which are somewhat enlarged. They lie rather close in between the retinula-cells. Smaller pigment cells lie outside these. Their nuclei can be fairly easily made out.

The whole layer of pigment described above is known as the *iris*, or *iris tapetum*.

Below each pseudocone there lies the excessively elongated portion of the ommatidium known as the **retinula**. This is formed from four *visual* or *retinula-cells* (*rt*), placed close together, with their long axes parallel. These cells are considered to be highly-specialized unipolar nerve-cells, placed with the pole or nervous end in contact with the fenestrate membrane. The opposite or distal end is drawn out into an extremely long border, carrying a row of excessively fine fibrils, arranged like the teeth of a fine comb. The retinula-cells of each ommatidium are so arranged that their fibrils pass inwards, and unite to form the peculiar structure known as the great rod or *rhabdome* (*rh*). The formation could be very roughly imitated by taking four ordinary long hair-combs and standing them up on end with their teeth all meeting inwards. The rhabdome itself appears to have a pale central axis surrounded by a dark outer portion. The fibrils which form it pass backwards right through the cell to which they belong, finally emerging at the pole in the form of the nerve-fibre process. The rhabdome is supposed to receive a stimulus from the rays of light focussed on it by the dioptric apparatus of the eye (lens and pseudocone), and to transmute this into the form of a nervous stimulus which is transmitted to the brain.

The retinula-cells themselves are somewhat flattened, sub-triangular or trapezoidal in cross-section, and stand out from the rhabdome in the manner seen in fig. 62 D–G. They contain pigment-granules throughout their entire length, but more especially thickly for about one-fifth of their length, at the level shewn in fig. 62, *rp*. This is called the *retinal pigment*.

The bases of the retinulae rest upon the *fenestrate* or *basement membrane* (*fm*), which forms the inner boundary of the eye proper.

This membrane is formed of a peculiar fibrous connective tissue. It is not only perforated, at regular intervals, to allow of the passage of the terminations of the nerve-fibres to the retinulae, but it is also pierced with regular rows of larger circular holes, through which tracheae of considerable size pass into the eye. Thus the fenestrate membrane resembles in appearance the bottom of a colander pierced with holes of two sizes. The tracheae, on entering the eye, occupy the large spaces or *lumina* (*lm*) which separate the slender bases of the ommatidia. It is usually stated [10, 152] that four of these tracheae surround the base of each ommatidium, though Calvert [30] shews six, arranged in a hexagon. I find that Calvert's statement is correct for the arrangement of the tracheae close to the fenestrate membrane (fig. 62, *f*). As they pass further into the eye, however, they quickly tend to become arranged in parallel rows, with only a slight indication of the original hexagonal arrangement left.

It is generally agreed that the eyes of insects possess no power of accommodation. There are no eye-muscles, and the lens is absolutely fixed, both as regards position and shape. Vigier[1], however, has described certain fibrillae in the eye of the Dragonfly which, he believes, enable the insect to shorten or lengthen its ommatidia. As the Dragonfly uses its eyes chiefly during flight, it seems obvious that nothing short of a very complete accommodation-apparatus would be of any value to it, owing to the rapid change of position of the objects to be viewed. It seems much more likely that the fixed focal length without accommodation is quite satisfactory, and that this is supplemented by the use of the ocelli for very near objects, which would otherwise be quite out of focus.

Variations in the form of the ommatidia. In the Zygoptera the ommatidia are all alike, and all deeply pigmented. In the *Libellulidae*, however, they are sharply marked off into two groups. In the upper group, the facets are larger, the ommatidia consequently further apart, the cones larger, and pigmentation is absent (fig. 63 A). In the lower group, the facets are smaller, the pigmentation very dense, and the ommatidia closely resemble those of Zygoptera. The line of separation between the two

[1] *C. R. Acad. Sci.*, Paris, 1904, CXXXVIII, pp. 775–777.

groups is marked in the living insect by a change of colour, the upper part of the eye being usually of a reddish hue, the lower greenish or yellowish. In the *Aeschninae* the two groups of ommatidia are present, and can be separated by the difference of colour, the upper portion being usually a brilliant green or blue, the lower more greyish. But there is an intermediate zone in which the size of the facets is graded, and the amount of pigmentation gradually decreases from below upwards. Ommatidia of both groups appear to be all of equal length in the *Aeschnidae*.

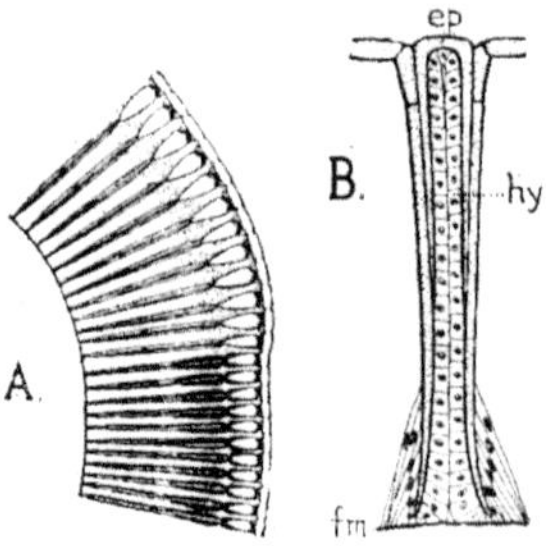

Fig. 63. A. T.S. through part of compound eye of *Diplacodes haematodes* Burm. (× 30) to shew transition from pigmented to unpigmented ommatidia. B. T.S. through median eye-line of *Aeschna brevistyla* Ramb. (×45). *ep* epicranium; *fm* fenestrate membrane; *hy* hypodermis. Original.

The differentiation of two groups of ommatidia is clearly correlated with a difference in the use of the two portions of the eye. During flight, the lateral and lower portions of the eye receive the impression of the surrounding objects, while the upper portion of the eye receives only rays of direct sunlight. It is not, therefore, necessary for this latter portion to be able to distinguish clearly the *form* of objects, provided that it is able to note the change in the intensity of light from above, produced by the veiling of the sun by clouds, or by the shadow of a possible enemy. While the absence of pigment undoubtedly renders this part of the eye unable to perceive definite forms clearly, it must also tend to mitigate the direct glare of the sun, by allowing the rays of light to diffuse equally in all directions.

The Zygoptera, on the other hand, never hawk about in the way that the larger Dragonflies do. It is essential for them, whether at rest or flying close to the surface of the water, to see surrounding objects clearly in all directions. Hence no differentiation of the upper ommatidia has ever been needed in their case, and none has arisen.

The Median Eye-line. We have already mentioned the fact (p. 11) that the large eyes of most *Libellulidae* and *Aeschninae* appear to meet along a median eye-line. A transverse section of this region (fig. 63 B) shews the true nature of the junction. The eyes are actually separated by an excessively narrow space, occupied by two layers of hypoderm-cells, and bridged over above by a narrow arch of chitin. At the borders of the eye, the chitin is turned inwards on each side of this arch, and extends downwards almost to the level of the fenestrate membrane. This formation is clearly the last remnant of the epicranium, which is therefore seen to be still actually continuous from vertex to occiput, even in the highest forms.

Function of the Compound Eye.

We have already stated that the compound eye of the Dragonfly is used for far vision; that is to say, far in comparison with the range of the ocelli. To the ordinary observer, a Dragonfly would appear very short-sighted, since it sees objects best at distances ranging from a few inches to five or six feet. Compared with other insects, however, the Dragonfly may be reckoned long-sighted. It has been stated that no insect can see beyond a distance of six feet. This may be so as regards perception of form. But in the case of the larger Dragonflies, perception of movement extends to a much greater distance. I have found that *Petalura gigantea* can be frightened away by waving a net in the air at a distance of twenty yards; while, in the case of *Hemicordulia tau* hovering motionless over a pool, a sharp movement at a distance of ten yards always produces a sudden dodge sideways on the part of the Dragonfly. On the other hand, by moving slowly and stealthily, without any jerks, it is possible, though not easy, to stalk a large Dragonfly sitting on a twig, and to capture it with the hand. I have caught *Anax papuensis* in this manner.

Theories of Vision of the Compound Eye.

Space will not permit us to go fully into this fascinating question. We shall simply state the two chief rival theories, and leave our readers to choose for themselves.

The **Theory of Mosaic Vision** was formulated by J. Müller in 1826 [95]. Its principal supporters have been Huxley, Lubbock and Exner. It is easily understood from Lubbock's well-known diagram (fig. 64). If a number of transparent tubes be arranged side by side in front of a retinal screen, and separated from one another by a coating of black, the only light which will traverse each will be those rays parallel to the long axis of the tube (e.g. *aa′*, *bb′*, *cc′*). All oblique rays (e.g. *aw*, *ay*, *bx*, *bz*) fall on the sides of the tube, and are absorbed by the black pigment. Now if the tubes be made very numerous, long and narrow, like the ommatidia of the compound eye, each separate unit would only perceive a very small portion of the total field of vision. The impression received through such an eye with 20,000 ommatidia would be in the form of a landscape made up of 20,000 little pieces, each accurate as to colour, distance and intensity. Anyone who has seen pictures constructed of mosaics will readily admit that such an impression would be an excellent representation of the scene, though not so perfect as the complete picture formed by the vertebrate eye.

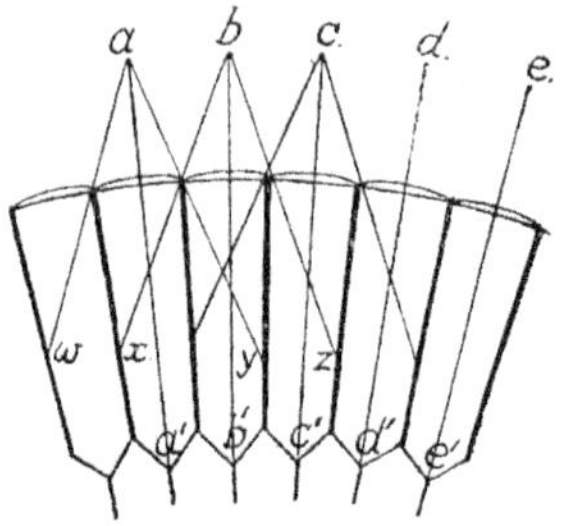

Fig. 64. Diagram to explain method of vision of the compound eye. After Lubbock.

Now, let us suppose that a very small object, so small as only to be represented on *one* of the 20,000 mosaic pieces, were *suddenly* to move. The nerve-ending receiving the impression through the corresponding ommatidium would at once respond to the movement; the mosaic would be altered *by a sudden alteration in one of its pieces.* Such a small movement might well pass unnoticed by the vertebrate eye. But the compound eye would be quick to register it, and the brain would act accordingly. If however the change were made very slowly and steadily, it would probably pass unnoticed. What we know of the power and method of vision of the Dragonfly exactly agrees with this. The quiver of an eyelid on the part of the collector is enough to frighten the coveted prize away, though the stealthy approach of the collector's whole bulk may not even be noticed.

Huxley accepted the Mosaic Theory, with the important addition that he considered the rhabdomes to be not merely *light-transmitters*, but *light-transmuters*. That is to say, they change the form of energy which we call light into that other form known as nervous energy. In this form, the theory has found wide acceptance.

Exner [53] propounded what amounts to a modification of the Mosaic Theory. Through a freshly cut-out compound eye of a Fire-fly, he obtained a photomicrograph of a distant church steeple, together with a window-pane on which the letter R had been pasted in black paper. The image was erect and complete. He explained this by supposing that the images formed by the separate ommatidia overlap and merge, so as to give the impression of a complete whole. He termed these "superposition" images, as opposed to non-overlapping or "apposition" images. It can be easily understood that an alteration in the position of the pigment layers might convert a set of apposition images (forming a mosaic) into a continuous image, by admitting just enough light to cause the boundaries to merge, and so obliterate the darkened borders of the mosaic pieces in a hexagonal pattern. It is probable that all insects with strong visual powers can modify the position of their pigment layers sufficiently to allow of such accommodation in various degrees of light-intensity.

The **Theory of Dioptric Vision** was enunciated by Lowne [86] in 1884. It differs entirely from the Mosaic Theory, both as to the method of vision and the limits of the eye. The supporters of the Mosaic Theory consider the fenestrate membrane to be the inner limit of the eye proper. Thus the corneal lens and crystalline cone form the *dioptric* portion of the eye, while the retinulae are the *receptive* portion or *retina*, in direct communication with the optic ganglion. All structures lying below the fenestrate membrane are considered to belong to the optic ganglion. But, according to Lowne, the whole of the portion lying above the fenestrate membrane is dioptric in function, while the true retina is the pigmented portion lying below the fenestrate membrane. The rhabdomes must then act as another set of lenses, in such a manner that all the rays of light received by the whole eye are brought to a focus on the retina, yielding a complete image like that formed by the vertebrate eye. This image would be an erect one. Hickson [73] supports Lowne's view of the position of the retina.

Optical Phenomena of the Compound Eye.

When a Dragonfly is caught and held in the hand, the eyes are seen to glow with a most beautiful light, generally of a semi-metallic green or blue colour, sometimes red, brown or grey. This is a reflection of light from the interior of the eye, and is called the *Internal Light.*

Again, if we peer into the eye, there will be seen, apparently deep down in the eye, a very distinct black spot, surrounded by six less distinct but similar black spots. These are the *pseudo-pupils.* The central pseudo-pupil is always exactly opposite the eye of the observer, and moves with great rapidity as his position changes. Its formation is due to the almost complete absorption of the direct light passing straight in through the corneal lens, while part at least of the oblique light is reflected back. Thus the central area appears black, the surrounding parts bright. The six secondary pseudo-pupils are probably diffraction-images corresponding with the six sides of the hexagon surrounding the central pseudo-pupil. If the curvature of that part of the eye gazed into be not approximately symmetrical (as, for instance, near the border of the eye), the pseudo-pupils will not appear rounded, but distorted and elongated.

Further, we can easily remark, in the eye of a live Dragonfly, two or three irregular greyish areas, which, like the pseudo-pupils, move when the observer moves. These are pseudo-pupils of the third order, derived from the secondary pseudo-pupils in the same way that the latter are derived from the central one. They are actually twelve in number, but only two or three can fall within the limits of the cornea, for any given position of observation.

The Eyes in the Larva.

The compound eyes are functional, and of considerable size, in the newly-hatched larva. The ocelli, on the other hand, only appear at metamorphosis in the Anisoptera, and late in larval life in the Zygoptera. In the case of the compound eye, the number of ommatidia, as well as the length of the retinulae, increases at each ecdysis, while the optic ganglion becomes gradually more separated off from the brain proper, and increases in complexity. It is interesting to note that the smallness of the larval eye in the *Libellulidae* is due to the fact that only the smaller (pigmented) ommatidia are functional in the larva. These form the definitely projecting eye of the larva. Between these and the epicranium proper there is a rather flat unpigmented area, in which the elements of the larger (unpigmented) ommatidia are gradually formed during later larval life. At metamorphosis the functional eye of the larva becomes the lower pigmented portion of the imaginal eye, while the nearly flat unpigmented part becomes convexly curved to form the upper portion. In the larva, the optic ganglion is actually at first only in connection with the pigmented part, the connection with the unpigmented ommatidia being a late outgrowth of lateral nerve-elements from the ganglion.

The Antennae (figs. 2, 65).

The external form of the antennae has already been dealt with in chap. II. They are of small size, both in the larva and the imago. They are, however, comparatively larger in the larva, and appear also to be more functionally active than in the imago. Their reduction is clearly correlated with the increase in power of the compound eyes. The largest antennae are found in some of the older Zygoptera, such as *Synlestes* (fig. 2 F, G), while the smallest occur in the *Libellulidae* and *Aeschninae*, where the eyes reach their maximum development. The sense of smell, which has been proved to be located in the antennae of insects, has in the Dragonflies become entirely subordinated to that of sight. It is doubtful whether it was ever at all well-developed in the Order. The scent of carrion, which so powerfully attracts many insects, would be a matter of no moment to the Dragonfly, which feeds exclusively on living insects. On the other hand, their well-developed sense of taste is sufficient to enable them to reject any obnoxious morsel caught on the wing.

In the antennae of the larva of *Aeschna*, Zawarsin [204] has demonstrated the existence of a *Johnston's organ* (fig. 65). This is undoubtedly a sense-organ. It may have an auditory function, but is more probably connected with testing the purity of the water. It lies in the second segment or pedicel. It is not visible in chitin preparations, being purely a nervous formation. It consists essentially of a ring of large bipolar nerve-cells (*J*), whose fibres run back to form two bundles passing into the two main divisions of the antennary nerve. It is difficult to understand how such an organ could respond to anything except the vibration or movement of the antennae as a whole, or, more particularly, to the movements of the distalia inserted into the pedicel just above it.

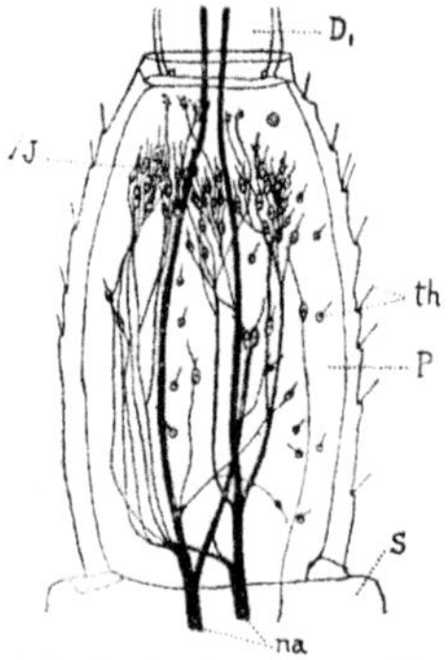

Fig. 65. Pedicel of antenna of larva of *Aeschna* (× 70). D_1 first distale; *J* Johnston's organ; *na* the two main branches of the antennary nerve; *P* pedicel; *S* scape; *th* sensillae. Methylen blue preparation, after Zawarsin.

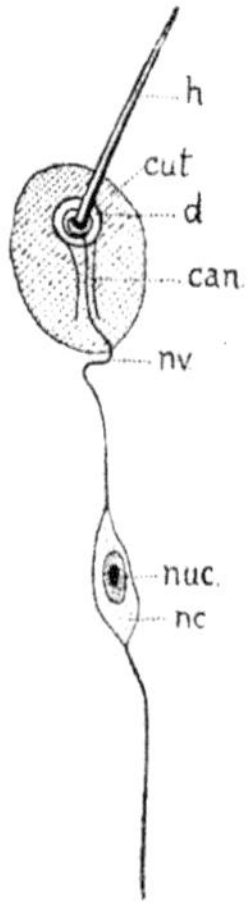

Fig. 66. Sensilla (tactile hair) from antenna of larva of *Aeschna* (× 140). *can* canal; *cut* cuticle; *d* circular disc of hard chitin; *h* hair; *nc* nerve-cell; *nuc* its nucleus; *nv* nerve-ending. Methylen blue preparation, after Zawarsin.

Each segment of the antenna carries a few *sensillae* or sensory hairs (fig. 66). These are organs of touch. They are most numerous, though shortest, on the scape and pedicel. Each hair is freely movable, set in a rounded depression (*d*), from which a canal (*can*) runs inwards through the cuticle. This canal carries the fine nerve-ending (*nv*) of a large bipolar sensory nerve-cell (*nc*).

Zawarsin also found *tympanulae* on the scape and pedicel. These are tiny circular grooves, with a raised wall covered over by a chitinous membrane. A canal lying in the cuticle, in close proximity, carries the nerve-ending. Such an organ is responsive to waves of sound, and may be regarded as a very elementary auditory organ.

The extreme tip of the antenna was also found to be sensitive. Here the chitin becomes very thin and delicate, forming a broad funnel-like canal with a short hair placed on each side. This seems to be an organ of touch. It is innervated by two or three elongated bipolar cells.

In the imago, the only sense-organs so far observed in the antennae are *tactile hairs* and *tympanulae*. The hairs resemble those of the larva, but are less numerous. The tympanulae were discovered by Lespès [85]. A group of four, arranged in a line, were found on the third joint in *Sympetrum*, and on the fifth joint in *Agrion*. Lespès states that they contain *statoliths*, and are convex in profile view. This suggests a balancing or orientating function. Lespès also found, in *Agrion* only, on the third joint, a peculiar tympanula closed by a stretched whitish membrane. Possibly this is a feeble auditory organ.

Experiments with Antennae.

It has been often asserted that Dragonflies deprived of one or both antennae are quite unable to direct their flight. In order to test this assertion, I have several times taken into the field with me a pair of fine dissecting scissors, with which I carefully cut off one or both of the antennae of various Dragonflies captured. I have treated in this way a number of specimens belonging to the genera *Hemicordulia*, *Anax*, *Aeschna*, *Ischnura*, *Austrolestes* and *Synlestes*. In no single instance did the operation affect the insect in the least. *Synlestes* has the largest antennae of any Dragonfly known to me; yet the loss of them caused it no inconvenience. While returning late one afternoon, I captured a specimen of *Anax*, which, on examination, proved to be one whose antennae I had removed in the morning. It was hawking about in the ordinary manner. I can only suggest that it is not loss of the antennae, but shock caused by damage to the nerve, if the operation is clumsily done, that causes them any temporary inconvenience.

The Labrum and Epipharynx (figs. 67, 68).

The *epipharynx* (fig. 67, *ep*) is a soft chitinous outgrowth from the under surface of the labrum. It is undoubtedly an organ of taste. The sensory portion consists of two curved rows of tiny circular organs, in the form of yellow specks on the surface. One of these taste-organs is shewn, highly magnified, in fig. 68. It consists of a minute chitinous cone (*c*) set in the middle of a thick

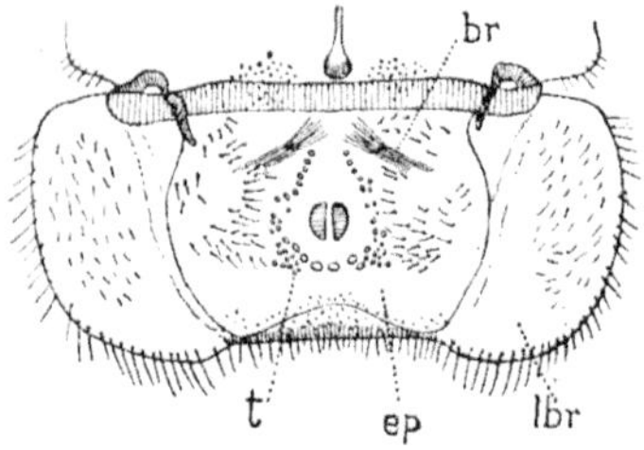

Fig. 67. Labrum (*lbr*) and epipharynx (*ep*) of *Aeschna brevistyla* Ramb. (× 20). Underside, shewing brushes of hairs (*br*), taste-organs (*t*) and sensillae. Original, chitin preparation.

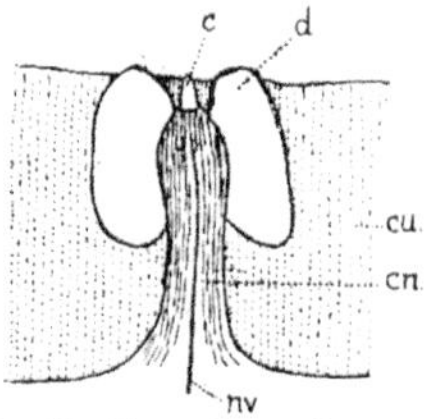

Fig. 68. Section through one of the taste-organs on the labrum of the larva of *Aeschna* (× 1000). *c* cone; *cn* canal; *cu* cuticle; *d* disc of hard chitin; *nv* nerve-ending. Methylen blue preparation, after Zawarsin.

disc of strong chitin of yellow colour (*d*), and connected with a canal (*cn*) carrying a fine nerve-ending (*nv*). Some of these organs are also found on the surface of the labrum itself. The epipharynx also carries two stiff brushes of hairs (fig. 67, *br*) pointing inwards. Numerous small scattered tactile hairs occur both on the labrum and epipharynx, and a row of strong hairs along the border of the former organ.

The Hypopharynx (fig. 69).

The hypopharynx (fig. 69) is well-developed in Dragonflies. It is a soft chitinous outgrowth from the base of the labium, projecting into the buccal cavity in the form of a distinct tongue. It is usually regarded as an organ of taste. It is, however, covered only with numerous long yellow hairs. Zawarsin [204] has shewn that the nerve-supply of these hairs is of the ordinary kind. He therefore concludes that the hypopharynx is merely an organ of touch in the larva. We think the same holds good for the imago.

The opening of the salivary ducts lies at the base of the hypopharynx (fig. 45 E).

Chordotonal Organs (fig. 70).

The peculiar organs known as *chordotonal organs* (fig. 70), and supposed to be organs of hearing, occur on all the legs of Dragonfly larvae, and probably also in the imago. The larva of *Aeschna* was found by Zawarsin [204] to have three sets of these organs on each leg, two on the trochanter and one on the tibia close to the knee-joint. Each set consists of a number of minute organs called *otaria*. Each otarium (fig. 70 B) is a small slit covered by smooth

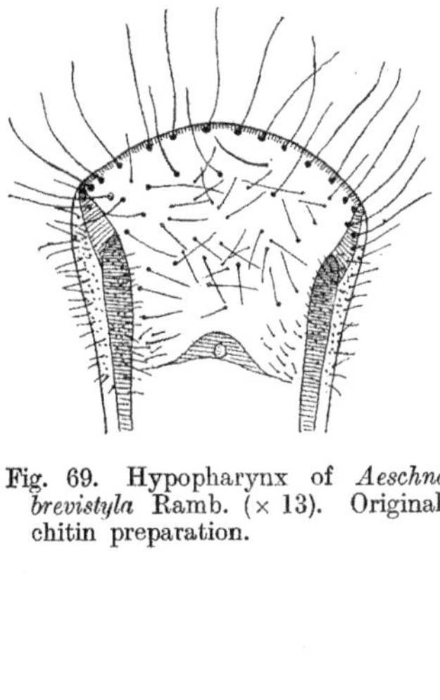

Fig. 69. Hypopharynx of *Aeschna brevistyla* Ramb. (× 13). Original, chitin preparation.

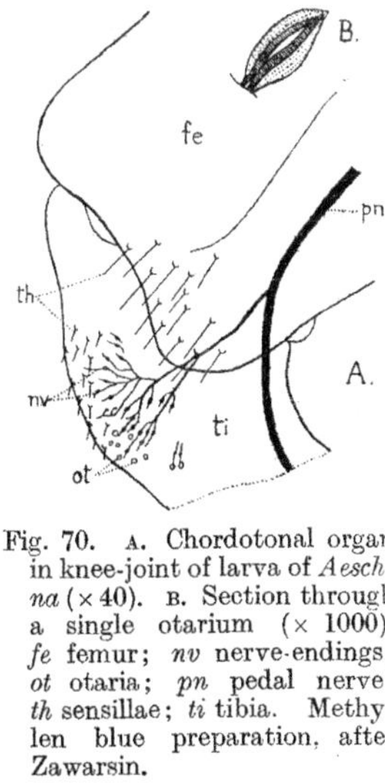

Fig. 70. A. Chordotonal organ in knee-joint of larva of *Aeschna* (× 40). B. Section through a single otarium (× 1000). *fe* femur; *nv* nerve-endings; *ot* otaria; *pn* pedal nerve; *th* sensillae; *ti* tibia. Methylen blue preparation, after Zawarsin.

chitin, under which is found a tiny rod or *style*. This rod is hollow, and carries a bundle of excessively fine nerve-fibrils. The total length of such an organ is only 0·01 mm.

Chordotonal organs differ from tympanulae in that the auditory vibration is received on a rod instead of on a stretched membrane or tympanum. Possibly such an organ only registers vibrations of much lower frequency than those which produce sounds. Their presence on the legs of the larva suggests that they may respond to currents set in motion by other animals in the water, and thus warn the Dragonfly of the approach either of an enemy or a victim. They would be particularly useful at night-time, when the eyes are probably of little use.

Stridulating Organs.

Berlèse [10, p. 174] has figured two small processes (*laminae stridulatoriae*) situated one on either side of the prothorax anteriorly, above the prosternum, in *Aeschna*. These he described as stridulating organs, which produce a sound by rubbing against two roughened patches behind the eyes, on either side of the posterior foramen. I know of no other observations on stridulation in Dragonflies, nor have I ever heard them emit any sound, except a sharp click of the wings occasionally when flying. In view, however, of the proved presence of minute auditory organs on the antennae, it is clear that more observations are required on this interesting point.

Other Sense Organs.

Minute organs of taste and touch occur scattered all over the mouth parts. They are however nowhere numerous, except on the labrum and epipharynx. Sensory hairs, with a tactile function, occur on many parts of the body, such as the antennae, the edge of the labrum, the legs, the abdomen, the anal appendages, and on the wing-sheaths of many larvae.

To sum up, the Dragonfly shares with the Bird the distinction of possessing the keenest vision of any animal in its Phylum. In both cases, the exceptional development of the sense of sight is correlated with an aërial existence. The senses of taste and touch are normally developed. But the senses of smell and hearing are very poor, the former being almost, if not quite, absent, while the latter is confined to organs of excessively minute size. To these five senses, as ordinarily understood, we must add the sense of direction or *orientation*, which is probably located in the tympanulae. The larvae may, perhaps, also possess a kind of *chemical* sense for gauging the purity of the water.

CHAPTER VIII

THE CIRCULATORY SYSTEM

The Circulatory or Blood-vascular System of the Dragonfly has been very little investigated, chiefly owing to the excessive delicacy and fineness of the structures forming it. In the larva, the broader and shorter abdomen, and the more abundant blood supply, make the study of the blood-system more easily carried out. We owe to Zawarsin [203] a very complete account of the heart in the larva of *Aeschna*.

It will be found convenient to treat the Circulatory System in three parts, (1) the closed system, (2) the open system or haemocoele, (3) the blood itself.

(i) *The Closed System.*

The Closed System in the larva is entirely dorsal, and consists only of the Dorsal Vessel. In the imago there is, in addition, a partly closed Ventral Vessel. The Dorsal Vessel is a slender tube lying dorsally above the alimentary canal. It begins in the ninth abdominal segment, where it is widest, and runs forward to the second segment, where it narrows sharply to a very fine tube. This latter penetrates through the thorax into the head, opening into the haemocoele above the brain. The broader portion of the Dorsal Vessel is called the *Heart*, the narrower portion the *Aorta*.

The Heart (figs. 53, 71).

The Heart proper differs from the Aorta, not only in its greater width, but also in possessing definitely contractile walls. It consists of eight *chambers*, corresponding to the eight abdominal segments (viz. numbers 9 to 2) in which it lies. These chambers will be spoken of as the first, second, etc., beginning from the

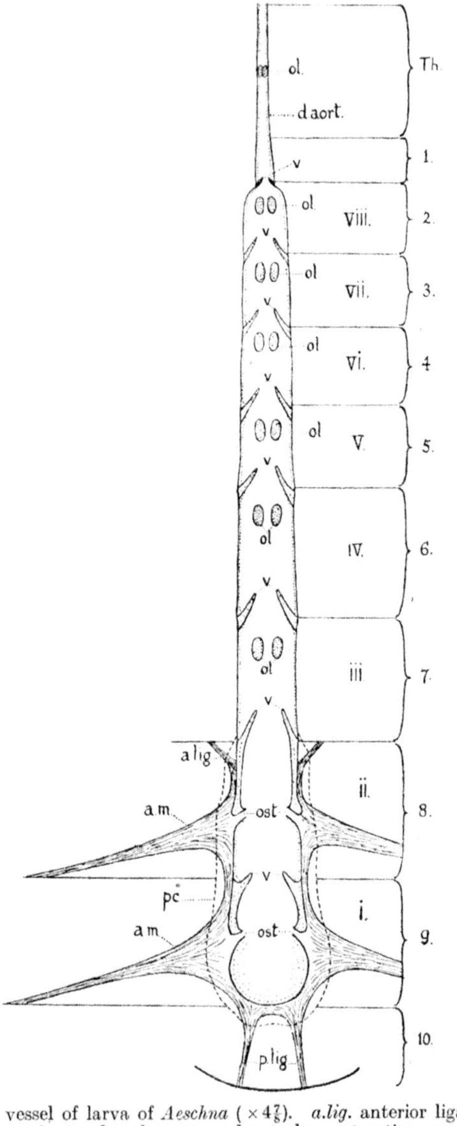

Fig. 71. Dorsal vessel of larva of *Aeschna* (× $4\frac{7}{8}$). *a.lig.* anterior ligament; *a.m.* alary muscle; *d.aort.* dorsal aorta; *ol* osteole; *ost* ostia; *pc* pericardium (dotted outline); *p.lig.* posterior ligament; *Th* thorax; *v* valves of heart; 1–10 abdominal segments; i–viii chambers of heart. After Zawarsin.

posterior end. Hence the eighth chamber lies in the second abdominal segment.

In most insects, each separate chamber can not only propel the blood forward, but can also receive blood from the haemocoele by openings, or *ostia*, in its walls. The heart of the Dragonfly larva is, however, differentiated into two very distinct portions, which we shall call the *Receptive* or *Hind Heart*, and the *Conductive* or *Fore Heart* respectively.

The *Receptive Heart* consists of the first two chambers only, viz. those in the ninth and eighth abdominal segments (fig. 71, i and ii). Each chamber not only performs the *ventricular* function of pumping the blood forward (into the chamber in front of it), but it also carries out the *auricular* function of receiving the blood from the haemocoele by means of *ostia*. Each chamber has a pair of these openings (*ost*) situated dorso-laterally near the middle of its wall. The wall is more muscular, and its powers of contraction and expansion greater, than in the case of the chambers of the Fore Heart.

The *first chamber* (i) is firmly suspended above the posterior end of the rectum. Its walls are held in position by two strong *alary muscles* (*a.m.*), attached to the side walls of the ninth segment, and also by two *posterior ligaments* (*p.lig.*) passing backwards, and attached to the wall of the tenth segment above the anus. Posteriorly the chamber is rounded off blindly; the anterior half is almost cylindrical. The *ostia* (*ost*) open, one on either side, above the alary muscles. Their openings are guarded by valves directed slantingly forwards. Each valve is continued forwards along the side wall of the heart to end in a larger flap (*v*) which forms the valve separating this chamber from the second.

The *second chamber* (ii) resembles the first, except that it is cylindrical throughout. It is supported by two strong *alary muscles* (*a.m.*) attached to the side walls of the eighth segment, and in front by two *anterior ligaments* (*a.lig.*) attached to the partition between the seventh and eighth segments. Its ostia and valves resemble those of i.

Close to each ostium of the Receptive Heart lies a small *phagocyte organ* (not shewn in the figure). This is simply a mass of corpuscles enclosed in a skeletal network of fibres. These organs lie just at the points where the blood-current is most intense. Their functions appear to be (1) the manufacture of fresh blood corpuscles, (2) the extraction of certain impurities from the blood stream. A pair of *lymphoid organs* also lie between the heart and the pericardium, opposite the ostia, and appear to pulsate with the heart.

The *Conductive Heart* consists of six chambers (iii to viii) lying in segments 7 to 2. All six are very closely similar. Each is cylindrical, and opens into the one before it by means of a pair of valves (*v*). There are no alary muscles, and no ostia. Hence the function of the Conductive Heart is purely *ventricular*; and even in this its action is not so powerful as that of the Hind Heart.

Replacing the absent ostia, on the wall of each chamber of the Conductive Heart, is a pair of peculiar oval organs, called by their discoverer Zawarsin "ostia-organs." As they are almost certainly reduced and closed-up ostia, we shall use the term *osteoles* for them. These *osteoles* (*ol*) lie close together on the dorsal wall of each chamber, a little in front of the middle. There is also a single pair on the aorta. Zawarsin believes that they have been formed by the fusion of an original phagocyte organ with an ostium proper. Consequently their chief function must be the manufacture of fresh blood corpuscles.

The Conductive Heart narrows slightly from behind forwards. In the second segment it ends abruptly, and passes by a pair of valves into the very narrow *aorta* (*d.aort.*).

Surrounding the whole heart, but only distinctly separated from it around the Hind Heart, is the **Pericardium** (*pc*), a space bounded by a delicate cellular wall supported by fibres of the alary muscles. On reaching the region of the heart, the muscle proper becomes differentiated into the so-called "elastic tissue." This, by interweaving of its fibres with the pericardial epithelium, forms a kind of net-bag in which the hind-heart lies free, except ventrally, where the fibres pass on into the wall of the heart itself. Specially thickened portions of this elastic tissue form the anterior and posterior ligaments.

In the Fore Heart, the pericardial wall fuses completely with the wall of the heart, so that no distinct pericardial cavity is recognizable.

Histology of the Heart. The wall of the heart is composed of three layers, an outer *adventitia*, a middle *muscular layer*, and an internal *intima*.

The *adventitia* consists of a special fibrous connective tissue resembling that of the fenestrate membrane of the compound eye. It also encloses peculiar spindle-shaped cells and phagocytes. It is separated from the muscular layer by a fine membrane.

The *muscular layer*, which is the contractile portion of the wall, is formed of complicated muscle-cells, partly fused into syncytia. The cell-masses are so arranged that the heart is distinctly divided into two lateral halves, by longitudinal dorsal and ventral zig-zig lines, or *sutures*, and also by *cross-sutures*, in the form of half-rings lying alternately to right and left of the longitudinal sutures. The combined result is to divide the heart into hexagonal areas. The muscle-fibres are striped, but there is no sarcolemma.

The *intima* is a fine internal membrane lining the lumen of the heart. It is almost certainly formed from the original sarcolemma of the muscular layer.

Innervation of the Heart. On either side of the heart there runs a fine longitudinal nerve-cord, the *alary nerve*, derived probably from the eighth ganglion of the ventral nerve-cord. Each alary nerve gives off strong branches to the alary muscles. Besides these, motor nerves derived from the nerve-supply to the intersegmental muscles, and hence connected with each of the ganglia of the ventral nerve-cord, enter the alary nerves at different levels, and give off a rich network of nerves all over the wall of the heart.

Functions of the Heart. Besides the principal function of pumping the blood forward in a closed stream, the heart has been shewn to perform other duties. Thus, Zawarsin [203], by injection of coloured substances into the blood, found that this foreign matter was quickly removed by the phagocyte organs of the hind-heart. Voinov [187] found that the entire wall of the heart was capable of removing certain soluble impurities from the blood, during its passage through the organ, and fixing it temporarily in the adventitia, by means of the phagocytes. Thus the heart is capable of performing a partial excretory function (see p. 118).

The Aorta.

The *dorsal aorta* (*d.aort.*) is a very delicate, straight, non-contractile tube, which carries the blood stream forward through the thorax to the head. It is not divided into chambers. Its chief peculiarity, in Dragonfly larvae, is the presence of a pair of small but distinct osteoles (*ol*) on its dorsal wall, in the middle of the thorax.

In the imago the dorsal vessel is excessively attenuated and elongated, owing to the corresponding form of the abdomen. No detailed studies have been made of it, but its structure seems to be essentially similar to that of the larva.

The Ventral Vessel.

The *ventral vessel* or *ventral sinus* of the imago is not really a closed tube, but a kind of groove overlying the ventral nerve-cord. A membrane placed on either side of the nerve-cord serves to separate it from the haemocoele. During the respiratory movements of the long slender abdomen, each contraction, by narrowing the abdomen transversely, forces the edges of these membranes up into a set of curved ridges, one pair in each segment. The blood is thus constrained to flow between these ridges, from before backwards. Thus an almost completely closed circulation is set up in the imago.

(ii) *The Haemocoele.*

As in all insects, the general body-cavity is not the same as the embryonic coelome. The latter becomes reduced to the narrow pericardial space, and the lumina of the developing reproductive organs. A secondary enlargement of the blood spaces, on the other hand, forms the general body-cavity. This is termed the *haemocoele*. Owing to the position of the various organs in the body, the haemocoele forms a kind of imperfect sinus along either side of the middle line, chiefly ventrally. It extends into the legs and antennae. It is filled with blood plasma, in which a weak and irregular circulation of corpuscles, from before backwards, is easily noticeable.

(iii) *The Blood.*

The blood of Dragonflies is a lymph-like fluid, in which corpuscles of various kinds live and float. The fluid or *plasma* is not absolutely colourless, but is usually of a delicate yellowish or greenish tint. In the larvae of some *Gomphinae* it is a very bright green.

The **blood corpuscles** appear to be of only two kinds, *miocytes* and *amoebocytes*. The former are the more numerous. In a transparent larva they can easily be seen travelling up the dorsal vessel, or passing backwards round the bases of the legs, and along the edges of the abdominal segments. They are of an elongate oat-shape. In the newly-hatched larva of *Anax papuensis* they vary in length from 13 to 15 μ. The probable limits of size, in larvae of different genera, may be placed at from 10 μ to 16 μ. In the very young larva, the number of corpuscles present in a single chamber of the heart at any given moment is usually only from six to ten; while, in the whole animal, there do not appear to be more than two or three hundred of them. There can be little doubt that, after each ecdysis, their number is largely added to by the manufacture of new corpuscles in the osteoles and phagocyte-organs. Both miocytes and amoebocytes are often spoken of as *phagocytes*, since they are supposed to play a part closely similar to that of the leucocytes, or white corpuscles, of the blood of vertebrates.

The *amoebocytes* differ from the miocytes chiefly in their irregular amoeboid form. Voinov [187] states that it is the amoebocytes alone which undertake the function of phagocytes. But he also denies the presence of the phagocyte-organs described by Zawarsin. It seems possible that both the oat-shaped corpuscles and the amoebocytes are only variations of a single form of corpuscle. Both have been seen to undergo binary division. Under certain circumstances the shape of the elongated corpuscle is known to alter. The distinction between them may therefore be only an arbitrary one.

There are no corpuscles in the blood of insects corresponding with the red corpuscles of vertebrates. The presence of a complicated tracheal system renders oxygen-carrying cells unnecessary.

Wandering *fat-cells* are often to be seen in the blood of Dragonflies. The large cells known as *oenocytes* also appear to be present, at any rate in the abdomen of the larva; but I know of no definite observations concerning them.

One of the principal functions of the blood is undoubtedly to receive from the mid-gut the products of digestion, and to carry them to the various organs, or to the fat-body for storage, until such time as they may be needed. The blood also dissolves the carbonic acid gas formed in the body of the insect. In the larva, it almost certainly plays some direct part in the process of respiration, by carrying this gas to points where it may be exchanged for oxygen from the water, by the ordinary processes of diffusion.

The Circulation.

In the young larva, directly after hatching or ecdysis, the whole course of the blood can be followed with ease. The blood enters in two streams on either side of the hind-heart, the current passing quite suddenly from sluggish to rapid as the blood approaches the ostia. The pulsations of the hind-heart are fairly regular, and usually from 60–70 per minute. Each pulsation drives a considerable quantity of blood forward into the next chamber. Between pulsations, the corpuscles remain almost stationary in the chambers of the heart. On reaching the aorta the corpuscles are forced forwards in jerks, so that an irregular stream of them is seen to enter the haemocoele in the upper part of the head. This stream at once spreads out, passing at first slantingly downwards beneath the brain. The main stream then circles forwards and upwards over the brain, and finally divides into two separate streams passing slantingly downwards and backwards towards the neck, through which the blood passes in two closely contiguous streams, separated by the nerve-cord. Some of the corpuscles pass into the bases of the antennae, where a circulation can be seen in the scape and pedicel.

In the thorax, a circulation along either side of the nerve-cord includes the basal joints of the legs (coxa and trochanter). A weaker current also passes backwards more dorsally. In the older nymphs, blood circulates in the spaces around the tracheae of the developing wing.

On entering the abdomen, the circulation becomes rather sluggish. The weak dorsal stream passes slowly backwards along either side of the heart. Most of its corpuscles pass into the heart in the eighth segment. Ventrally, a somewhat stronger stream passes along either side of the abdomen, as well as medially above the nerve-cord. This stream passes right back to the anus, and, in the Zygoptera, bathes the bases of the caudal gills. It then turns upwards and forwards, entering the heart through both chambers of the hind-heart.

As the larva grows, the haemocoele extends further into all the appendages, so that blood can circulate for a considerable distance into the antennae, the legs, and the caudal gills of the Zygoptera.

CHAPTER IX

THE RESPIRATORY SYSTEM

All insects breathe by means of *tracheae*, i.e. tubes designed to carry the oxygen supply direct to all the organs of the body. These tracheae are of ectodermal origin, being developed as paired segmental invaginations in the embryo. The mouth of the invagination becomes the *spiracle* or *stigma*. The tubes unite in the body of the insect to form a complete *tracheal system* consisting of paired longitudinal trunks, from which numerous branches arise. These branches divide again and again, until they end in exceedingly fine tracheae known as *capillaries*.

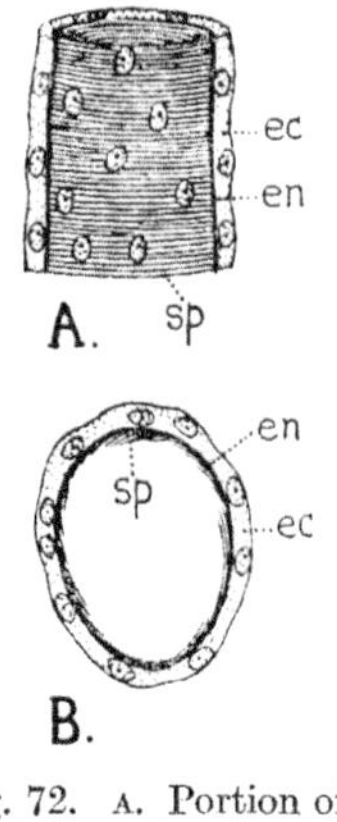

Fig. 72. A. Portion of a small trachea from larva of *Synlestes weyersi* Selys. B. T.S. through same (×800). *ec* ectotrachea; *en* endotrachea; *sp* spiral thread. Original.

The structure of a trachea (fig. 72) is easily understood, if we bear in mind its ectodermal origin. In transverse section it is seen to be formed of two layers, an outer *ectotrachea* (*ec*), and an inner *endotrachea* (*en*). The ectotrachea is a cellular layer corresponding with the hypodermis. Its cells are flattened, hexagonal, with distinct broadly oval nuclei, which are often seen bulging out from the contour of the smaller tracheae. The endotrachea is a chitinous non-cellular layer secreted by the cells of the ectotrachea, and corresponding with the cuticle. On its inner side it is strengthened by a peculiar thickening known as the *spiral thread* (*sp*). This is not a single complete spiral, but consists of a number of short coils, each the product of a single cell. These are so closely interwoven as to produce the appearance of a complete whole. The cross-section of the spiral thread in one of the larger tracheae of a Dragonfly larva is easily seen to be quadrilateral.

The tracheal system of the Dragonfly imago is of the type known as an "open" system, i.e. the spiracles are functional, and air is drawn into the system directly through them. It seems to be quite clear that the larva had originally an exactly similar system. With the adoption of the aquatic habit, however, the spiracles ceased to be useful, except on such special occasions (e.g. metamorphosis) as the larva was able to leave the water and breathe air directly. Hence the spiracles have persisted in the larva, although not generally functional, and often indeed quite closed up. The requisite oxygen is obtained by the agency of a system of specially developed *tracheal gills*, from which special branches known as *efferents* pass to the main tracheal trunks. Thus the tracheal system in the larva becomes more complicated than that of the imago.

Tracheal gills are developed in the larvae of Dragonflies in three places:

(i) In all Anisoptera, within the anterior portion of the rectum, by specialization of the rectal epithelium and pads. The whole formation makes up the beautiful structure known as the *branchial basket*.

(ii) In most Zygoptera, on the three anal appendages of the larvae, i.e. the two cerci and the appendix dorsalis. These are known as *caudal gills*.

(iii) Laterally, on certain of the abdominal segments, in a very few archaic *Calopterygidae*. These paired gills are remnants of original paired abdominal appendages, of which, in other Dragonflies, the only pair left are the cerci. They are homologous with the gills of the May-fly larva. The larvae possessing them also have caudal gills.

Besides these recognized forms of gills, there are the three rectal pads or folds of Zygopterid larvae, which probably function as blood-gills. These were dealt with on p. 116, and need not be further discussed.

We shall deal with the respiratory system under five headings:

(*a*) the tracheal system of the imago, (*b*) modifications in the general tracheal plan in the larva, (*c*) the branchial basket of Anisoptera, (*d*) the caudal and lateral gills of Zygoptera, (*e*) the movements of respiration.

A. The Tracheal System of the Imago.

1. *The Spiracles* (figs. 73, 74).

There are ten pairs of spiracles in the imago, two belonging to the thorax, and eight to the abdomen. They are situated on the last two thoracic and the first eight abdominal segments respectively.

The **mesostigmata** (fig. 73 A, B), or anterior thoracic spiracles, are situated on the mesostigmatic laminae, as described on p. 24. Each is a rather large transversely elongated opening, guarded by two thickened chitinous lips, one of which usually carries a number of stiff hairs arranged in sets. Within this lip there projects the peculiar organ called the *comb* (B), a stiff chitinous framework formed by a number of regularly arranged projecting spines or teeth, supporting between them a kind of honey-comb network of weaker chitinous "cells." The size and arrangement of teeth and cells varies in different genera. Landois [83], who first discovered this comb, believed that it was a sound-producing organ. It seems to be simply a specialized portion of the closing apparatus, probably derived from an original single row of strong bristles.

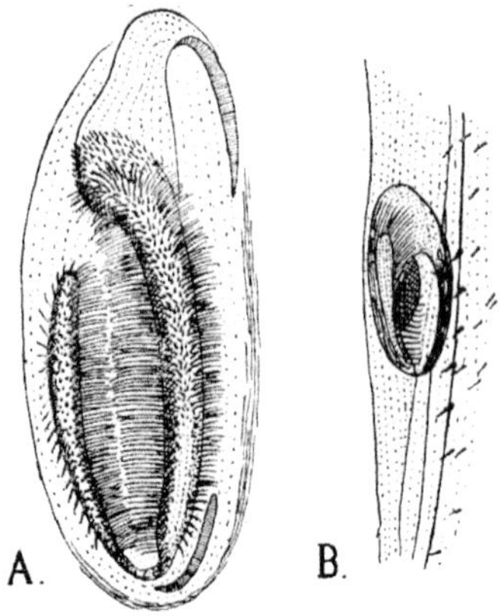

Fig. 74. Abdominal spiracles (×63). A. *Petalura gigantea* Leach, 7th segment. B. *Orthetrum caledonicum* Br., 7th segment. Original, chitin preparation.

The **metastigmata** (fig. 73 C), or posterior thoracic spiracles, lie in the lower portion of the metepisterna, as described on p. 24. Each is a small but distinct oval opening, about half as long as the mesostigma, but much wider by comparison, and more regular in shape. It carries no comb, but a narrow chitinous flap or ridge lies below one lip. On this flap is a regularly arranged row of stiff curved bristles, projecting inwards.

The **abdominal spiracles** (fig. 74) lie in the anterior portions of the pleural membrane of the first eight abdominal segments. They increase in size from the first segment to the eighth, the seventh pair being fairly large, the eighth pair much larger than any of the others. They are of an elongated oval form. In the older groups they lie obliquely to the body axis; in the higher groups they become longitudinally placed. In structure they are usually not unlike the metastigma (fig. 74 B, cf. fig. 73 C). In some of the older Anisoptera, e.g. *Petalura* (fig. 74 A), they become highly developed, with an extraordinary array of bristles on both lips, reminding one of the cylinder of a musical box. Some of these bristles are elongated, and become curved inwards in two opposing rows, which completely close the opening when the lips become approximated towards one another. In the *Libellulidae* (fig. 74 B) the sternum becomes armed with spines in the region of the stigma. In some cases the whole sternum becomes densely spinulose. In the Zygoptera, the internal flap seems to be absent.

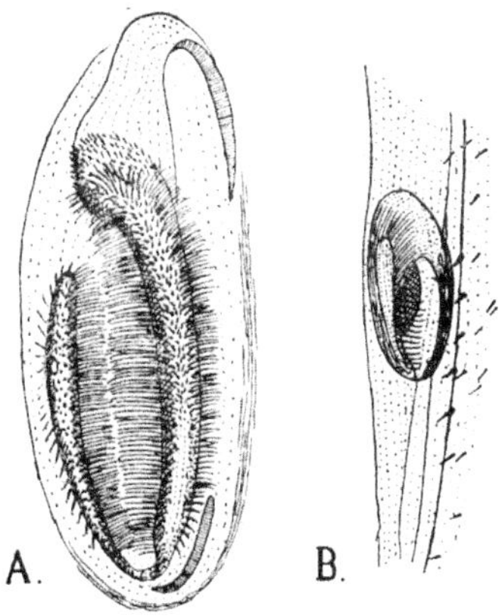

Fig. 74. Abdominal spiracles (×63). A. *Petalura gigantea* Leach, 7th segment. B. *Orthetrum caledonicum* Br., 7th segment. Original, chitin preparation.

2. *The Tracheae* (fig. 75).

The tracheal system of the imago consists of three pairs of main longitudinal trunks, with their numerous branches and capillaries. These trunks are named the *dorsal*, *visceral*, and *ventral trunks* respectively, from their positions [22].

The Dorsal Trunks (*DT*). These are the largest of the three, and the only pair to extend into the head. They run from the eighth to the first abdominal segment, on either side of the heart, above the alimentary canal. They are not directly connected

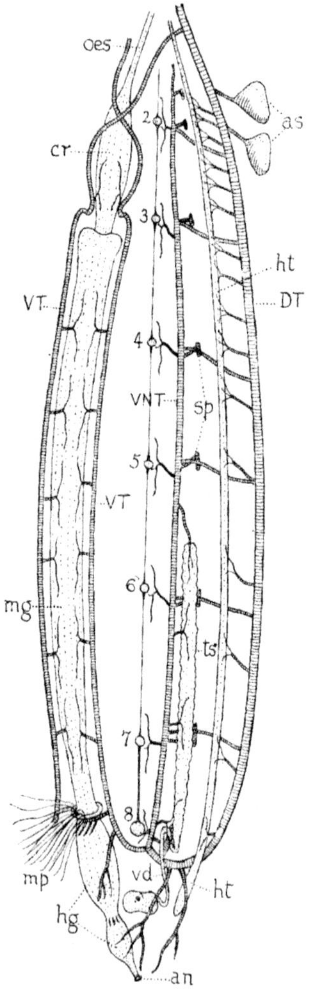

Fig. 75. Tracheal system of abdomen of *Aeschna brevistyla* Ramb. ♂ (×2½). Heart displaced to right, alimentary canal to left; left dorsal and ventral trunks and left testis removed. *an* anus; *as* air-sacs; *cr* crop; *DT* dorsal trunk; *hg* hind-gut; *ht* heart; *mg* mid-gut; *mp* Malpighian tubules; *oes* oesophagus; *sp* abdominal spiracles; *ts* testis; *vd* vas deferens; *VNT* ventral trunk; *VT* visceral trunk; 2–8 ganglia of ventral nerve-cord. Original, from a dorsal dissection.

with the abdominal spiracles, but each gives off six cross-branches to the corresponding ventral trunk, close to the levels of the second to the seventh of these spiracles. On entering the thorax, the dorsal trunks dip ventrally downwards, and run forwards to below the mesostigmata, from which they receive short strong branches. They also receive longer branches from the meta-stigmata. On entering the head, each divides into two. As the tracheal supply of the head is practically the same in imago and larva, and as it can only be followed satisfactorily in the latter, we have postponed a general account of this part to the next section.

Each dorsal trunk gives off the following principal branches:

(i) Branches to the three legs (*pedal tracheae*).

(ii) Branches to the two wings (*alar trunks*, best developed in the larva).

(iii) Branches to the thoracic muscles.

(iv) Short branches with blind ends, capable of being distended into air-sacs (*as*). These air-sacs are of very irregular occurrence (i.e. they are probably not noticed unless they are distended). Usually at least two pairs can be found in the second segment, lying on either side of the usually more distended crop. I have also found small air-sacs in other segments, particularly from the sixth to the eighth. Along this region, the dorsal trunk gives off branches which ordinarily lie close alongside the main trunk, directed backwards. In some cases the posterior ends are found to be distended, forming long sausage-shaped air-sacs. I have noticed this particularly in *Orthetrum* and *Hemicordulia*, where the abdomen is dilated.

(v) The six cross-branches to the ventral trunks, already mentioned. The most posterior of these is very slender, the others strong and thick.

(vi) Numerous branches to the heart, particularly along the anterior half. These branches, on reaching the heart, bifurcate into two widely diverging portions, one running anteriorly, the other posteriorly. From each of these, numerous smaller branches arise, so that the heart is covered by a complete net-work of tracheae.

At its posterior end, each dorsal trunk turns inward to enter a strong transverse branch lying in the eighth segment, by means of which it becomes connected with the ventral trunk of the same side. This cross-trunk is probably the highly developed homologue of series (v); its greater size being correlated with the great enlargement of the corresponding spiracle. The genital duct crosses it on the inner side. From it, two strong tracheae are given off to supply the last two segments and the anal muscles.

The dorsal trunks are easily recognized by their dark reddish-purple or brownish colour, as well as by their large size.

The Visceral Trunks (*VT*). These are smaller than the dorsal trunks, and of a much paler colour, either whitish or lightly pigmented in brown or pink. They run closely alongside the whole extent of the mid-gut, laterally. Anteriorly, each visceral trunk bends inwards in a kind of "kink" in the region of the gizzard, then bends outwards along the surface of the crop, which it *crosses dorsally*, finally entering the dorsal trunk of the opposite side close up to the thorax. The right visceral trunk usually passes over the left, towards the anterior end of the crop. This point is called the *visceral crossing*. Posteriorly, each visceral trunk turns downwards in the eighth segment to join the ventral trunk, at its point of union with the large cross-trunk mentioned above. Each visceral trunk gives off the following branches:

(i) A branch to the crop and gizzard.

(ii) Six short branches to the mid-gut. Each of these, on reaching the gut, bifurcates into two diverging portions, which spread out anteriorly and posteriorly, ramifying over the surface.

(iii) A strong branch to the zone of entry of the Malpighian tubules. This divides into a number of short branches corresponding with the separate Malpighian conduits. Each short branch quickly divides into five or six fine tracheae, one of which accompanies each Malpighian tubule along its entire length.

(iv) A strong branch to the hind-gut.

The Ventral Trunks (*VNT*). These are well developed and darkly pigmented, but not so large as the dorsal trunks. They lie ventrally in the abdomen, above the pleural membrane, one on either side of the ventral nerve-cord. Anteriorly they become very attenuated, and divide up into two or three small branches, one of which appears to reach the metastigma, or a branch arising from it. Another small branch joins a branch from the dorsal trunk. Posteriorly the ventral trunk is connected with the dorsal and visceral trunks of the same side, in the manner already described. Each ventral trunk gives off the following branches:

(i) Eight short branches to the corresponding abdominal spiracles on the same side.

(ii) Six branches to the dorsal trunk, as already described.

(iii) Seven branches to the seven abdominal ganglia of the ventral nerve-cord. Each of these runs inwards transversely, close to the corresponding ganglion. It then divides into two strongly diverging branches, of which the posterior is the largest. The anterior runs forwards to supply the nerve-cord; the posterior gives off a branch to the ganglion, and then runs backwards to supply the nerve-cord. The branches are so arranged on either side of the nerve-cord that the latter appears at first sight to be supplied from two continuous longitudinal tracheae.

(iv) Branches to the gonads. In the male, the testis is supplied by four branches, arising from the ventral trunk in segments 5–8 respectively. In the female, an immense number of branches supply the ovary, particularly the anterior portion. These penetrate the fat-body, and then branch and re-branch in a complicated manner, forming a strong net, in which the ovary and its enveloping fat-sheath are firmly held.

B. **Modifications in the Larva** (figs. 76, 77).

More is known about the tracheal system of the larva than about that of the imago, not only because the broader and shor er abdomen, and the greater thickness of the tracheal trunks, render dissection easier, but also because, just after ecdysis, almost the whole of the tracheal system of the larva can be followed under the microscope. The general plan is evidently the same, the three pairs of longitudinal trunks having the same relative positions, and giving off much the same branches [111, 147].

Modifications in the Abdomen. The dorsal trunks are of great thickness, and of a peculiar semi-metallic reddish or coppery colour. They are strongly bent outwards in the region of the mid-gut in segments 3–4. Posteriorly, about the end of segment 7, each divides into two branches. One of these turns inwards to supply the posterior part of the rectum, the anal region, and, in Zygoptera, the median caudal gill. The other turns sharply outwards to receive a short branch from the junction of the

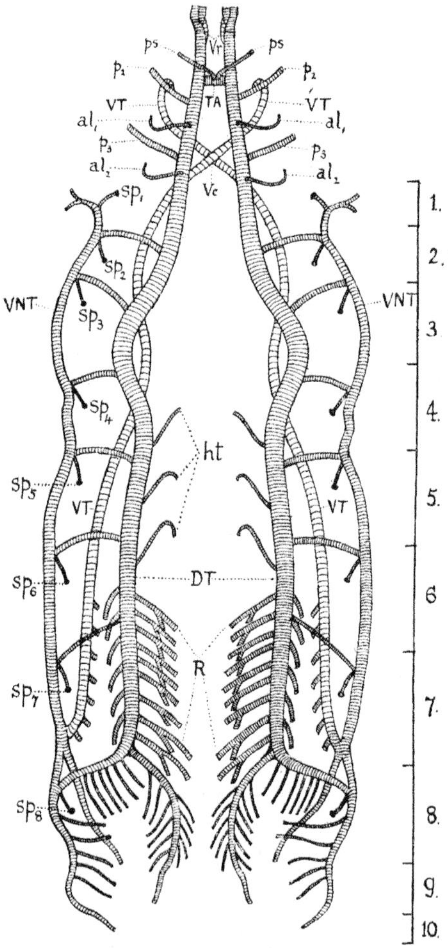

Fig. 76. Tracheal system of larva of *Dendroaeschna conspersa* Tillyard, exclusive of head (× 6). al_1, al_2 alar trunks; *ht* trachea to heart; p_2, p_3 pedal tracheae; *ps* branch to metastigma; *R* rectal efferents; *TA* thoracic anastomosis; *Vc* visceral crossing; *Vr* visceral remnant; 1–10 abdominal segments. Other references as in fig. 75. Original, from a dorsal dissection.

visceral and ventral trunks, and then runs backwards to supply the sides of the ninth and tenth segments, and, in Zygoptera, the lateral caudal gill. The tracheae developed in connection with the branchial basket in Anisoptera are dealt with in the next section. The visceral and ventral trunks unite in segment 7; from their junction a short branch runs to the dorsal trunk; a longer branch runs backwards through segments 8–9. The visceral trunks are very poorly developed in Zygoptera.

The abdominal spiracles are small and non-functional. The branches connecting them with the ventral trunks are short and thin. They appear to be solid except just before and after ecdysis, when the endotrachea is withdrawn through the temporarily open spiracle. The visceral trunks are shining white in colour; they cross one another at the visceral crossing (*Vc*) as in the imago. Each ends up anteriorly on the second pedal trunk of the opposite side, close under the dorsal trunk.

Modifications in the Thorax. In the larval thorax, the dorsal trunks do not bend ventrally downwards, but pass forward in the synthorax dorsally, close under the pleural ridges. Each gives off two *alar trunks*, which enter the wings at the anal end of the base, forming a loop from which the wing-tracheae arise, and leaving it by the costal side. The alar trunk of the hind-wing then curves downwards, externally to the dorsal trunk, and finally enters the third pedal trachea close to the region of the metastigma. The alar trunk of the fore-wing follows a similar course, ending up on the second pedal trachea. Just in front of the level of this latter trachea, the two dorsal trunks converge closely, and become united by a short thick *thoracic anastomosis* (*TA*), from which small branches pass back to the metastigmata. On reaching the level of the mesostigma, each dorsal trunk opens into a kind of chamber below that organ, formed by enlargement of a short cross-branch. In front of this, each dorsal trachea divides into two branches, which pass forwards to the head. In the Zygoptera, the second pedal trachea joins the dorsal trunk just under the mesostigma.

A peculiar formation connecting the anterior and posterior portions of the dorsal trunk, just in front of and behind the mesostigmatic chamber respectively, is the *visceral remnant* (figs.

76, 77, *Vr*), a short white loop which appears to be the severed anterior end of the visceral trunk. If that be so, the connection of the visceral trunk with the pedal tracheae is a secondary formation.

Tracheation of the Head (fig. 77). The labium is supplied by a pair of strong tracheae, which arise from a transverse loop, derived from the two first pedal tracheae close to their bases. All the rest of the head is supplied by the two pairs of large dorsal

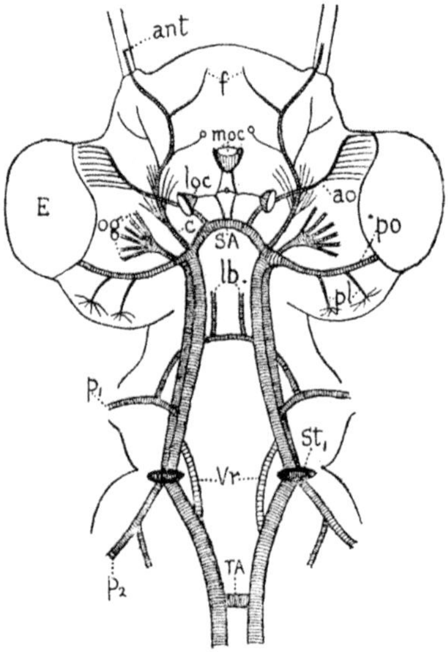

Fig. 77. Tracheal system of head and part of thorax of larva of *Synlestes weyersi* Selys (× 13). *ant* antennary trachea; *ao* anterior optic trachea; *c* cerebral trachea; *E* eye; *f* frontal branches; *lb* labial tracheae; *loc* lateral ocellus; *moc* median ocellus; *og* trachea to optic ganglion; p_1, p_2 pedal tracheae; *pl* branches to postocular lobe; *po* posterior optic trachea; *SA* superior anastomosis; St_1 mesostigma; *TA* thoracic anastomosis; *Vr* visceral remnant. Original, cedar oil preparation.

trunks. Of these, the two uppermost, soon after entering the head, are united by a semicircular arch, the *superior anastomosis* (*SA*). From the two bases of this arch, large tracheae pass transversely outwards to supply the eyes, giving off branches to the postocular lobes by the way. Higher up on each side of the arch, smaller branches pass outwards and forwards to the anterior region of the eye. At the summit of the arch, in Zygoptera, two small tracheae come off, each of which soon branches to supply the median ocellus, and the lateral ocellus of its own side. Thus the median ocellus receives a double tracheal supply. The lower branches of the dorsal trunks divide into two branches below the bases of the arch. The more posterior of these two branches is very thick. It soon divides up in a fan-like manner to supply the optic ganglion. The more anterior branch is slenderer. It runs forwards in a double curve to enter the antenna, giving off branches to the brain, the mandibles and maxillae, and the frontal region, on the way.

The above description was made from a nearly full-grown larva of *Synlestes* (fig. 77). In *Aeschna*, the superior anastomosis is comparatively weak, but the other main branches are much stouter than in *Synlestes*. There is no branch corresponding to the one from the arch to the anterior part of the eye. The branches to the optic ganglion and the brain are united for a short distance. They then separate, and spread out as two immense fans, deeply coloured with brown pigment.

Modifications in the Spiracles. The mesostigma in the larva is well formed and easily seen. It is shorter and broader than in the imago, somewhat resembling the metastigma of the latter, though not so rounded. It is open, and provided with a row of bristles beneath one lip. It is clear that this spiracle can become functional whenever the larva has need to leave the water for a time. Many *Aeschnine* larvae wander about on rocks or other damp places at night, using these spiracles for breathing. The metastigma is very small, being little more than a point. In the Anisoptera it is just open, by a slit; in the Zygoptera, it seems to be quite closed. With few exceptions, it appears to be non-functional. The abdominal spiracles have been already mentioned.

An excellent account of the thoracic spiracles and abdominal tracheation of Anisopterid larvae has been given by Oustalet[1] [111]. I know of no detailed account of the tracheae in the imago. An excellent short account is, however, given by Calvert [22].

[1] He, however, denies the existence of abdominal spiracles.

C. **The Branchial Basket.**

This unique, highly efficient, and very beautiful structure is only found in the larvae of Anisoptera. The "basket" is formed by expansion of the anterior two-thirds or more of the rectum, which becomes a wide barrel-like chamber. From its walls the gills project inwards in six longitudinal sets, developed originally, as I have recently shewn [179], from the six rectal pads, which remain undifferentiated in the posterior part of the rectum. The whole structure of the branchial basket is so complicated, that we shall divide our account into five sections: (1) the efferent tracheal system, (2) the types of gill-system, (3) the ontogeny and phylogeny of the types, (4) the histology of the gills, and (5) the physiology of respiration.

1. *The Efferent Tracheal System* (fig. 78).

Originally developed as a series of branches from the main trunks to the rectum, this tracheal system, on the birth of the larva, at once takes on a definite *efferent* function, conveying the gas received by the gills from the water to the main longitudinal trunks. Thus, while it is most convenient to trace the system inwards from the large trunks to the fine capillaries, we shall use an efferent notation in describing them. The arrangement may be gathered from figs. 76, 78.

There are six series of principal branches, or *primary efferents* (ef_1), to the gill-basket, arranged in three pairs. Two series come off from the dorsal trunk of each side, one from the visceral trunk. The ventral trunks have no part in the formation. Each primary efferent, on approaching the gill-basket, divides into two *secondary efferents* (ef_2), which pass through the muscular tunic to the bases of the gills. Here they break up in various manners, according to the type of gill. These branches enter the gill-base, and give off a very large number of tracheal *capillaries*, fine tubes without any spiral thread, destined to receive the gas from the water in the rectum. Each capillary forms a *complete loop* along the free border of the gill, finally returning to the same secondary efferent from which it originally arose. The different systems of branching, and the courses taken by the capillaries, will be best understood by reference to fig. 80.

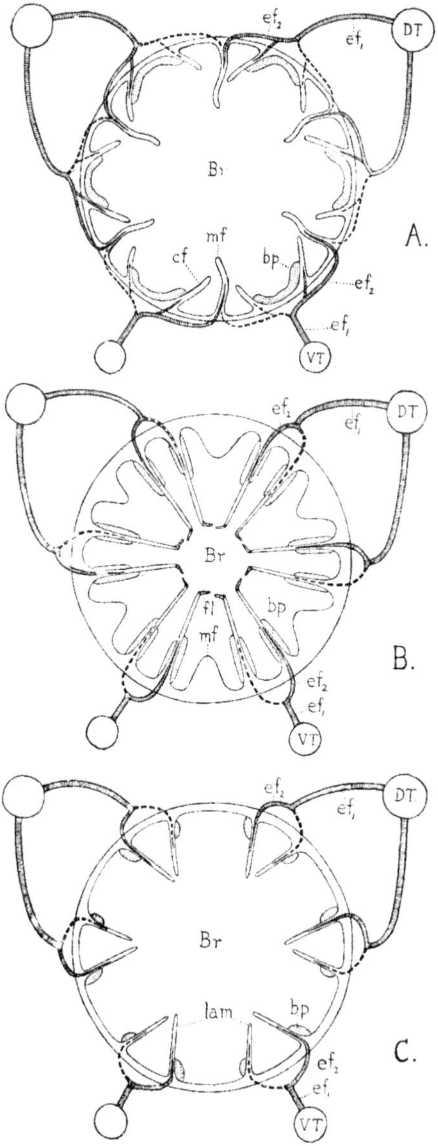

Fig. 78. Diagrammatic T.S. through branchial basket of Anisopterid larvae, to shew types of gill-system. A. Undulate Simplex. B Foliate Duplex. C. Lamellate Duplex. *bp* basal pad; *Br* branchial basket; *cf* cross-fold; *DT* dorsal trunk; ef_1 primary, ef_2 secondary efferent; *fl* foliae; *lam* lamellae; *mf* main-fold; *VT* visceral trunk. In C, only one lamella is shewn cut in each hemibranch, although six or more are really cut (see fig. 83 B). The dotted tracheae lie outside the true level of the section. Original.

2. *The Types of Rectal Gill-System* (figs. 78–80).

The gill-system may be either *Simplex* or *Duplex*. Distinct types of gill are found in both systems, and have been named according to their form [138, 179].

The Simplex System (figs. 78–79 A, 80, 1).

In the Simplex System there are six principal longitudinal gill-folds (the *main-folds*, *mf*) running the whole length of the basket, and arranged in the positions denoted by the even numbers on a clock-face diagram, i.e. in the same positions as the rectal pads of the imago. These are supported, to right and left alternately, by a double series of *cross-folds* (*cf*). A complete main-fold with its cross-folds is called a *holobranch*. The twelve sets of secondary efferent tracheae correspond with the twelve sets of cross-folds. Each, however, after supplying a branch to its cross-fold, continues on to enter the main-fold. Thus each main-fold receives a *double tracheal supply*, from right and left alternately. We see, from fig. 78 A, that each main-fold, therefore, is in connection, by alternate tracheae, with *two* sets of primary efferents. Or, in other words, the two sets of secondary efferents, arising from one series of primary efferents, do not enter a single main-fold, but pass, one to the left side of one main-fold, the other to the right side of the next main-fold. This complicated arrangement becomes much simplified in the Duplex System.

The main-folds are crinkled and grooved, projecting obliquely backwards into the rectum in the manner seen in fig. 79 A. The cross-folds are also crinkled, but do not project so far inwards.

There are two very distinct types of Simplex System known:

(1) *The Undulate Type* (figs. 78–79 A, fig. 80, 1). This occurs in the *Cordulegastrinae*, *Petalurinae* and in the genus *Austrogomphus*. The gill-folds remain entire, and are undulated along their whole length. The free edge of the gill carries minute spines at short intervals. The tracheal branches approach fairly near to the free edge, giving off capillaries along their whole length. These capillaries run nearly straight to the free border, and then turn round to form loops, returning to a near-by branch of the same secondary efferent. Pigment is seldom developed in these gills.

(2) *The Papillate Type* (fig. 80, 2, 3). This occurs in most of the *Gomphinae*. The edges of both main and cross-folds are broken up into long slender papillae, along each of which a strong *axial trachea* runs almost to the tip. A small number of capillaries (about twenty) arise from this trachea, each forming a complete loop within the papilla. The papilla is strengthened externally by frequent short transverse chitinous rods. Purple or mauve pigment is developed below the bases of the papillae. This is clearly a specialization from type (1), and of greater efficiency, since the water bathes each papilla on all sides.

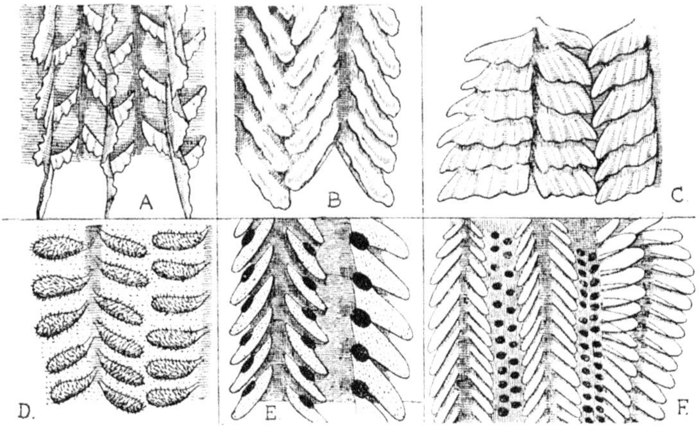

Fig. 79. Portions of the freshly opened gill-basket, to shew form of gills. A. Parts of three holobranchs from *Austrogomphus ochraceus* Selys (undulate simplex). B. The same from *Austroaeschna multipunctata* Martin (implicate). C. The same from anterior portion of basket of *Aeschna brevistyla* Ramb. (normal foliate). D. The same from *Anax papuensis* Burm. (papillo-foliate). E. The same from *Synthemis macrostigma* Selys (archi-lamellate). F. Anterior halves of three holobranchs from *Diplacodes haematodes* Burm. (neo-lamellate). Original.

The Duplex System (figs. 78 B, C; 79 B–F; 80, 4–11).

In this system, the main-folds may be present, but not functioning as gills (*Aeschninae*), or entirely absent (*Libellulidae*). The gills are formed entirely from the double series of cross-folds, which become enlarged to form *twelve series*, or *hemibranchs*, *of separate obliquely-placed or transverse gills*, arranged in six double series, or *holobranchs*. As will be seen from fig. 78, each holobranch

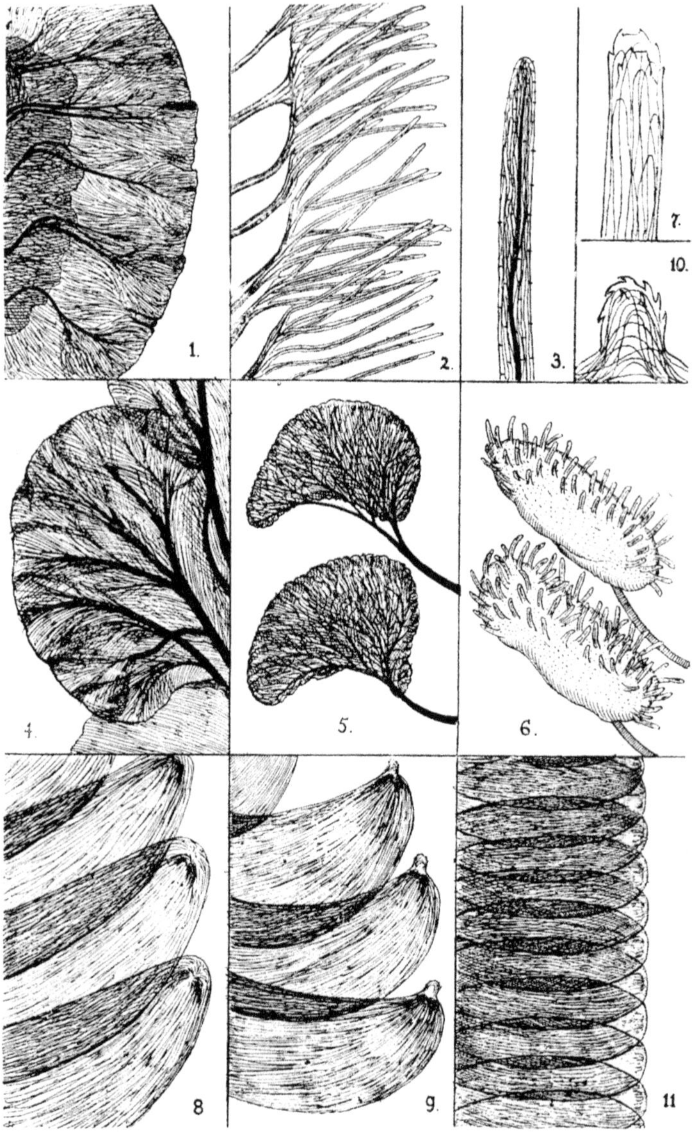

Fig. 80. Structure of rectal gills. 1. From main-fold in fig. 79 A. 2. From main-fold of *Hemigomphus heteroclitus* Selys (papillate simplex). 3. A single papilla of same, apical third. 4. From a gill in fig. 79 B. 5. Two foliae from posterior portion of gill-basket shewn in fig. 79 C. 6. Two papillo-foliae from fig. 79 D. 7. A single papilla of same, apical half. 8. Three lamellae from fig. 79 E. 9. Three lamellae of *Austrocordulia refracta* Tillyard. 10. Supra-vortical papilla of same. 11. Ten lamellae from fig. 79 F. (1, 2, 4, 8–11, ×75; 5, 6, ×40; 3, ×460; 7, ×680; 10, ×570.) Original, from photomicrographs.

lies with its axis of symmetry in a position corresponding with one of the *odd* numbers on the clock-face diagram. This is because the holobranch is formed, not from the right and left sets of cross-folds originally supporting one main-fold, but from the right set of one together with the left set of the next. Now we have already shewn that two such sets have their trachea connected with a single series of primary efferents (fig. 78 A). Hence the Duplex System combines a definite increase in the complexity of the gill-system, with a very useful simplification of the efferent tracheal system. As the tracheae to the main-folds are aborted, each separate gill now has its own secondary efferent, and each holobranch its own single series of primary efferents.

Three main types are easily recognized:

(1) *The Implicate Type* (fig. 79 B; fig. 80, 4). This occurs only in the *Brachytronini*. The gills form a series of obliquely-placed projecting concave tiles, slightly overlapping one another from before backwards. The true axis of symmetry of a holobranch lies above the letter B in fig. 79. Owing to the main-folds being still present, but non-functional, the cross-fold character is not yet lost, and the overlapping ends of the gills from two consecutive holobranchs (to the left of B) are connected with the underlying main-fold. Thus the gills can be spread out on a slide so as to resemble very closely the Undulate Simplex Type, from which they have been clearly derived (cf. fig. 80, 4, 1). Pigmentation is very rarely present.

(2) *The Foliate Type* (fig. 79 C, D; fig. 80, 5–7). This is found only in the *Aeschnini*. Here the gills are constricted off so as to form a series of separate leaves or *foliae*, the secondary efferent representing the stalk. The free ends of the gills are much enlarged, with capillary loops spreading all over them in immense numbers. There are two sub-types, (*a*) the *normal foliate* sub-type, found in *Aeschna* (fig. 79 C; fig. 80, 5), with the foliae resembling crinkled rhubarb or cabbage-leaves, and (*b*) the *papillo-foliate* sub-type, found in *Anax* (fig. 79 D; fig. 80, 6, 7), with the foliae swollen up into oval humps on short stalks. All over the free surface of these humps a number of minute papillae arise. These differ from the larger papillae of the Papillate Simplex Type in not containing any axial trachea, but simply five or six drawn-out

capillary loops. The free end of the papilla is armed with strong hooks. Pigment of a deep brown, purplish or even blackish colour is strongly developed in the foliate type. The total number of capillaries in the gill-basket of *Anax* reaches 100,000 or more.

(3) *The Lamellate Type* (fig. 79 E, F; fig. 80, 8–11). This beautiful and unique type occurs only in the *Libellulidae*, with little variation throughout this immense family. Each gill is in the form of a separate flat lamella, set on a broad obliquely-placed base (figs. 81, 82). The number of gills in a hemibranch is from twelve to thirty. Near the anterior outer corner of the base of each gill is a prominent rounded disc of an orange or brown colour,

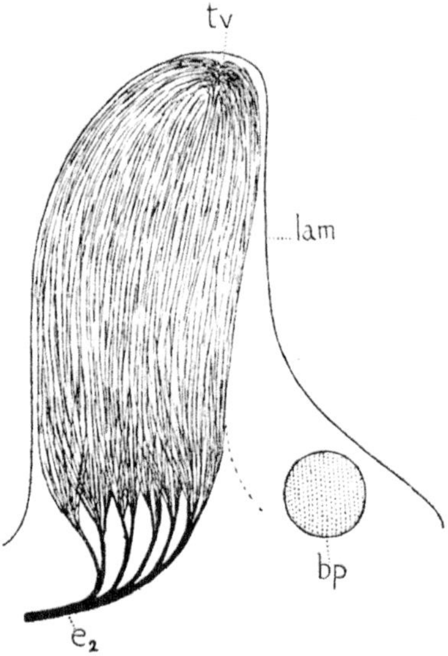

Fig. 81. A single lamella from the gill-basket of *Hemicordulia tau* Selys ($\times$ 87). *bp* basal pad; e_2 secondary efferent; *lam* blade of lamella; *tv* tracheal vortex. Original.

the *basal pad* (*bp*). The secondary efferent trachea of the gill breaks up, beneath the base, into a number of strong tracheae close together. From each of these there comes off a whole brush or pencil of capillaries, which run up to the oval tip of the gill, and then turn inwards to form the very characteristic appearance called the *tracheal vortex* (*tv*), finally returning to join other pencils of capillaries.

The lamellae in a hemibranch overlap from before backwards (cf. fig. 79 B, E), and can be raised or lowered. They are prevented from damaging one another by the presence of three *lamellar tubercles* (fig. 82), one on the posterior surface (*tp*) towards the tip, and two on the anterior surface (*ta*) near the middle and base respectively. Fig. 82 shews three consecutive gills of *Austrocordulia refracta*, viewed in profile.

The secondary modifications in type may be briefly noted and named:

(i) *Archi-lamellate sub-type.* This occurs only in the *Synthemini* (fig. 79 E). There are only twelve widely-separated lamellae in a hemibranch, and the basal pads are of enormous size.

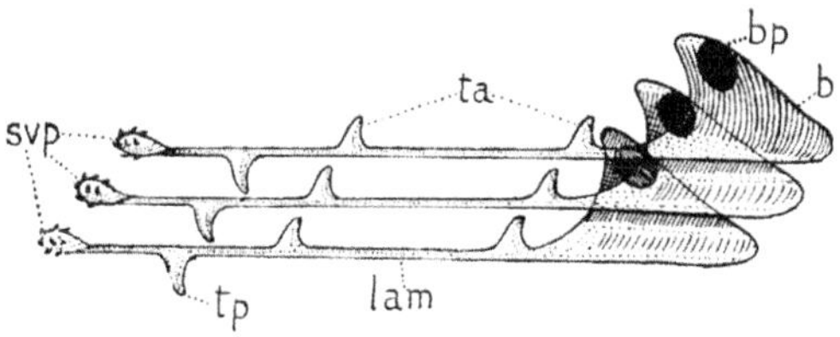

Fig. 82. Three lamellae from *Austrocordulia refracta* Tillyard (×87) seen in profile. *b* base; *bp* basal pad; *lam* blade of lamella; *svp* supra-vortical papilla; *ta* anterior tubercles; *tp* posterior tubercle. Original.

(ii) *Neo-lamellate sub-type* (fig. 79 F). Here the number of gills is greatly increased; they are placed close together, and the basal pads are much smaller.

(*a*) *Corduline form.* The lamella is irregular in outline, somewhat hump-backed or nodding at the tip (fig. 81). Each gill only overlaps the one following it. A very lovely mauve or pale purple pigment is nearly always present.

(*b*) *Libelluline form.* The lamella becomes nearly regular in form, with broadly rounded tip (fig. 79 F). Each one overlaps two or more following it. The tracheal vortex is weak or nearly absent. The gills may be pure white (*Orthetrum*), or pigmented as in (*a*), or even of a deep shining black colour (*Diplacodes*).

In *Austrocordulia* (fig. 80, 9) there is an extraordinary development of a strong short terminal or *supra-vortical papilla*, armed with curved spines (fig. 80, 10; fig. 82, *svp*). It is clearly an evagination comparable with the papillae of *Anax*.

Main-folds are completely absent in the Lamellate Type.

3. *Ontogeny and Phylogeny of the Types.*

All newly-hatched Anisopterid larvae have a gill-system closely resembling the Undulate Simplex Type. This persists throughout the larval life of the more archaic groups (*Petalurinae, Cordulegastrinae* and the genus *Austrogomphus*). In the *Gomphinae* the Papillate Type develops by evaginations of the gill surface, separated by constrictions. In the *Aeschninae* the larvae are born with only the main-folds present. The cross-folds appear as small buds close to the secondary efferents, at the third instar. They increase in size, much resembling the *Libellulid* lamellae at first, only smaller. Later they overgrow the main-folds, and the latter lose their tracheae. The foliate type is established in *Aeschna* by the seventh instar. In *Anax* the humps develop earlier, and the papillae form as groups of tiny evaginations during the sixth and seventh instars. In the *Libellulidae* the larvae are born with six main-folds, each very short, and with only two main undulations. These are supported by only one or two cross-folds on either side. The cross-folds are already as large as at the fourth instar of *Aeschna*; and, except for a slight crinkling, are clearly destined to become lamellae. At the next ecdysis, the main-folds lose their pigment, and become non-functional; the cross-folds become definite lamellae. At the fourth instar, the main-folds are quite gone; the cross-folds number from five to six in a hemibranch. Lamellae continue to develop at each instar, until the full number is attained.

We may summarize this evidence as follows: The Undulate Simplex Type is the basic form of gill. The Papillate Simplex Type is evolved directly from it. The Implicate Duplex Type (ontogeny not yet examined) appears to be another simple modification in a different direction, the cross-folds alone forming the gills. The Foliate Duplex Type is a higher development from the Implicate Type, by constricting-off of the separate cross-folds. This line culminates in *Anax*.

The Lamellate Type, though necessarily classified with the other Duplex Types, is a distinct line of development from the very base of the series. It appears to have arisen at a time when only a small region of the rectum had become specialized for a respiratory function. The main-folds were quickly lost, and the beautiful and very regular formation, which we now see, has been formed by constant addition of extra lamellae. In the *Synthemini*, the most archaic of *Libellulidae*, these still only number twelve in a row.

Fig. 78 shews diagrammatically the connection between the three main types.

4. *Histology of the Rectal Gills* (fig. 83).

All forms of gill have the same histological structure, with

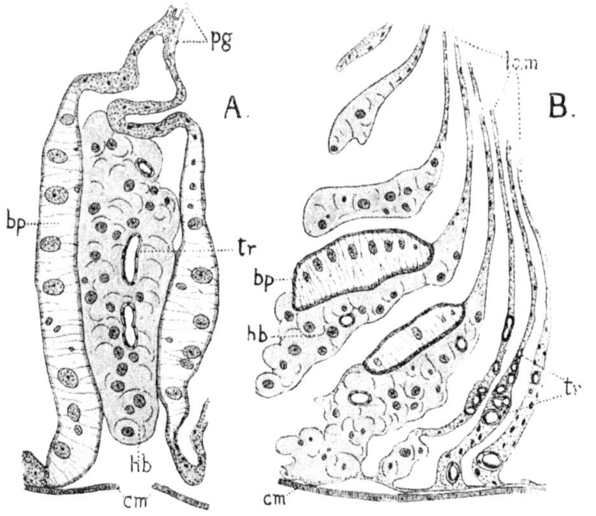

Fig. 83. Histology of the rectal gill. A. T.S. through base of gill from *Aeschna brevistyla* Ramb. B. The same through several lamellae of *Cordulephya pygmaea* Selys (× 250). *bp* basal pad; *cm* circular muscle; *hb* hypobranchial tissue; *lam* blades of lamellae, cut at different levels in the one section; *pg* pigmented epithelial syncytium of gill; *tr* tracheae. Original.

only unimportant differences of detail. The gill consists of the following parts:

(1) *An external cuticle.* This is continuous with the general intima of the hind-gut. It is excessively fine, and covers the gill on both sides. Where it turns over at the free edge of the gill, it is armed with small spines. It is quite colourless.

(2) *An epithelial syncytium.* This is a very flat layer of completely fused cells, underlying the cuticle of the gill. It is a specialized portion of the epithelium of the hind-gut, and may be termed the "respiratory epithelium." The nuclei are oval, numerous, and irregularly placed. Except at the base, the syncytia of the two walls of the gill fuse together, blocking out the haemocoele, and only allowing of the passage of capillaries. Pigment granules are frequently scattered in the epithelium.

(3) *Tracheal capillaries.* These run in the fused syncytial mass. They usually pass along into the gill closer to one wall, and return closer to the other. The lumen is minute, and collapses soon after death. There is no spiral thread. The nuclei of the ectotrachea are few, and jut out strongly at the sides of the lumen. The structure suggests great stretching of an originally shorter and wider tracheal loop.

(4) *Basal pads* (*bp*). At the base of the gill the two walls are not fused. A prolongation of the haemocoele separates them. In this cavity the larger tracheal branches (*tr*) run. Either on one side, or on both, the epithelium of the gill is seen to be enormously swollen, forming a huge *pad*, tensely filled with liquid. These pads are only found on the anterior side of the gill in most forms; in the *Aeschninae* (A), they occur on both sides. In all forms except the *Libellulidae* they are diffuse and non-localized, so that they are not noticeable on opening the gill-basket. In the *Libellulidae*, however, in correlation with the extreme separation of the gill-system into definite lamellae, the basal pads have become restricted and localized into rounded pigmented discs, which are very conspicuous objects in the opened gill-basket. The pads possess huge oval nuclei, from which a fibrillar network radiates out. No separate cell-boundaries can be seen. The protoplasm is restricted to the fibrils and the extreme borders of the pad; the spaces between the fibrils are filled with liquid, apparently water. The pads appear to form strong basal supports for the gills.

Just as the thin epithelium of the gill is clearly homologous with the undifferentiated rectal epithelium of other insects, so it seems highly probable that these basal pads are derived from the original swollen portions, known as the rectal pads.

(5) *The hypobranchial tissue* (*hb*). The space at the base of the gill, between the two walls, is more or less completely filled with a peculiar tissue surrounding the larger tracheae. This tissue consists of large rounded lobules, each with one or more rounded nuclei. The cell-boundaries are not very distinct. It has been called "connective tissue," and also "adipose tissue," but most resembles the latter. It appears to be a specialized development from the fat-body [145, 179].

5. *Physiology of Respiration.*

We cannot here enter into the details of the extremely problematical and difficult subject of insect respiration in general, and aquatic larvae in particular. No really acceptable explanation of the action of the rectal gills has yet been put forward. The solution, however, appears to lie along the simple line of explanation indicated by Ris [138], following Lowne.

It seems clear that the primary factor in rectal respiration is the *epithelial syncytium*, and that its action is exceedingly simple. The presence of a layer of separate cells, in the unspecialized rectum, formed, I believe, a definite barrier to the rapid absorption of oxygen, when the larvae first took to the water. By fusion of the cells into a single protoplasmic syncytium, and by extensive stretching of the same to the necessary tenuity, there is developed between the circulating water and the underlying capillaries a structure just strong enough to prevent injury to the latter, but sufficiently permeable to act as a diffusion membrane. The action of such a membrane would be as follows:

When the larva hatches, the tracheal system is quickly filled with carbonic acid gas, derived from the region of the mid-gut, as already described on p. 70. By this means, *all the capillary loops* become filled with gas. As soon as this is accomplished, water is drawn into the rectum, and regular respiratory movements begin. At the start, the pressure of CO_2 in the capillaries is about 760 mm. In the water drawn into the rectum, the partial pressure of this gas is less than 1 mm., while that of O is about 160 mm., and that of N about 600 mm. It follows, therefore, that CO_2 must diffuse rapidly outwards from the capillaries into the water, and that it will be replaced, in the course of time, by a mixture of N and O differing little, if at all, from atmospheric air. Experiments which I have made on newly-hatched larvae shew that equilibrium is practically established in from one to three hours after hatching.

Now the nitrogen is not required for the growing larva, but the oxygen is being continually drawn upon. Thus the partial pressure of the O in the tracheal system is always *tending to diminish*.

This is the *driving force* for the extraction of further oxygen from the water. The oxygen pressure must be kept up. For every cubic millimetre of oxygen extracted from the tracheal system for general use by the larvae, an equivalent volume passes in through the syncytium to the capillaries. There is no taking up of oxygen into chemical combination, as in the case of red blood-corpuscles. The rectal syncytium is simply a diffusion membrane, preserving the equilibrium of oxygen and nitrogen on either side of it, and hence continually allowing the passage of oxygen inwards from the water to the capillaries, while preventing any increase in the amount of nitrogen.

As regards the basal pads, these, I believe, act as supports for the gills, and nothing else. In the Implicate Type of gill-basket, where the arrangement is such that the gills support one another, no basal pads are developed. In the case of the basal pad, the advantage of fusion of the originally separate cells is that the whole organ is enabled to act as a single cell. Its outer wall becomes an osmotic membrane, allowing water to pass into the pad until the requisite condition of turgescence is obtained.

In conclusion, it is of interest to notice that the rectum of Anisopterid larvae has a three-fold function (*a*) that of respiration, (*b*) that of excreting the faecal pellets (as part of the hind-gut) and (*c*) that of propulsion. By squirting the water out of the branchial basket forcibly backwards, the larva propels itself forwards in a series of jerks. This is its usual method of progress.

D. **Caudal and Lateral Gills** (figs. 35, 40–42, 84–86).

Nearly all Zygopterid larvae possess three external caudal gills. Of these, the *median caudal gill* is dorsal, unpaired, and developed from the appendix dorsalis, while the *lateral caudal gills* are placed latero-ventrally, one on either side of the anus, and are developed from the cerci. In a few cases (some species of *Argia*, and some Hawaiian species of *Agrion*) the gills are much reduced, and somewhat resemble the appendages of Anisoptera.

It is not possible yet to determine definitely the actual value of these organs for respiration. Each is connected with its base (i.e. tergite or sternite of segment 11) by a definite breaking-joint, so that it is easily cast off if seized by an enemy. The larvae frequently lose one or more of their gills, and continue to live without them until the next ecdysis, when they are reproduced in a shorter form. It seems quite clear that the larvae do possess an auxiliary means of breathing in water, viz. by means of rectal respiration. There is, however, no development of true rectal *tracheal gills* of the kind seen in the Anisoptera, and hence the rectum must be a much less efficient organ of respiration than in that suborder.

In the young larva, the gills are filamentous and hairy. They soon acquire a triquetral form (i.e. triangular in cross-section). This triquetral form is retained in a few larvae, with slight modifications, throughout larval life. In most cases, however, it either tends to swell up into a *saccus*, or to become flattened into a *lamella*. The histology of all these forms is essentially similar; their external forms, and the distribution of their tracheae, vary very greatly.

Further, the gill may be from the first either a simple complete whole (*simple gill*) or it may have an extra distal joint (*constricted gill*). Gills developed along the latter lines shew a very definite constriction into two parts, in all except the most highly developed forms. These tend to obliterate the constriction, which may become reduced to a *node* (*nodate form*), i.e. simply the remains of the original joint on one side of the lamella; or the node may be lost, and its original position only marked by a spine, or by a difference in the consistency of the basal and distal portions of the gill. Also, it is possible for the whole distal part to be lost, so that an originally constricted gill becomes secondarily a simple gill. Such cases can only be determined by careful ontogenetic studies.

The Triquetral Gill.

The best-known example of a triquetral gill is the lateral gill of the *Calopteryginae* (fig. 40). In cross-section this is an isosceles triangle with its longest side (base) placed internally, facing the median gill. The thickened mid-rib of the gill lies at the vertex of this triangle, externally. There are two main longitudinal tracheae. The edges of the gill may be spiny. The shape is evidently suitable for resting either on the ground or against any moderately flat surface, while still exposing a large area to the water.

The Saccus, or Saccoid Gill (figs. 35, 84 c, 87).

Saccoid gills appear to be confined to the *Epallaginae, Thorinae* and *Protoneurinae*. In the first two cases, the saccus is a complete or simple one; in the last, it is very distinctly constricted, at a point somewhat distally placed from the middle. That the saccus is a development from the triquetral gill appears clear from a study of the genus *Diphlebia*. Here the young larva has triquetral gills. These gradually swell up, as the larva grows, until they assume a nearly rounded saccoid form (fig. 35).

In *Pseudophaea* (fig. 87) the gills are so swollen up as to appear quite rounded, except for a small tapering tip, which occurs also in *Diphlebia*. In *Cora* [31] the gills are also swollen and rounded, but the apex is broad and irregularly broken off.

In the *Protoneurinae* (fig. 84 c) beautiful constricted sacci occur in *Neosticta* and *Nososticta*. The basal portion is a short thick sausage-shaped bladder; the distal portion arises from a very constricted base, swells out considerably, and then tapers to a point.

All saccoid gills are very dark and opaque, so that it is not easy to make out the course of the tracheae. There are from two to four large longitudinal tracheae in each saccus.

The Lamellar Gill (figs. 41, 42, 84).

Lamellar gills are typical of the *Lestidae* and the *Agrionidae*. The median gill of the *Calopteryginae* is also a lamella.

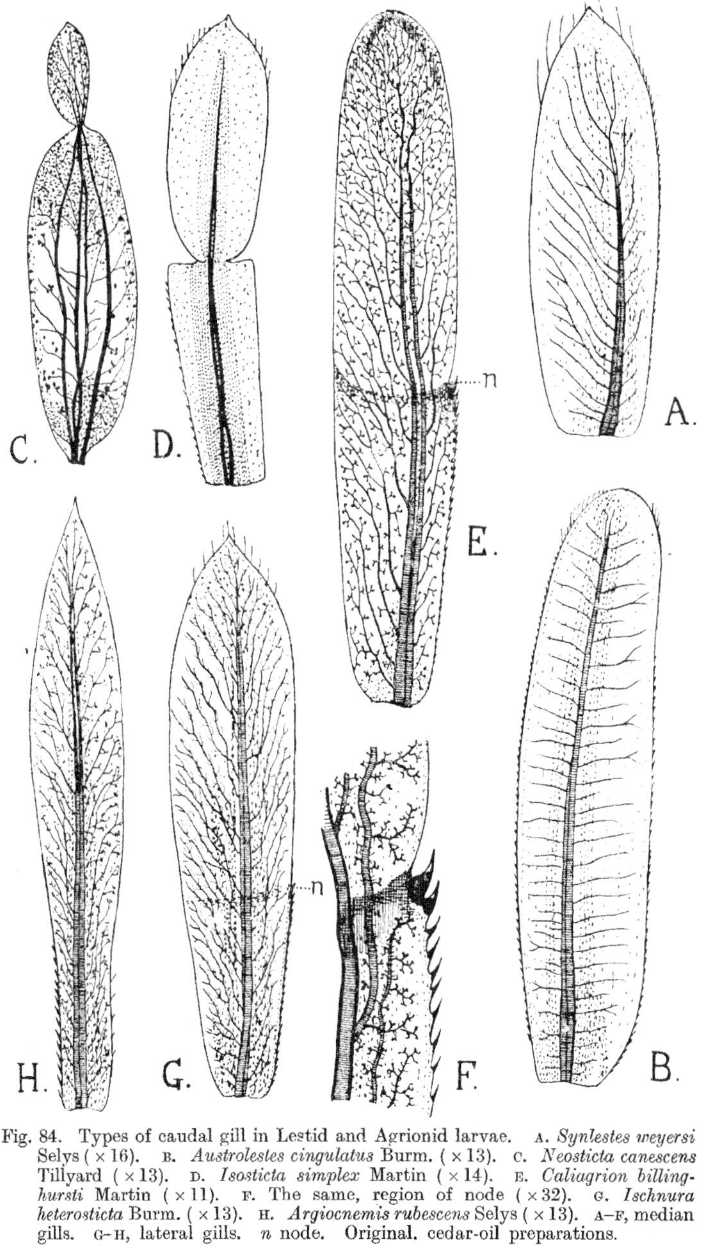

Fig. 84. Types of caudal gill in Lestid and Agrionid larvae. A. *Synlestes weyersi* Selys (×16). B. *Austrolestes cingulatus* Burm. (×13). C. *Neosticta canescens* Tillyard (×13). D. *Isosticta simplex* Martin (×14). E. *Caliagrion billinghursti* Martin (×11). F. The same, region of node (×32). G. *Ischnura heterosticta* Burm. (×13). H. *Argiocnemis rubescens* Selys (×13). A–F, median gills. G–H, lateral gills. *n* node. Original, cedar-oil preparations.

Lamellae may be developed from an original triquetral form in two ways:

(1) As a *horizontal* lamella, by a gradual widening of the base of the triangle representing the cross-section of the triquetral gill, and a corresponding flattening of the other two sides below it. Thus the mid-rib, originally placed externally, becomes mid-ventral in position, with the broad horizontal base lying dorsally above it[1]. This is a very rare formation, found only in *Argiolestes* (fig. 42 A). I do not know whether it is typical for the *Megapodagrioninae* or not, as we know little of the larvae of this subfamily. It is clearly a specialization to enable the larva to flatten its gills close against a rock, and still expose a large surface to the water. *Argiolestes*, when at rest, usually raises these gills, like a fan, slightly away from the rock surface. When disturbed, they are at once flattened closely down, so as to escape attention.

(2) As a *vertical* lamella, by elongation of the base, together with approximation of the other two sides towards it, in a vertical plane[2] (fig. 86 C). This process is easily understood by comparing cross-sections of the lateral and median gills of *Calopteryx*. It clearly first arose as a specialization in the case of gills which no longer rested on the surface. Nearly all the forms which possess lamellar gills live freely in masses of water-weed, matted roots, etc. The vertical lamella not only exposes its whole surface to the water, but the three gills together form an efficient parachute, used by the larva when descending through the water.

We can only briefly mention the various types of vertical lamellae known, together with the types from which they have been derived:

A. *Lestidae* (figs. 41, 84 A, B). In all known larvae of this family the lamella is of the *simple* form, elongated oval in shape, often with a broadly rounded end, and usually of exceptional length. In the *Lestinae* (figs. 41, 84 B) the secondary or branch tracheae come off from the main trachea at right angles, and run nearly to the border of the lamella before branching. The finer

[1] This applies only to the lateral gills. The median gill has been formed by horizontal flattening of a gill originally diamond-shaped in cross-section.

[2] This is the process for the lateral gill. The median caudal lamella appears to be formed generally by symmetrical narrowing, in a vertical plane, of a gill in which the cross-section was originally diamond-shaped.

branches and capillaries run along the edges much like the finer veins in a leaf. Bands of beautiful dark brown or purple pigment often cross these lamellae, and give them a striking appearance. In the *Synlestinae* (figs. 41, 84 A) the gill is much shorter, and the tip slightly pointed. The branch tracheae come off somewhat obliquely from the main stem. [The median gill of *Calopteryx* also belongs to this simple type.]

B. *Agrionidae* (figs. 42, 84 D–H). There is considerable variety in the type of gill found in this family. It is probable that most of the forms have been descended from a two-jointed or *constricted* type, though many highly specialized forms shew little sign of this at present. The following types occur:

(i) *Constricted Saccus* (fig. 84 C). This is found in *Nososticta* and *Neosticta*, of the *Protoneurinae*. The larvae are rock-dwellers. The shape of the gill suggests a candle surmounted by a flame. There are either two, three or four main tracheae, all of which penetrate through the constriction.

(ii) *Constricted Lamella* (fig. 84 D). This very beautiful opaque gill occurs in *Isosticta* (*Protoneurinae*). It is clearly derived from (i) by flattening. The larva dwells freely in weeds, roots, or débris. The change from saccus to lamella seems to be correlated with the change from the rock-dwelling to the free-living habit of the larva.

(iii) *Nodate Lamella* (fig. 84 E, F). This is a fairly common form of gill, found in many *Agrioninae* (*Caliagrion*, *Pseudagrion*, *Austroagrion*, *Erythromma*, *Agrion*). The node is a reduction from an originally complete constriction. The prenodal border is always armed with spines, which stop short at the node. The largest spine borders the node. The joint may be traced more or less completely across the gill, but is usually only effective close to the node itself. The tracheation is often extremely beautiful, and of a dendritic character. The tip is usually rounded.

(iv) *Subnodate Lamella* (fig. 84 G). This common form also occurs in many *Agrioninae* (*Ischnura*, *Enallagma*). The position of the node is now only marked by the termination of the row of prenodal spines, together with a more or less faint trace of the transverse joint. The basal (prenodal) part of the gill is usually slightly thicker than the rest. The tip tends to become pointed.

(v) *Denodate Lamella* (fig. 84 H). When all trace of the node is lost, except for a slight basal thickening, edged by a short row of minute spines, the gill may be termed *denodate*. The very narrow, sharply pointed denodate gill of *Argiocnemis* is clearly a step onwards in reduction from the type seen in *Ischnura*. The gills of *Pyrrhosoma*, *Agrion mercuriale*, and possibly *Platycnemis* appear to be denodate.

(vi) The form of gill found in *Mecistogaster* (*Pseudostigmatinae*) is very peculiar [31]. The basal portion is stalk-like, the distal foliate and pointed. Possibly the pointed gill of *Platycnemis* is related to this form.

(vii) The gills in the *Argiini* appear to be simple, and either triquetral or saccoid. They seem to be of a more archaic type than those mentioned above, and suggest that the larvae are very sluggish creatures. Some species of *Argia* have very reduced caudal gills, the anal appendages differing little from the form seen in the Anisoptera.

Types (i) to (v) appear to form a phylogenetic series. A more complete study of these organs might help us towards a natural classification of the *Agrioninae*. They suggest, at any rate, that the tribal divisions at present adopted are not natural, but simply aggregations of a number of convergent forms.

Arrangement of the Main Tracheae. In the typical median lamella (fig. 85) there are two main tracheae or *stems* running close together beneath the mid-rib, which is more or less unsymmetrically placed. In the lateral lamellae, only a single trunk trachea (the *lower* anal branch of the dorsal trunk) enters the base of the gill. In the *Agrioninae* this divides into two almost at once. In the median gill, the base is broader transversely, and receives *both upper* anal branches of the dorsal trunk, which run close together under the mid-rib. Thus each lateral gill is only connected with the dorsal trunk of its own side, whereas the median gill is connected equally with both. The bifurcations of the dorsal trunks occur in the ninth segment.

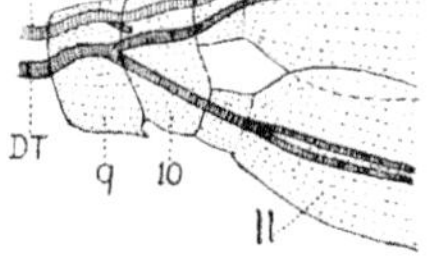

Fig. 85. Arrangement of main gill tracheae in an *Agrionid* larva, seen from left side. *DT* left, *DT'* right dorsal trunk; *ll* lateral, *lm* median caudal gill; 9, 10 abdominal segments. Original.

Structure and Histology of the Caudal Gills (fig. 86).

The Triquetral Gill possesses a thick cuticle, considerably swollen along the mid-rib. Beneath this lies a layer of pigmented hypoderm cells. In the interior of the gill there are two main longitudinal tracheae, with a number of obliquely-placed branches. There are also two main longitudinal blood-channels, one dorsal (efferent), the other ventral (afferent); these become confluent

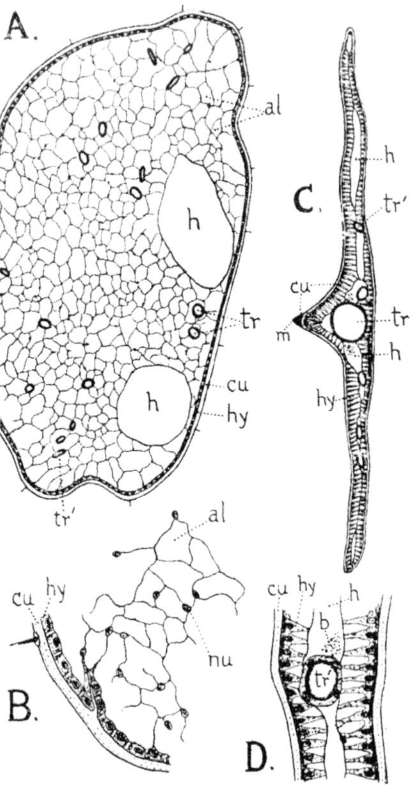

Fig. 86. Histology of the caudal gill. A. T.S. through basal third of saccus of *Diphlebia lestoïdes* Selys (×47). B. Small portion of same (×200). C. T.S. through basal third of lateral lamellar gill of *Synlestes weyersi* Selys (×47). D. Small portion of same (×200). *al* alveoli; *b* blood-plasma; *cu* cuticle; *h* haemocoele; *hy* hypoderm; *nu* nucleus of alveolus; *tr* main longitudinal trachea; *tr′* branch trachea. Original.

distally. The tracheae and blood-spaces are enclosed in a mass of reticulated tissue, which divides the interior of the gill into a large number of separate *alveoli*. This internal arrangement of parts is more fully explained in the account given for the saccus (see below), the histology of the two forms of gill being closely similar.

The saccus (fig. 86 A, B) is covered by an unpigmented cuticle (*cu*), which usually carries small hairs at intervals. Beneath the cuticle is a continuous layer of polygonal *hypoderm* cells (*hy*) with large oval darkly staining nuclei. These cells are filled with a purplish-brown pigment. The boundaries between the cells are pale. At certain points the hypoderm-layer is turned inwards (B), so as to form a kind of delicate *internal lamina*, which probably helps to support the very flimsy internal structure of the gill. From the hypodermis of these laminae, as well as, apparently, from a number of points on the main periphery of the hypodermis, delicate strands of tissue pass inwards to form an immense number of more or less spheroidal chambers or *alveoli* (*al*) which almost completely fill up the internal cavity of the gill. Each alveolus has, at some point on its wall, a single large nucleus (*nu*) not unlike the nucleus of a hypoderm cell. Most of the alveoli appear to be quite empty, but some clearly contain blood-plasma.

Near the base of the saccus there are usually four main longitudinal tracheae (*tr*), two of which remain symmetrically placed and unbranched for some distance. On either side of these lies a very large prolongation of the haemocoele, in which both blood-plasma and blood-corpuscles can be clearly seen. One of these large blood-channels is placed dorsally, the other ventrally, as in the triquetral gill. The two channels become confluent distally. In the median gill, the dorsal channel is afferent, the ventral efferent; in the laterals, the circulation is reversed. The other two tracheae branch out into the cavity of the gill, the branches (*tr′*) travelling obliquely towards tke gill-border. Further along the gill, all four tracheae give off branches. These branches in their turn divide, until there are seen very numerous minute tracheae at all points of the periphery. These, on reaching the hypodermis, turn so as to run parallel to and close alongside it, and in some places appear to press in so as to be almost contiguous with the cuticle.

The above description applies to the sacci of *Diphlebia* and *Neosticta*, which I have examined. Ris [137] has made a very careful study of the saccus of *Pseudophaea*, which differs, apparently, from the above description in possessing within its alveoli peculiar branching tufts of fibrils, shewing beautiful moss-like designs.

The lamellar gill (fig. 86 C, D) is of simpler structure. The cuticle (*cu*) resembles that of the saccus in being unpigmented, but it does not usually carry hairs. The cuticle is thickened along the *mid-rib* (*m*), which is developed only on the outer side of the lateral gill, but on both sides of the median gill. Beneath the cuticle is a layer of hypoderm cells (*hy*). These, in the most flattened portions of the lamella, resemble the hypoderm cells of the saccus. Where the gill is thickened, the hypoderm cells become elongated, and tend to assume a pyramidal form, with clear spaces interposed along their inner ends. The pigment is nearly all deposited in the layer of protoplasm lying closest to the cuticle, where also the nuclei are situated. The alveoli are much reduced in number, but can still be seen to be present, especially in the thickened part of the gills, under the mid-rib.

The inner ends of the hypoderm cells rest on a definite basement membrane. The basement membranes of the two opposite walls of the lamella are contiguous in many places. In others they are separated by blood-channels or extensions of the haemocoele (*h*), or to allow of the passage of tracheae (*tr*, *tr'*). Either one or two main longitudinal tracheae (*tr*) run close under the mid-rib; in the median gill there are always two. The various types of branching of the tracheae from this main stem have been already described and figured.

It is instructive to compare the structure of a cross-section of a lamellar gill with that of a larval wing-sheath. The two organs are, histologically, somewhat similar, yet they serve two very different purposes.

Our knowledge of these caudal gills is insufficient to allow of any speculations as to their physiology. Some authors, indeed, deny their respiratory function, and will only allow that they act as a kind of tail-fan, after the manner of the uropods and telson in many Crustacea. We know that many larvae use their caudal lamellae as a kind of parachute when descending through the water, and that most, if not all, Zygoptera can exist without them. The problem as it stands offers an immense field for interesting original research.

Lateral Abdominal Gills (fig. 87).

These occur on either side of abdominal segments 2–8 in the larvae of *Pseudophaea*, *Bayadera*, and *Anisopleura*, and on segments 2–7 of the larva of *Cora* [31, 62, 104, 137]. Their position is distinctly ventral, and they appear to be remnants of true paired abdominal appendages, serially homologous with the thoracic legs. In shape they are slender filaments. Ris [137] states that their internal structure in *Pseudophaca* resembles that of the saccus, except for the extreme attenuation of the organ. The alveoli are few in number, and moss-like fibrils are absent.

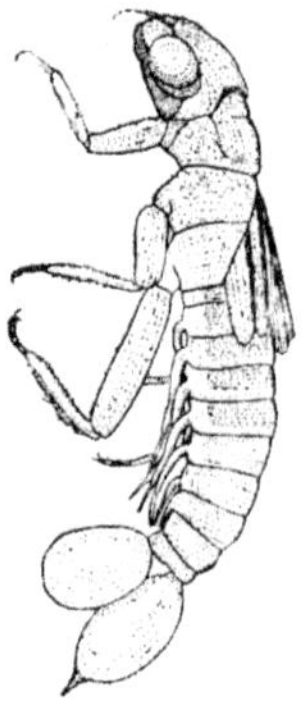

Fig. 87. Larva of *Pseudophaea* sp. (× $1\frac{1}{2}$) to shew caudal sacci and lateral abdominal gills. Drawn from a photograph sent by Dr F Ris.

E. **Movements of Respiration.**

In the imago of the Dragonfly, the respiratory movements can be followed with ease, by holding the live insect in the hand by means of its wings. The broad-bodied *Libellulinae* are the best for this purpose. *Inspiration* is effected by an increase in the volume of the abdomen, the sternites being lowered and the tergites expanded latero-ventrally. At the same time the whole abdomen becomes slightly lowered. By this movement, air is drawn into the spiracles. The act of inspiration is brought about simply by the relaxation of the transverse or dorso-ventral muscles of the abdomen.

The converse movement, that of *expiration*, is brought about by the contraction of the transverse muscles. By it the abdomen is contracted, the sternites being raised and the tergites narrowed, while the abdomen as a whole is slightly raised. The movement, too, is quicker than that of inspiration. It seems, however, clear that air can, if necessary, be prevented from escaping from the tracheae by the action of the combs of bristles, which lie just inside the spiracles, and close them from within, whenever the pressure is in that direction. Also, we know that the tracheae in the imago maintain an internal pressure greater than that of the atmosphere, since they collapse almost at once when the animal is dissected, whereas those of the larvae do not. Such a pressure could only be maintained, provided that the spiracles can remain closed during the movement of expiration.

We do not know for certain how insects get rid of their carbon dioxide. It is supposed that a great deal escapes through the integument. In the larva, the act of ecdysis may be connected with the removal of this and other impurities, which may help in the formation of the cuticle. In the larva also, rectal respiration must remove a great deal of this gas, by the ordinary processes of diffusion.

In the Anisopterid larva, the respiratory movements of the rectum are easily watched. A large volume of water is drawn in rather quickly. This can be let out slowly, with no apparent effort, or it may be shot out with great force so as to propel the larva forward.

In the Zygopterid larva, no definite respiratory movements can be seen with the naked eye, beyond an occasional waving of the gills to and fro. In larvae which have lost their gills, the abdomen is often strongly arched; or it may be jerked quickly to and fro. Rectal respiration in these larvae can be watched by a careful use of carmine particles. The vigorous rectal respiration which takes place in newly-hatched Zygopterid larvae can be followed with ease under a low power, owing to the transparency of the integument. The whole problem of the respiration of these larvae is, at the present time, much in need of further investigation.

The degree to which the spiracles are functional in Dragonfly larvae was investigated by Dewitz [49] in a deeply interesting and well-known paper. (See also Miall [94].)

CHAPTER X

THE BODY-WALL AND MUSCLES

The Body-Wall (fig. 88).

The Body-wall in the Dragonfly, as in other insects, is formed of a strong external non-cellular chitinous *cuticle,* with an underlying cellular layer, the *hypodermis.*

The Cuticle is not a single layer, but consists of several parallel laminae, of different staining intensities. It is formed as a secretion

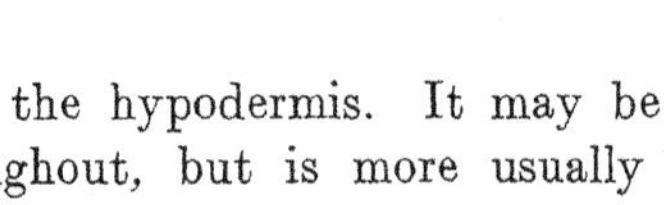

Fig. 88. Section of body-wall of abdomen of larva of *Synlestes weyersi* Selys ($\times 1000$). cut_1 darkened outer layers of cuticle; cut_2 clear inner layer of same; *hy* hypoderm; *m* muscle-fibres; *n* nucleus of hypoderm cell. Original.

from the cells of the hypodermis. It may be quite clear and transparent throughout, but is more usually pigmented with black, brown or orange. The pigment is always densest in the outer layers (cut_1). The layer next the hypoderm (cut_2) remains unpigmented. The outermost layer may be either smooth or sculptured, frequently by close and delicate striae. Often it is covered with minute hairs or bristles, each set in a tiny depression surrounded by a raised circular wall. In certain regions numerous minute pores occur in it [39].

At various points the cuticle is turned inwards into the body of the animal, forming structures designated as *endoskeleton.* Such are the *tentorium,* the *posterior foramen,* the *apodemes* of the thoracic and abdominal segments. At the spiracles it is also turned in and thickened. The internal chitinous linings of the stomodaeum and proctodaeum are also directly continuous with the external cuticle, just as the cells secreting them are continuous with the hypodermis.

The Hypodermis (*hy*) is a continuous single layer of regular hexagonal flattened cells, closely underlying the cuticle throughout. It serves not only for the production of the latter, but also for the attachment of the body-muscles. The muscle-fibril is inserted into the inner border of the hypoderm cell, which shews in the region of attachment a change from a granular to a fibrillar structure of the cytoplasm (fig. 88, cells above *m*). Here and there, enlarged hypoderm cells are met with having larger nuclei than usual. These receive the terminations of sensory nerve-fibres, and serve to give a perception of general sensation to the body-wall, as distinct from the sense-organs. By this arrangement of muscles and nerves it becomes clear how, at ecdysis, the cuticle is shed without any local damage to the underlying parts.

The Haemocoele. The large open space below the hypodermis is filled with blood, and is called the *haemocoele* (p. 162). It is quite distinct from the *coelome* or body-cavity of Vertebrates. This latter, in the Dragonfly as in all Arthropods, practically disappears after embryonic development is completed. Its only remnants are the pericardial cavity and the lumina of the developing gonads.

The Muscles.

In the Dragonfly all the muscles are striated, without exception. Zawarsin [203] has conclusively demonstrated the striated character of the alary muscles in the Dragonfly larva.

The muscle fibres are bound together into bundles or fasciculi, with or without an enveloping sarcolemma. In the larva, the muscles are all of one kind, closely resembling the striated muscles of Vertebrates. In the imago, we meet with two main classes of muscle, known as *leg-muscle* and *wing-muscle* respectively (Weissmann). The former class includes the head muscles and the segmental muscles of the abdomen. In these there is only a single bundle, with a definite sarcolemma. In the wing-muscles, each separate muscle is formed of a number of bundles, bound together by a continuous inter-fibrillar substance, but without a sarcolemma. Most of the muscles are of a white. creamy, or pale greyish colour. The wing-muscles are exceptional in being dark yellow, inclining to orange. The cross-striation of

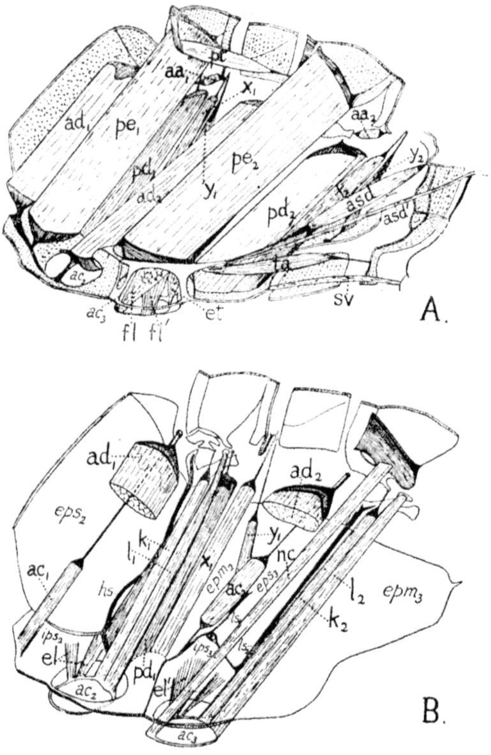

Fig. 89. Synthoracic musculature of *Sympetrum* (× 10). A. Inner or proximal view. B. Outer or distal view, with *pe* removed and *ad* cut short. Explanation of references for muscles in text. References to parts of thorax (in italics) in legend of fig. 1. *ac* acetabulum. Adapted from Berlèse.

the wing-muscles differs from that of the other muscles. In the former the striae are thick, and light and dark alternately. In the latter the striae are all alike, not darkened, but carrying rows of small dots. In both forms of muscle the striae are broader and more easily seen than in the striated muscles of Vertebrates. Further, whereas the nuclei of the leg-muscle fibres are situated peripherally, those of the wing-muscle fibres are placed centrally, and are much larger and more elongated.

The muscles may be attached, both at origin and insertion, either directly (as in the case of the body-muscles to the hypodermis) or by means of a *tendon*, i.e. a chitinous ingrowth from the cuticle. The ordinary tendon is conical, receiving the muscle-fibres into its broad hollow end; it is of a pale colour. Special strong tendons, of a dark colour and peculiar shape, occur in connection with certain muscles, notably most of the wing-muscles. Such tendons are known as *cup-* or *cap-tendons*, from their shape, which is either in the form of an inverted cup, a skull-cap, a flattened pyramid or wedge. The stalk of the tendon may be either long or short.

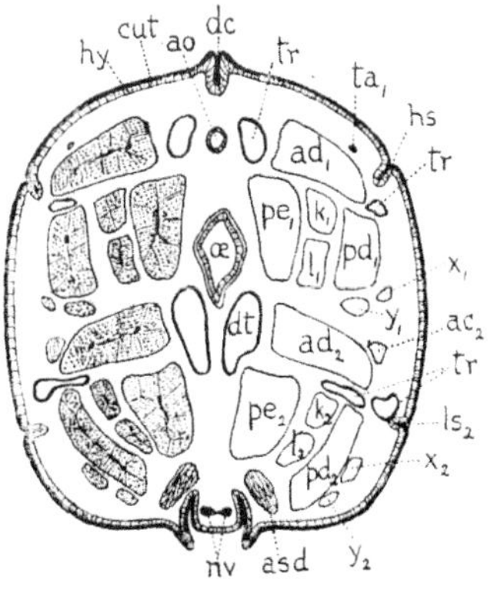

Fig. 90. Section parallel to notal plane through synthorax of *Ischnura aurora* Br. (× 20). Muscles as in fig. 89, except ta_1 = tendon of ac_1. *ao* aorta; *cut* cuticle; *dc* dorsal carina; *dt* dorsal tracheal trunk; *hs* humeral suture; *hy* hypoderm; ls_2 second lateral suture; *nv* nerve-cord; *œ* oesophagus; *tr* trachea. Original.

The development of the muscular system in the Dragonfly is very great. Even a brief account of all the muscles in its body would take up far more space than a volume of this size can afford. It will be sufficient here to tabulate the principal muscles of the various regions of the body, giving their names, their places of origin and insertion, and any further details that may be of interest to the student. The use of Latin names has been avoided [10, 22, 84, 120, 121, 184].

Muscles of the Head.

Number	Name	Origin	Insertion	Remarks
(1) (2) (3)	Elevators, Depressors, Rotators } of the Head	Prothorax and interior part of mesothorax	Margins of posterior foramen	Small muscles
(4)	Elevators of the labrum	By tendon from suture between vertex and frons	Base of labrum	Two, right and left
(5)	Depressors of the labrum	Do., outside (4)	Roof of mouth behind labrum	Two, right and left
(6)	Flexor of the mandible	Postocular lobe	Base of mandible, inner side	Large, fan-shaped
(7)	Extensor of the mandible	Do., outside (6)	By tendon to base of mandible, outer side	Smaller, less powerful than (6)
(8)	Primary flexor of the maxilla	By tendon from back of head	Cardo and stipes, inner margin	
(9)	Secondary flexor of the maxilla	Back of head, behind (8)	By small tendon into inner lobe of max., inner side	Strap-shaped, smaller than (8)
(10)	Extensor of the maxilla	Back of head, between (8) and (9)	Cardo, outer side	
(11)	Flexor of the inner lobe of the maxilla	Stipes, inner side	Base of inner lobe, near inner border	
(12)	Extensor of do.	Stipes, outer side	Do., near outer border	
(13)	Flexor of the palp	Stipes, inner side	Base of palp, near inner border	
(14)	Extensor of do.	Stipes, outer side	Do., on outer border	
(15)	Adductors of the labium	Below posterior foramen	End of submentum, on either side	Two, united at origin (Imago)
(16)	Submental	Submentum	Squame, on either side	Do. (Imago)
(17)	Adductor of the median lobe	Do.	Base of mentum	(Imago)
(18)	Abductor of the lateral lobe	Do.	Base of lateral lobe, on outer side	Larva
(19)	Adductor of do.	Do.	Do., inwards from (18)	Larva
(20)	Extensor of the mask	Behind hypopharynx	Base of mentum, ventrally	Larva
(21)	Flexor of the mask	On tentorium	Lateral margin of mentum, more dorsally	Larva

N.B. Note the absence of muscles for moving the movable hooks (see p. 79).

Muscles of the Wings (Fig. 89)

(For abbreviations of thoracic sclerites, etc., see legends of figs. 1, 8)

Number	Name	Reference Letters	Homologue in other Insects	Origin	Insertion	Remarks
(22)	Principal Elevator	*pe*	First tergosternal	Squame of furca	*Fw*: By cap-tendon into wing-base below *C*. *Hw*: by short strong do. from *C* to $R+M$	Very large. Raises the wing
(23)	Anterior Accessory Elevator	*k*	Eleventh tergosternal	Upper outer margin of *ac*	Do.	Small
(24)	Posterior Accessory Elevator	*l*	Sixth lateral	Do., just behind (28)	By pale tendon into wing-base below *Sc*	Small
(25)	Anterior Depressor	*ad*	Third tergosternal	Squame of prefurca	By cap-tendon into *ap*	Large. Lowers the costal side of the wing
(26)	Anterior Accessory Depressor	*ac*	Fourth tergosternal	Sternoepisternal apodeme	By long cap-tendon, close to (25)	Short and small. Acts with (25)
(27)	Posterior Depressor	*pd*	Pleuroradial (Notopleural)	Sternoepimeral apodeme	By cap-tendon into pleural membrane at base of wing, below $R+M$	Large. Lowers the whole wing
(28)	First Accessory of Posterior Depressor	*x*	First Pleurometapteral	*Fw*: ventral part of epm_2. *Hw*: hind margin of S_3	By slender pale tendon into wing-base below *A′*	Small. Lowers the posterior border of the wing
(29)	Second Accessory of Posterior Depressor	*y*	Seventh Lateral	*Fw*: interpleural apodeme. *Hw*: posterior epimeral apodeme	By long tendon into wing-base close to (28)	Shorter and smaller than *x*. Same action as *x*
(30)	Adductor of the Secondary Anal Vein	*aa*	(Fulcro-alar)	Fulcrum of notum (opposite the level of *M*)	By short tendon into base of *A′*	Very short and small. Pulls the wing backwards

N.B. The names adopted are essentially those of Poletaieff [120] and Calvert [22], except (30). In the figures, muscles of the fore-wing are indicated by the suffix "1," those of the hind-wing by the suffix "2." The names denoting homology with the thoracic muscles in other insects are those given by Berlèse [10]. Lendenfeld's very verbose and involved paper [84] dealing with the thoracic skeleton and wing-muscles of the Odonata, is best disregarded; not only because his naming of the muscles is very fanciful, and based on false analogies with the muscles of the limbs of Vertebrates, but also because his interpretation of the thoracic sclerites is fundamentally erroneous.

Muscles of the Legs (figs. 34, 89).

Number	Name	Reference Letters	Origin	Insertion	Remarks
(31)	First Flexor of Fore-leg		Pronotum	Procoxa	= Notocoxal of prothorax
(32)	Second Flexor of Fore-leg		,,	Trochanter, lower wall	Noto-subcoxals
(33)	Extensor of Fore-leg		,,	Trochanter, upper wall	
(34)	First Flexor of middle or hind-leg	*fl*	Midventral ridge of furca	Coxa, internal lateral wall	Sternopedals
(35)	Second Flexor of middle or hind-leg	*fl*		Coxa, posterior external wall	
(36)	Third Flexor of middle or hind-leg	*fl*		Coxa, anterior external wall	
(37)	Extensor of middle or hind-leg	*el*	,, ,,	Coxa, posterior internal wall	= Third pedal. Has two separate fasciae inserted together
(38)	Flexor of trochanter	*ftr*	Base of coxal wall	Base of trochanter, inferior	
(39)	Extensor of trochanter	*etr*	,, ,,	Base of trochanter, superior	
(40)	Femoro-trochanteric		Inferior wall of trochanter	Basal membrane of femur	
(41)	Flexor of tibia	*ftb*	Whole length of inferior wall of femur	By strong tendon to base of tibia, inferior	Long base of attachment, short tendon of insertion
(42)	Extensor of tibia	*etb*	Do., superior wall	Do., superior	
(43)	Flexor of tarsus	*fts*	Whole length of inferior wall of tibia	By short tendon to base of tarsus, inferior	,, ,,
(44)	Extensor of tarsus	*ets*	Do., superior wall	Do., superior	,, ,,
(45)	Long Flexor of claws	*fcl*	Whole length of tibia	By very long tendon, branching distally into two slips, one for each claw	
(46)	Notocoxal	*nc*	Metanotum	Wall of coxa, anteriorly	Has no homologue in mesothorax

Other Thoracic Muscles (fig. 89).

Number	Name	Reference Letters	Origin	Insertion	Remarks
(47)	Prosternal		Furca of mesothorax	Furca of prothorax	Only one pair
(48)	Protergal	*pt*	Anterior part of mesonotum	Anterior part of metanotum	,, ,,
(49)	Tergosternals of prothorax		Pronotum	Prosternum (prefurca)	Two pairs
(50)	Lateral Thoracico-abdominal	*ta*	Furca of metathorax	1st abdominal sternite, anteriorly	One pair
(51)	Submedian Ventral Thoracico-abdominal	*sv*	Posterior epimeral apodeme, ventrally	2nd abdominal sternite, anteriorly	One pair, with common tendon of origin
(52)	First Auxiliary Sternodorsal	*asd*	Posterior process of furca of metathorax	1st abdominal tergite, anteriorly	These help to straighten the abdomen out, acting with (54) and (55)
(53)	Second Auxiliary Sternodorsal	*asd′*	,, ,,	2nd abdominal tergite, anteriorly	

Segmental Muscles of the Abdomen.

Number	Name	Origin	Insertion	Remarks
(54)	Superior Longitudinal Tergal (Two sets)	Anterior dorsal part of tergite	Posterior dorsal part of tergite of segment next in front	Straighten abdomen out
(55)	Lateral Longitudinal Tergal (Two sets)	Inferior lateral margin of tergite, and pleural membrane	Do., but placed higher up in each segment	Do.
(56)	Longitudinal Sternal (Two sets)	Anterior part of sternite	Posterior part of sternite of segment next in front	Small. Bend the abdomen
(57)	Vertical or Dorso-ventral (= Vertical Expiratory of Plateau)	Tergites of segs. 1–8	Second lateral sternal processes of segs. 1–8	Contract the segments
(58)	Transverse Genital	Tergite of seg. 2. Male only	Sternite of seg. 2	Specially developed in connection with genitalia

The muscles connected with the Internal Organs are dealt with in connection with those organs.

Nature and Mechanism of the Wing-Muscles.

The Dragonfly differs from all other insects in the fact that the great muscles of the synthorax are directly connected with the wing-bases by means of tendons. In all other insects the muscles by means of which the action of the wings is controlled are not directly connected with those organs, but are connected with the notum. Those that pass from the notum to the sternum are called *tergo-sternals* (or dorso-ventrals). Those that pass from the notum to the pleurum are called *laterals* (or notopleurals). From the table on p. 207 it will be seen that all the wing-muscles of the Dragonfly, except one, are homologues of either tergo-sternals or laterals. The direct attachment to the wing-bases has been brought about by the development of special sclerites, derived from the notum, in the formation of those bases.

The *adductor of the secondary anal vein* is the exception mentioned above. It has no homologue in any other insect (except perhaps in the *Locustidae*). Its development is correlated with the peculiar formation of the vein to which it is attached (p. 64).

Another peculiarity in which the Dragonfly differs from the more highly specialized insects, and resembles the lower insects of much less powerful flight, is the fact that fore and hind-wings are completely independent. Each has exactly the same muscle-supply, and each works entirely on its own. The fore-wings can be destroyed without affecting the working of the hind, and vice versa. A number of authors, chiefly in text-books, have asserted that a strong ligament connects the bases of fore and hind-wings together. The error seems to have arisen from a misinterpretation of a well-known figure by Lendenfeld (see Sharp, *Camb. Nat. Hist.*, Vol. v, p. 416; Packard, *Text-Book of Entomology*, p. 158) in which what appears to be the cut border of the thorax looks like a ligament. In his original figure, Lendenfeld left this piece unnamed.

A third point of interest is the presence of only one true *axillary* or wing-pivot in the Dragonfly. This acts as the principal fulcrum

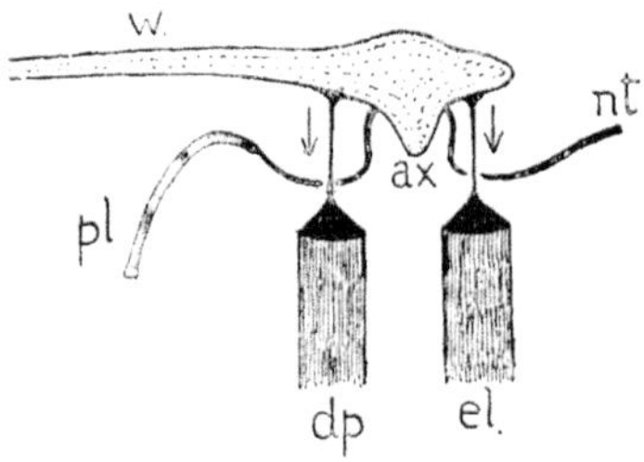

Fig. 91. Diagram to illustrate antagonism of two parallel muscles acting on wing of Dragonfly. *ax* axillary; *dp* depressor; *el* elevator; *nt* notum; *pl* pleurum; *w* wing. Original.

of the wing. An examination of the position of this fulcrum shews us that it lies closer to the middle line than do the tendons of the depressor muscles. The tendons of the latter are inserted at a level *external* to it (i.e. further from the middle line). Those of the elevators, on the other hand, are inserted at a level internal to it (i.e. nearer the middle line). Thus the wing acts as a *lever of the first order* (i.e. the fulcrum lies between the power and the weight) for the elevators, but as a *lever of the third order* (i.e. the power between the fulcrum and the weight) for the depressors. By referring to fig. 91, we thus see plainly how two sets of muscles which are antagonistic, but pull very nearly in the same direction, can both elevate and depress the wing. (This diagram is not in any sense intended to shew the true shape and positions of the structures indicated, but merely to explain their action.)

The Diaphragm.

According to Amans [1], who discovered it, the *diaphragm* is an elliptical membrane stretched across the body-cavity of the larva, between the fourth and fifth abdominal segments, in front of the Malpighian tubules. It is inserted externally, above and below, on to the inter-segmental muscles, and internally on to the circular muscle-tunic of the mid-gut. It shews muscle-fibres stretching in different directions. Amans considers that it plays an important part in respiration, defaecation, and especially in the play of the mask. The diaphragm is slightly inclined to the vertical plane. A contraction of the abdominal segments causes it to become taut, and consequently the blood exerts a forward pressure on the narrow cavity of the mask. This causes the movable hooks to shoot out suddenly, as explained on p. 79.

CHAPTER XI

THE REPRODUCTIVE SYSTEM

The Reproductive System in the Dragonfly consists of two very distinct parts (1) the internal organs, or *gonads,* with their ducts and accessories, (2) the external organs (*external genitalia*), which have to do with the processes of copulation and oviposition.

THE MALE.

1. *Internal Organs* (figs. 92, 93).

These consist of a pair of *testes,* with their ducts or *vasa deferentia,* together with a median *sperm-sac* or *reservoir* [128].

The Testes (*ts*). These are very elongated subcylindrical organs, very often of a dark brownish colour, but sometimes much paler. They lie dorsally on either side of the alimentary canal. The anterior end of the mature testis reaches as far forward as the fourth segment, the posterior end being in the eighth segment, in which it usually makes a small bend or loop, forwards and inwards. The testis is supplied with reddish brown *genital tracheae* as branches from the ventral trunk. The fat-body forms a more or less complete whitish sheath to the organ, with which it is fairly closely bound up by means of the tracheae.

The surface of the testis is irregularly grooved and wrinkled. This is seen, in sections, to be due to the fact that the body of the organ is formed of a very large number of separate *lobules,* of a more or less spherical shape. In each lobule a very large number of male cells are being developed. Just after metamorphosis, the more dorsally placed lobules are filled with rounded *spermatidia* (fig. 93, s_1). Towards the middle, these are seen to have become divided up into much smaller rounded cells, the immature *spermatozoa* (s_2). On the ventral side, nearest the intestine, the lobules are filled with ripe or nearly ripe spermatozoa (s_3) of an elongated form. There do not seem to be any definite ducts to the lobules; we must therefore regard them as formed by foldings and constrictions of the original gonad-wall.

The process of **spermatogenesis** has not been studied in detail in the Dragonfly, though the later stages in the growth of the spermatozoa have been followed by Bütschli [20], in the case of

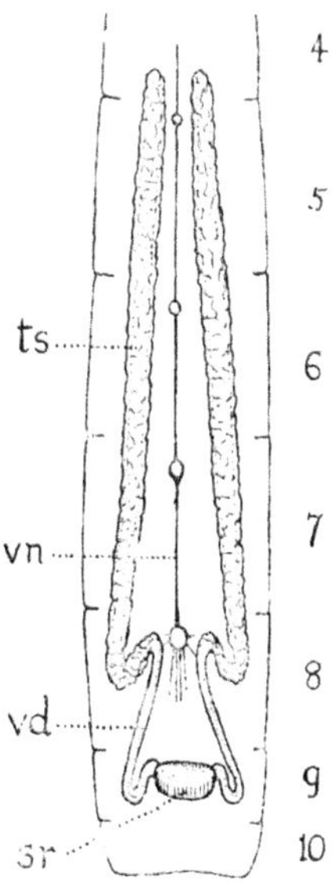

Fig. 92. Internal reproductive organs of *Aeschna brevistyla* Ramb., ♂, dorsal view (×2). *sr* sperm-reservoir; *ts* testis; *vd* vas deferens; *vn* ventral nerve-cord; 4–10 abdominal segments. Original, from a dissection.

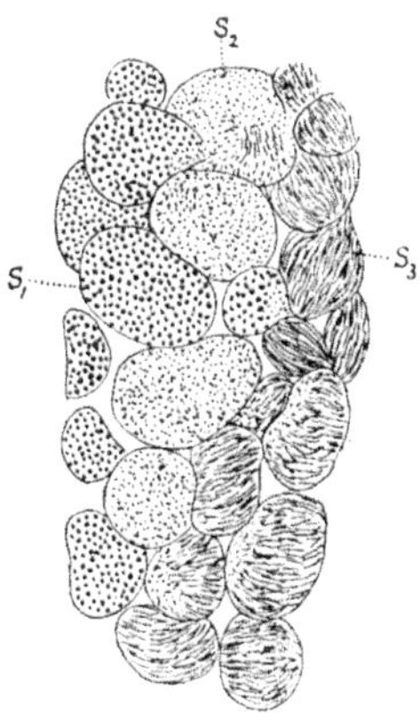

Fig. 93. L.S. through portion of testis of a freshly-emerged male of *Ischnura aurora* Br., to shew lobules; s_1 spermatidia; s_2 immature spermatozoa; s_3 ripe spermatozoa (×150). Original.

Agrion and *Calopteryx*. Fig. 94 shews a mature spermatozoon of the latter genus. The head (*h*) is very short, 0·01 mm., cylindrical, with elongated nucleus (*nu*). The middle piece (*m*) is five times as long as the head, and almost as wide. The tail (*t*) is of great length, nearly 0·2 mm., excessively thin, and undergoes violent lashings by means of which the organism is propelled forwards. In the *Libellulidae*, the tail is much shorter and less slender.

The vas deferens (*vd*) is a rather narrow tube, which runs along the middle of the inner side of the testis, being enclosed within that organ by the overgrowth of its lobes around it. It leaves the bent-up end of the testis in the eighth segment, and passes backwards to the ninth. There it makes a small bend or loop, turning forwards, inwards and downwards to enter a short *common duct* just above the genital pore. In the mature male, this common duct is dilated dorsally into a conspicuous *sperm-sac* (*sr*), lying above the points of entry of the two vasa deferentia.

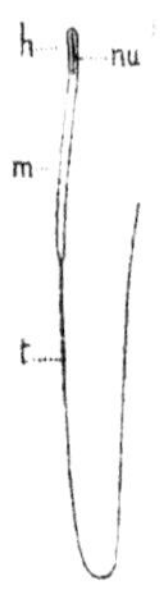

Fig. 94. Mature spermatozoon of *Calopteryx splendens* Harris (×250). *h* head; *m* middle-piece; *nu* nucleus; *t* tail. After Bütschli.

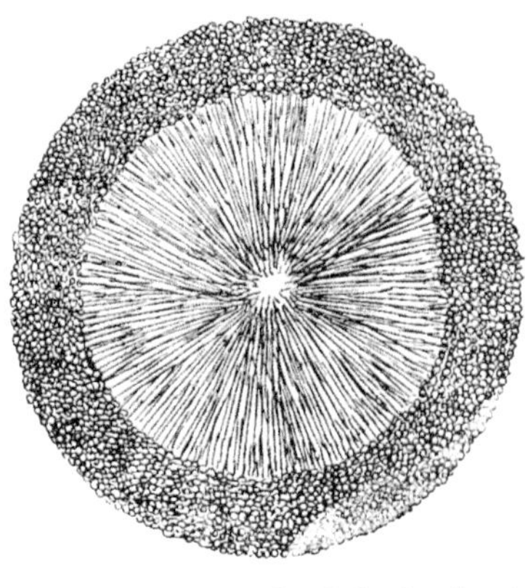

Fig. 95. Sperm-capsule of *Aeschna brevistyla* Ramb. (×330). Original.

If the vas deferens of a mature male be opened[1], it is found to contain, not separate sperms, but a large number of little round balls, or *sperm-capsules* (fig. 95). Each of these is about 0·15 mm. in diameter, and appears to contain the whole mass of sperms derived from a single testis-lobule. The sperms are all radially arranged, their inner ends (apparently the heads) forming a slight spiral (about a turn and a half) around the centre. Externally, the capsule is enclosed by a thick coat, apparently formed of minute gelatinous globules, one affixed to each sperm.

[1] The observation was made with *Aeschna brevistyla*.

The whole capsule is slightly sticky, and eminently suited for the difficult operation of the transference of the sperms externally from the ninth to the second segment.

The wall of the vas deferens is very thick, being formed of high columnar cells with large nuclei, each of which shews numerous chromatin granules and a conspicuous, deeply-staining nucleolus.

The genital pore opens mid-ventrally on the ninth sternite. It is guarded by two small chitinous flaps, the *valvules*, homologous with the valves or lateral processes of the ovipositor in the female.

2. *External Organs* (fig. 96).

The copulatory apparatus of the male Dragonfly is one of the most remarkable structures in the Animal Kingdom. The "palpal organ" on the pedipalp of the male Spider, and the hectocotylous arm of the Cephalopod Mollusc, extraordinary as they are, do not defy all explanation, since in each case they are modifications of an appendage already present. But the apparatus of the male Dragonfly is not homologous with any known organ in the Animal Kingdom; it is not derived from any pre-existing organ; and its origin, therefore, is as complete a mystery as it well could be. It is useless to talk vaguely, as some observers have done, about relationships with the Progoneate Myriapods, which have the genital opening far forward on the body. The true genital opening of the male Dragonfly remains far back, on the ninth segment. The copulatory apparatus of the second and third segments has nothing to do with that. It is an entirely new and unique development, for the reception of the sperms, and for their subsequent introduction into the female. All we can say with certainty is, that its formation is correlated with the great length of the abdomen, with the use of the anal appendages as claspers, and with the adoption of the "tandem" position during pairing. Backhoff [4] has shewn that this apparatus develops late in larval life (at the antepenultimate instar) from originally median unpaired hypodermal cell-masses, which later on assume a bilaterally symmetrical form. Thus it has nothing to do with the lost segmental appendages of the abdomen, which disappear in embryonal life.

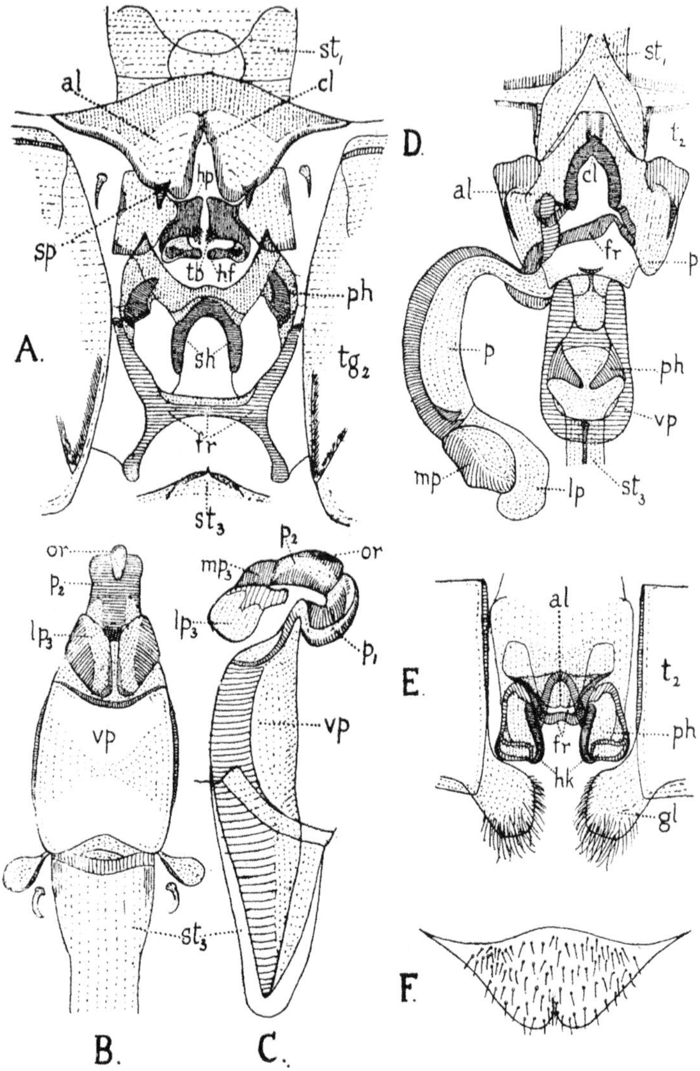

Fig. 96. Copulatory apparatus of male Dragonflies. A. *Aeschna brevistyla* Ramb., ventral view, with penis removed. B. Penis of same, ventral view. C. The same, lateral view. D. *Synlestes weyersi* Selys, ventral view, with the penis displaced to the left. E. *Orthetrum villosovittatum* Br., ventral view, to shew genital lobes (*gl*). F. Enlarged view of anterior lamina of same. (A–C × 11, D × 12, E × 15, F × 43). *al* anterior lamina; *cl* cleft of same; *fr* framework; *gl* genital lobe; *hf* hamular fold; *hk* hook (inner branch) of hamule; *hp* hamular process; *lp*, lp_3 lateral flaps of penis; mp_3 middle piece of distal joint of same; *or* orifice of same; *p* penis; p_1–p_3 its three joints; *ph* posterior hamule; *sh* sheath of penis; *sp* spine; st_1 first, st_3 third urosternite; *tb* tuft of bristles; t_2 or tg_2 second urotergite; *vp* vesicle of penis. Original, chitin preparations.

The apparatus is developed from the whole of the second sternite, together with the anterior part of the third. It consists of a *penis*, a *vesicle* for the storage of the sperm-capsules, and a number of complicated accessory structures arranged so as to protect the penis and also to guide and hold the ovipositor of the female during copulation. These are all lodged in a conspicuous depression of the second sternite, known as the *genital fossa*. This fossa opens posteriorly into a small sac, the *vesicle of the penis* (*vp*), developed from, and attached posteriorly to, the anterior border of the third sternite. From the distal end of this vesicle, which projects forwards, arises the strongly-chitinized median organ called the *penis* (*p*). Its lumen is continuous with that of the vesicle, and opens to the exterior near its distal end.

In the Anisoptera (A–C), the penis is complicated, three-jointed, and bent somewhat in the form of the conventional figure of the fish known as the "sea-horse." The basal joint (p_1) is large and strongly chitinized, and bent near its middle. It sometimes carries a fringe of hairs, or a long fleshy process. The second joint (p_2) is smaller, and generally more or less curved. The *orifice* of the penis (*or*) opens on its convex surface. The third joint (p_3) is small, wider than the second, and very variable in shape. It consists of a hard *middle piece* (mp_3) with two larger soft *lateral flaps* (lp_3), which are folded up when at rest, but expanded when the penis is erected. Long fleshy processes and other outgrowths may occur on this joint. When withdrawn, the penis is protected by a chitinous arch arising from the floor of the fossa, and called the *sheath of the penis* (A, *sh*).

In the Zygoptera (D), the form and disposition of the above parts is simpler. The penis (*p*) is not distinctly jointed, and occupies the same position as the sheath in Anisoptera (with which it is probably homologous). Its lumen opens basally into the body-cavity, not into the vesicle, and it does not open distally to the exterior. Also, whereas the penis of Anisoptera is supplied with muscles, nerves and tracheae, these all seem to be absent in connection with the penis of Zygoptera. It is difficult to say what part a poorly developed organ of this kind can play in copulation, unless it serves as a kind of hinge or link, by means of which the female pore is held in contact with the vesicle. The structure is primitive enough to suggest that the whole of the accessory apparatus originally arose from a slight depression in the second sternite, close under the anterior edge of the third. In this, perhaps, the male originally deposited sticky sperm-capsules, which the female sought for, during pairing, by bending her abdomen forward under that of the male.

The accessory portions of the apparatus consist of the supporting framework of the genital fossa, together with one or two pairs of *hamules* for the guidance of the ovipositor into position. The *framework* consists of the following parts:

(i) laterally, the two latero-ventral margins of the second tergite (tg_2),

(ii) anteriorly, a specialized bifid plate, the *anterior lamina* (*al*), developed from the anterior part of the second sternite, and

(iii) posteriorly, a less specialized and narrower plate of the same sternite, bordering segment 3 anteriorly, and called the *posterior lamina*.

The *hamules* consist of anterior and posterior pairs. The *anterior hamules* only occur in the *Aeschnidae*; they are absent alike in the Zygoptera and the rest of the Anisoptera. They are of complicated form, and function as claspers. In *Aeschna* (A) we distinguish two principal parts: the *hamular fold* (*hf*), and the *hamular process* (*hp*) carrying a distinct tuft of bristles (*tb*). *Posterior hamules* (*ph*) occur in all Dragonflies. They are of simple form in the Zygoptera and *Aeschnidae*. In the *Libellulidae*, however, they become highly differentiated, with outer and inner branches, the latter usually in the form of a sharp hook (*hk*). (See also fig. 176.)

In the *Libellulidae* (E) an additional complication is introduced by the projections of the postero-lateral angles of the second tergite, which form well-developed *genital lobes* (*gl*).

The *anterior lamina* (*al*) varies much in shape. In Zygoptera and *Aeschnidae* it has a deep median cleft (*cl*). In the former it is produced backwards to form two strong *posterior lobes* (*pl*), which probably play the same part as the anterior hamules of *Aeschnidae*. In the latter, a sharp spine (*sp*) often occurs on the lamina, on either side of the cleft. In the *Libellulidae*, the anterior lamina is hood-shaped, hairy, and either entire or slightly notched (F).

For further details of the almost endless variety of structure of the external genitalia, the reader should consult Selys [151], Thompson [161] and Backhoff [4]. The *Gomphinae* in particular shew the most extraordinary formations, especially in the penis.

The Female.

1. *Internal Organs* (figs. 97, 98).

These consist of a pair of *ovaries* with their *oviducts*, a *bursa* or *spermatheca*, and a pair of *accessory sacs* [128].

The ovaries (fig. 97, *ovy*) are very large, extending from the base of the abdomen down to the seventh segment, dorsally on either side of the alimentary canal. Anteriorly they are united by a median *dorsal ligament*, formed by fusion of the anterior ends of the egg-strings, and attached to the posterior wall of the thorax. Each ovary is enclosed in a thick sheath formed by the fat-body. Numerous tracheae given off from the ventral trunk pass to the lower ventral surface of the ovary, and ramify over its walls.

The ovary is formed of numerous separate *egg-strings*, arranged longitudinally, and held in place by the genital tracheae and the envelope of fat. Each egg-string is moniliform, and consists originally of three areas, the end-filament, the germinal area, and the maturation area. These areas are best examined in the well-grown larva, since at metamorphosis the whole ovary is fairly well advanced. The *end-filament* is the narrower anterior portion in which the germ-cells are first formed. It is formed of a single row of cells in a long thread. Most of these primitive germ-cells develop later into oogonia, but a few form follicle cells. Posterior to the filament is an area in which the egg-string is seen to be enlarged, owing to the fact that the germ-cells here become crowded together and are increasing in size. This is the *germinal area*, in which the large rounded oogonia are formed, each with a double nucleus, and in which also the formation of the yolk is first noticeable. In the third portion, or *maturation area*, forming the whole of the posterior part of the egg-string, the moniliform appearance becomes most marked, because the developing ova have become fairly large, and are separated out from the germinal mass as a single string of eggs. They lie with their long axes nearly parallel to the egg-string, and increase in size as they approach the oviduct. In this area, the oogonia ripen into mature ova, and the latter become surrounded by large masses of food-yolk. The whole mass of the egg becomes next enclosed in a delicate *vitelline membrane* formed from the egg itself. Round

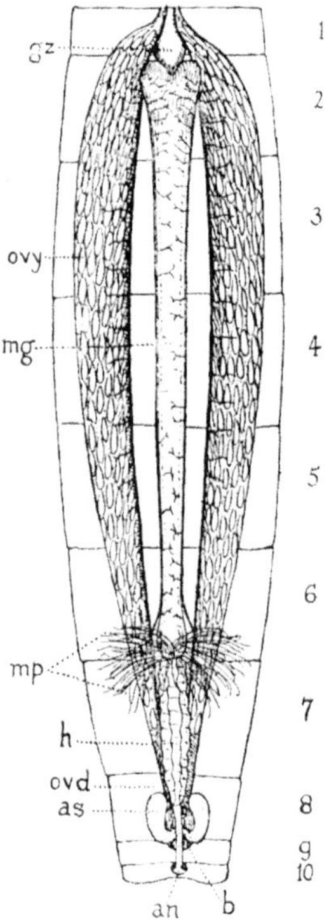

Fig. 97. Internal reproductive organs and alimentary canal of *Petalura gigantea* Leach, ♀, dorsal view (× $1\frac{1}{3}$). *an* anus; *as* accessory sac; *b* bursa; *gz* gizzard; *h* hind-gut; *mg* mid-gut; *mp* Malpighian tubules; *ovd* oviduct; *ovy* ovary; 1–10 abdominal segments. Original, from a dorsal dissection.

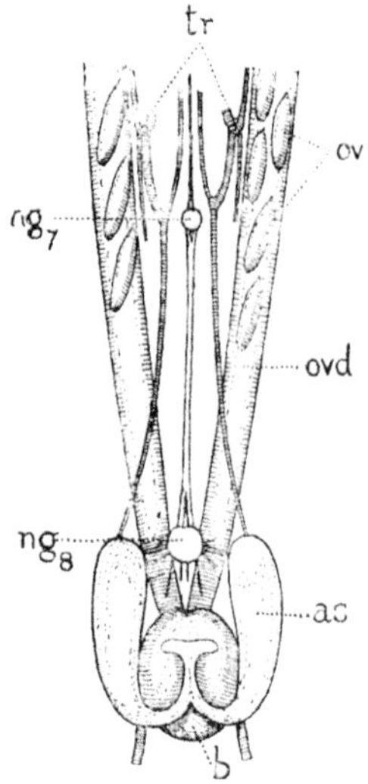

Fig. 98. Posterior part of fig. 97, enlarged (×4), and with alimentary canal removed. ng_7, ng_8 seventh and eighth ganglia; *ov* eggs; *tr* tracheae. Other references as in fig. 97. Original.

this, a thick chitinous outer shell or *chorion* is deposited by the follicle cells, which imprint their faint polygonal outline upon the structure. The eggs are always elongated when first formed, but may become rounded in some forms, when fully completed [91].

Posteriorly, the egg-strings open into the **oviduct** (*ovd*), a thin-walled cylindrical tube, down which the eggs pass one by one. Fertilization is supposed to take place here. The two oviducts open posteriorly in the eighth segment into a large pouch, the **spermatheca** or **bursa** (*b*). The walls of this pouch are very thick and muscular, and are further strengthened by an internally projecting ring or *collar* of hard chitin. Its cavity is often found to be distended with masses of spermatozoa. Opening dorsally into the bursa, by the union of their two short ducts, is a pair of elongated ovoid sacs, the **accessory sacs** (*as*). In mature females these sacs are tightly swollen and smooth, their rounded ends directed forwards and lying at the level of the eighth ganglion (ng_8). At other times they are wrinkled and collapsed. In some forms they appear as coiled tubes or *caeca*. The genital pore opens between the eighth and ninth segments, mid-ventrally.

2. *External Organs* (figs. 99–104).

The external genitalia of the female Dragonfly consist typically of three pairs of ventral processes or **gonapophyses**, outgrowths of the integument in the region of the genital pore. They begin to develop early in larval life, but have nothing to do with the primitive paired segmental appendages of the abdomen, which, as already stated above, are lost during embryonic life. The three pairs together form the **ovipositor**, or organ for use in the process of egg-laying. Van der Weele [192] has made an excellent study of the development and structure of this organ. As it is greatly reduced in a large number of Odonata (*Gomphinae, Libellulidae*), it is necessary to select a form having a well-developed ovipositor, if we would understand its structure correctly. For this purpose we have chosen *Synlestes weyersi* (fig. 99).

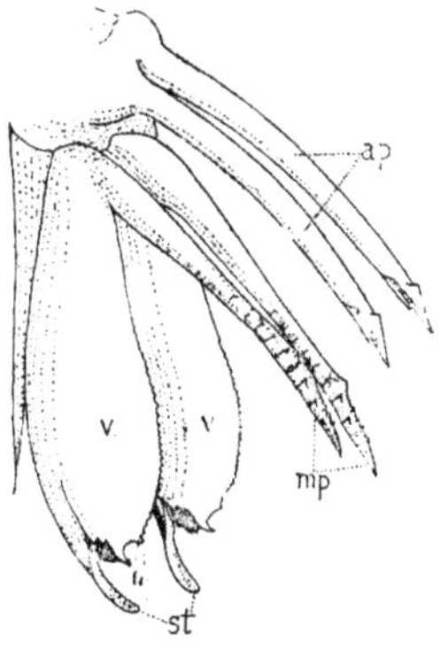

Fig. 99. Ovipositor of *Synlestes weyersi* Selys, ♀ (×10). *ap* anterior processes; *mp* median processes; *st* styles; *v* valve. Original, chitin preparation.

The three pairs of processes may be named *anterior processes* (*ap*), *lateral processes* or *valves* (*v*) and *median processes* (*mp*) respectively. The *anterior processes* (*ap*) are two sword-shaped structures, slightly curved, and with a short saw-like edge near the pointed distal end. Each process is fixed to a short base-piece, an outgrowth of the posterior part of the eighth sternite. Into these pieces are inserted the muscles whose contractions cause the ovipositor to emerge from its sheath. Together with the median processes, the anterior processes form that part of the ovipositor known as the *aculeus* or *terebra*.

The *lateral processes* or *valves* (*v*) are of a very different form, being rather broad, leaf-like, slightly convex lamellae. Their bases occupy the whole length of the ninth sternite, from which they are developed. The apex of each projects postero-ventrally under the tenth segment. At its end it carries a hard, pointed organ called the *style, styloid process,* or *sternorhabdite* (*st*). The structure of the style shews that it is a tactile organ. The outer border of the valve is finely denticulate, and there is a large tooth close to the style.

The *median processes* (*mp*) somewhat resemble the anterior ones, but the blade is narrowed in the middle, and the distal end carries a complicated set of saw-like teeth. The process is inserted into a short base-piece arising from the inner anterior side of the base of the corresponding valve. Into these base-pieces are inserted the muscles which retract the ovipositor back into its sheath.

From the above description, it is clear that the complete ovipositor of the Dragonfly is an archaic structure, homologous with the well-known form such as is found in the Sawflies. All the Zygoptera possess ovipositors closely resembling that of *Synlestes,* though seldom so highly developed for cutting. In the *Calopterygidae,* the organ is not so large in proportion to the size of the insect, but the parts are closely similar. In some *Agrioninae* (e.g. *Ischnura*) there is a ventral spine developed posteriorly on the eighth sternite, projecting under the base of the ovipositor. In many forms the valves are crossed by a diagonal ridge running to the style. A bunch of stiff hairs often takes the place of the large tooth seen in *Synlestes.*

The *Aeschninae* have a complete ovipositor, differing chiefly from that of the Zygoptera in the slenderness of the blades and valves, and in the fact that the style carries either a stiff bristle or a bunch of hairs distally (fig. 100 A, *br*). The action of the ovipositor is here supplemented by the development of a hard projecting process, the *dentigerous process* (fig. 100 A, *dp*), from the tenth sternite. In the *Brachytronini* (A) this forms a plate with a curved and strongly toothed distal border, projecting backwards below the styles. In most *Aeschnini* it is reduced to a rounded tubercle studded with fine denticles, the *dentigerous tubercle*. In the *Gynacantha* series it becomes a sharp fork, with two, three, or four prongs (B).

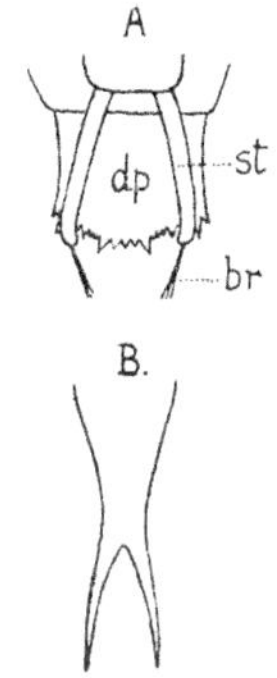

Fig. 100. A. Dentigerous plate and styles of *Austrophlebia costalis* Tillyard, ♀, ventral view. B. Fork of *Gynacantha rosenberghi* Selys, ♀. *br* bunch of hairs; *dp* dentigerous plate; *st* style. (×6.) Original.

The *Petalurinae* have an ovipositor almost as perfect as that of the *Aeschninae*, though the processes are reduced in length.

We may now briefly follow the stages in the reduction of the ovipositor, consequent on the loss of its original function:

(i) In the *Cordulegastrinae* (fig. 101) the valves are reduced to minute rudiments, and the saw-teeth of the median processes are absent. The terebra, however, is hypertrophied, forming a very hard and conspicuous organ projecting far beyond the end of segment 10. Ris [132] has observed these insects ovipositing in rough angular débris, formed by limestone deposits in small shallow streams in Switzerland. The insect flies in a vertical position, dipping its abdomen into the limestone deposit at intervals. Thus the hard point of the terebra protects the soft parts of the end segments from injury.

(ii) In the genus *Gomphomacromia* the valves are absent, and the anterior processes are blunt and basally united. The ovipositor appears to be at a reduction stage comparable with (i), minus the hypertrophy.

(iii) In the tribe *Synthemini* (fig. 102) a wide range in the reduction process is still to be seen. Forms such as *S. primigenia* and *S. macrostigma* appear to have long well-formed ovipositors; whereas others, e.g. *S. claviculata*, have none at all. An examination of the long ovipositors in this tribe shews us, however, that, as in (ii), the valves are absent, and the median processes reduced.

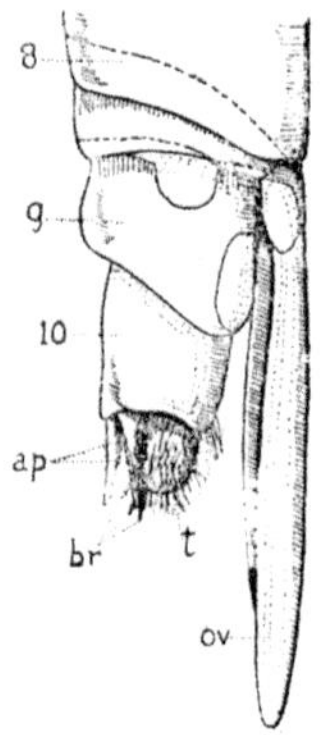

Fig. 101. Ovipositor of *Cordulegaster annulatus* Latr., ♀, lateral view (× 4½). *ap* anal appendages; *br* bristles; *ov* ovipositor; *t* tubercle. Original.

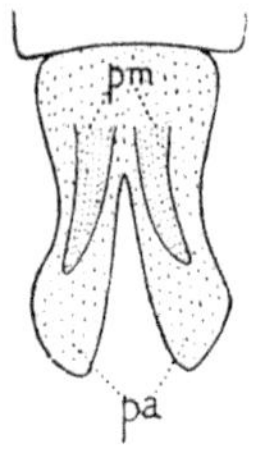

Fig. 102. Ovipositor of *Synthemis eustalacta* Burm., ♀ (× 6). *pa* anterior processes; *pm* median processes. Original, chitin preparation.

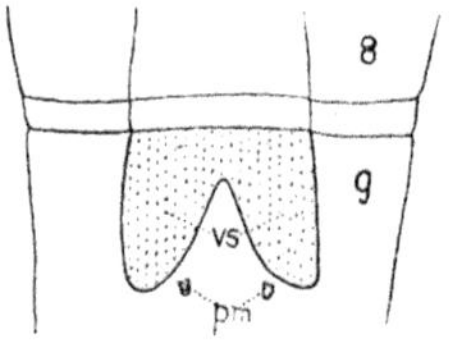

Fig. 103. Vulvar scale (*vs*) of *Tramea loëwii* Br., ♀ (× 4½). *pm* median processes; 8, 9 abdominal segments Original, chitin preparation.

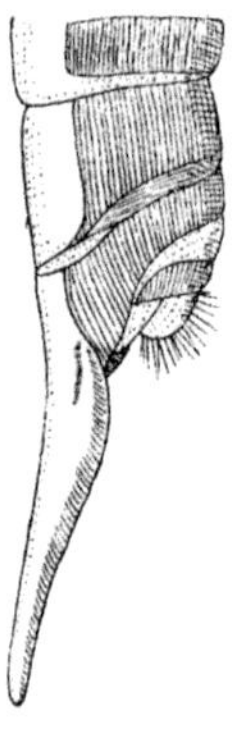

Fig. 104. End of abdomen of *Uracis ovipositrix* Selys, ♀, lateral view. After Ris.

It is the anterior processes which, by union of their bases, and widening of their distal ends, give the appearance of a large ovipositor. No species of this tribe uses its ovipositor for placing its eggs, and the latter are of the rounded form found in all *Libellulidae*.

(iv) By reduction of the median processes to mere vestiges, and a decrease of the anterior processes to small flaps, we reach the stage seen in many *Libellulinae* and *Gomphinae*. Finally, the former quite disappear, while the latter become either the small *valvula vulvae* or the bifid or complete *vulvar scale* (fig. 103, *vs*) situated on the anterior side of the genital pore.

A few *Libellulinae* (e.g. *Uracis*, fig. 104) have recently reacquired a distinct ovipositor, by elongation of the vulvar scale into a sheath-like process projecting from segment 8. A similar process is developed from segment 9, from near the basal rudiments of the original median processes; it is softer, and fringed with stiff hairs. The valves are absent.

We may sum up the position, as regards the gonapophyses, in the following table:

				Fate in ♀				
Name	Segment	Position	Order of Development	*Zygoptera Aeschninae* & *Petalurinae*	*Cordulegastrinae*	*Synthemini*	*Gomphinae* & *Libellulinae*	Fate in ♂
Anterior Processes	8	median ventral	3	terebra or aculeus	hypertrophied	basally fused or much reduced	vulvar scale	absent
Median Processes	9	,, ,,	2	terebra or aculeus	hypertrophied	reduced or absent	vestigial or absent	absent
Lateral Processes	9	latero-ventral	1	valves	vestigial	vestigial or absent	vestigial or absent	valvules

The Process of Oviposition (figs. 105–108).

As the form of the ovipositor is closely correlated, not only with the method of oviposition, but with the shape of the egg, it will be advisable to deal with both of these here.

We may classify the method of oviposition broadly into two divisions:

(1) **Endophytic Oviposition**, or the placing of the eggs in the stems or leaves of aquatic plants, or other objects close to or under the water. Correlated with this habit are (*a*) the complete form of ovipositor, and (*b*) the elongate form of egg.

(2) **Exophytic Oviposition**, or the dropping of the eggs freely into the water, or in other ways placing them outside the tissues of plants. Correlated with this are (*a*) reduction of the ovipositor, (*b*) the rounded form of egg.

It will be seen that the Zygoptera, *Aeschninae* and *Petalurinae* practise endophytic oviposition, while all the rest practise exophytic oviposition. Reduction, however, is not complete in the *Cordulegastrinae* and some *Synthemini*, so that these may be considered in a measure as intermediate or linking forms.

Bearing in mind the description given above of the ovipositor of *Synlestes*, we can now easily understand the process of oviposition in this genus. After alighting on a reed-stem above the water, the Dragonfly feels about with its styles until it finds a suitable place. It then everts the terebra, and makes a small incision with the sharp points of the anterior processes. The ends of the median processes slip automatically into this small slit, which is quickly enlarged by sawing with their teeth. As soon as the hole is made large enough to take several eggs, an egg is expressed through the genital pore, and guided along between the blades of the terebra into the tissues of the plant. Three or four eggs are placed in each hole. The whole body is then moved a short distance downwards, and the process repeated on a new spot. Further holes are made in a gradually descending series, not at all regularly placed (fig. 105 A). Finally, the median processes are retracted, the anterior ones fall back with them as their muscles relax, and the valves close up automatically. Other Zygoptera carry on the process in the same manner, but many kinds only place *one* egg in each incision. The eggs may be laid far above the water-level, at the water-level, or well beneath it. In the last case, the Dragonfly (either alone or accompanied by the male) descends down the stem under the water, where it appears as if silver-plated, owing to its being surrounded by a complete film of air.

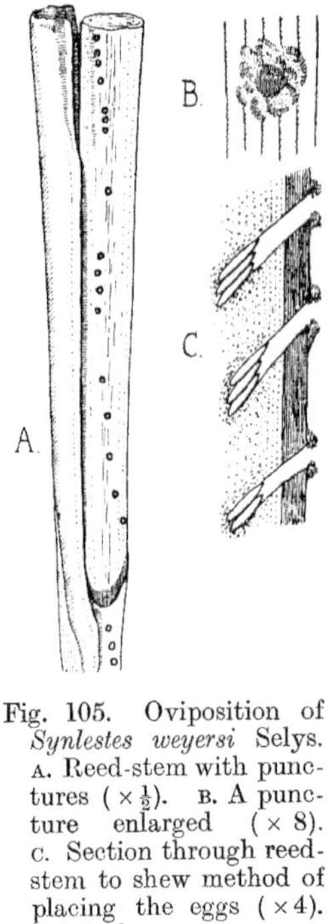

Fig. 105. Oviposition of *Synlestes weyersi* Selys. A. Reed-stem with punctures ($\times \frac{1}{2}$). B. A puncture enlarged ($\times$ 8). C. Section through reed-stem to shew method of placing the eggs ($\times$4). Original.

In the *Aeschninae*, some forms lay their eggs in soft tissues, as do the Zygoptera. The eggs are, however, generally more neatly and regularly arranged (fig. 106). This may be due to the increased efficiency of the styles, which perhaps measure off the distance between successive incisions. The terminal bristle or hairs would assist in this process. In the *Brachytronini*, the eggs are placed either in hard wood or in the matted roots of mosses or filmy fern growing on damp rocks. The hard wood is sawn across transversely by the dentigerous plate, before the egg is inserted. The process is slow, and attended with much danger to the Dragonfly, which shews great trepidation, and often leaves its sawing uncompleted. The oviposition of *Gynacantha* has not been observed.

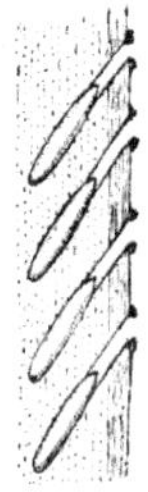

Fig. 106. Section of reed-stem to shew eggs of *Anax papuensis* Burm. in situ (× 5). Original.

Turning now to the exophytic forms, we have already noticed *Cordulegaster* (p. 223). In some *Synthemini*, the large fused anterior processes help to regulate the flow of eggs, and to prevent them from exuding too far. In the great majority of cases, the Dragonfly lays her eggs while flying over the surface of the water, merely by striking the tip of the abdomen from time to time against the water, and so washing off the steady flow of exuding egg-masses. The gelatinous substance surrounding these masses seems to dissolve on contact with water, so that the eggs spread out on the river bed. *Gomphinae* usually oviposit in the ripples of fast streams, *Libellulinae* over still water. This method of oviposition culminates in the genus *Cordulephya*, where the female performs extraordinary evolutions, with sudden dashes down to the water to lay its eggs. The movements are so swift and unexpected that the insect is almost invisible, and very difficult to capture.

In the genera *Tetragoneuria* (fig. 107) and *Sympetrum*, the eggs are laid in gelatinous ropes, attached to submerged twigs. A number of females help in adding to the length of rope, which may contain many thousands of eggs. It seems clear here that the gelatinous envelope does not dissolve, but rather tends to harden in the water [100].

Damage to the stems of the Blue Flag (*Iris versicolor*) has been observed by Needham [99] owing to the numerous punctures made by two species of Nearctic *Lestinae*. About one-fourth of the fruiting stems are killed annually.

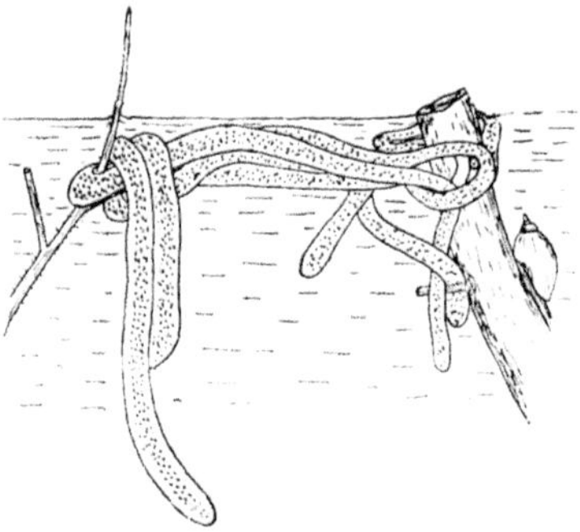

Fig. 107. Egg-rope of *Tetragoneuria cynosura* Say. After Needham.

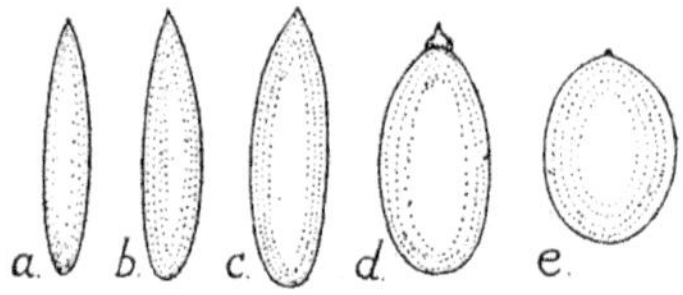

Fig. 108. Eggs of Dragonflies. *a*, *Austrolestes leda* Selys (×15); *b*, *Austropetalia patricia* Tillyard (×20). *c*, *Petalura gigantea* Leach (×20). *d*, *Hemicordulia tau* Selys (×30); *e*, *Diplacodes bipunctata* Br. (×35). Original.

Fig. 108 shews the forms of egg typical of the endophytic and exophytic groups.

CHAPTER XII

EMBRYOLOGY

Structure of the Egg.

The egg, when laid, consists of the following parts:

(i) The *germ-vesicle,* or nucleus of the mature germ-cell. This is a large vesicular nucleus shewing a fine network of chromatin. It is deeply embedded in the yolk, near the centre of the egg.

(ii) A fine network of *formative protoplasm,* ramifying through the yolk.

(iii) The *periplasm,* a very fine layer of formative protoplasm completely surrounding the yolk.

(iv) The *food-yolk,* an immense quantity of homogeneous, structureless yellow material, in the form of originally spherical globules compressed by mutual pressure. Drops of fat are also present.

(v) The *vitelline membrane,* quite structureless, colourless and exceedingly fine. It is secreted from the egg itself, and closely invests the yolk-mass.

(vi) The *chorion,* a tough outer covering, overlying (v), and formed of *chorionin.* This is secreted by the follicle cells of the ovary, which leave a more or less definite irregular hexagonal pattern upon it (representing the boundaries of the follicle cells) except at the two ends, where the same pattern falls upon the thickened gelatinous envelope. At the anterior pole of the egg (i.e. the pole of attachment to the egg-string), the chorion is strongly thickened. A delicate canal, placed obliquely a little to one side of the central axis, pierces the chorion at this end. This is the *micropyle,* or passage for the entry of the sperm. The chorion is also slightly thickened at the posterior pole, but in a less pointed fashion.

(vii) An outer layer of *gelatinous matter,* serving to protect the egg from contact with the air. This forms a more or less regular thin layer over the chorion. It is slightly thickened at the posterior pole. At the anterior pole it is very greatly thickened, forming a conical cap, known as the *pedicel*, by the apex of which attachment is made to the egg-string. The surface of this cap shews the impress of the follicle cells more plainly than the chorion does.

The egg has a *dorsal* and a *ventral* side, the former being distinctly convex, the latter much flatter. The terms dorsal and ventral, applied to these surfaces, indicate the future position of the *late* embryo, but not of the early embryo, which is reversed.

The processes of maturation, fertilization and extrusion of the polar bodies have not been followed in the Dragonfly's egg, the immense mass of yolk making the investigation very difficult.

Embryonic Development (figs. 109–112).

The present state of our knowledge of the embryology of the Dragonfly is due to Brandt [13], Heymons [71] and Tschuproff [181]. The work of Brandt on *Calopteryx* still remains the most valuable source of our knowledge. The other two authors have chiefly investigated problems concerning the germinal layers; their studies were made on *Libellula* and *Epitheca*. Much still remains to be done, particularly with regard to the early segmentation, and the formation of the various internal organs in later embryonic life.

Early Stages.

Cleavage or Segmentation of the Egg. The first segmentation-nucleus is formed, as in all insects, by union of the male and female pronuclei near the surface of the yolk. This nucleus then moves inwards, and sub-divides into a large number of daughter-nuclei. These nuclei then move outwards to the surface of the yolk, where each takes up a position surrounded by a small "island" of protoplasm. These islands are all actually connected together by fine threads. They finally fuse with the periplasm, and become marked off by constrictions between them, so that they soon take on the appearance of a single row of surface cells. This is the *blastoderm*. Since the cells are formed at the surface, this form of segmentation is known as *superficial*.

Vitellophags. While the blastoderm is being formed, certain cleavage cells migrate inwards. These increase by division to

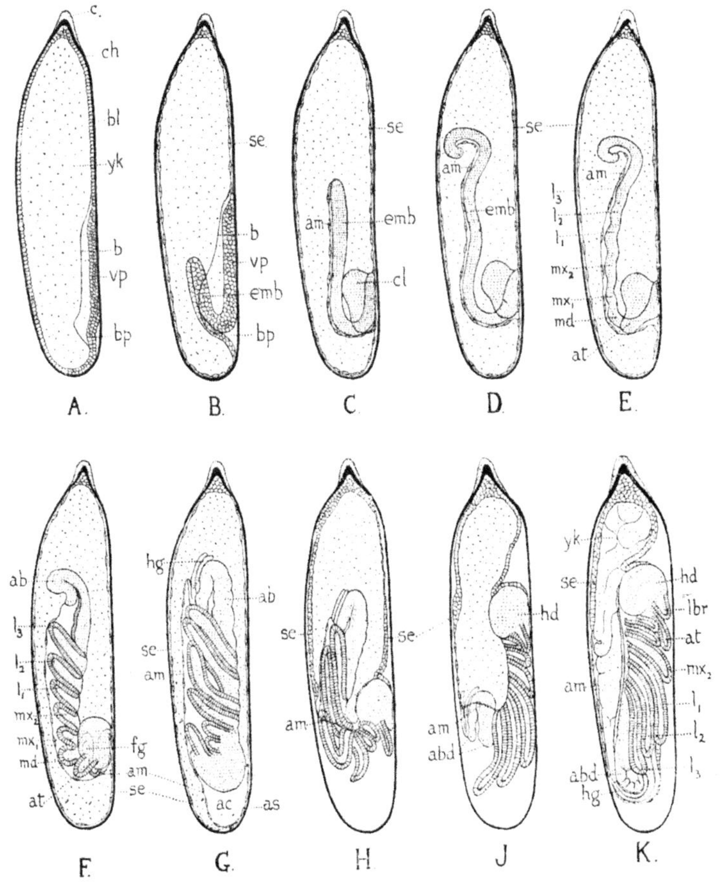

Fig. 109. Ten stages in the embryonic development of *Calopteryx*. Lateral view, dorsal surface to the left, ventral to the right. A. Formation of ventral plate. B–D. Stages in the invagination of the embryo. E–G. Formation of appendages. H–K. Rupture of amnion, and reversion of embryo towards its final position. *ab* abdomen; *abd* abdominal segment; *ac* amniotic cavity; *am* amnion; *as* union of amnion and serosa; *at* antenna; *b* lateral border of ventral plate; *bl* blastoderm; *bp* blastopore; *c* cap or pedicel; *ch* chorion; *cl* cephalic lobe; *emb* embryo; *fg* stomodaeum; *hd* head; *hg* proctodaeum; l_1–l_3 legs; *lbr* labrum; *md* mandible; mx_1 first maxilla; mx_2 second maxilla (labium); *se* serosa; *vp* ventral plate; *yk* yolk. After Brandt.

form large cells known as *vitellophags*, whose function it is to liquefy the yolk and carry it to the developing embryo. They grow to a great size, and possess enormous nuclei. They are not all used up by the time of hatching. Those that are left become finally enclosed, with their store of yolk, in the mid-gut of the young larva, where we have already met with them (p. 71).

The Ventral Plate (fig. 110, and fig. 109 A).

The first sign of the Dragonfly embryo is a slight thickening of the blastoderm on the posterior portion of the ventral surface of the egg. This thickening, known as the **ventral plate** (*vp*), is caused by proliferation of the blastoderm cells, which here become several layers deep. The ventral plate grows forwards, and quickly becomes a very distinct oval patch, occupying half or more of the ventral surface of the egg.

Near the posterior end of the plate—in that position, in fact, where the thickening first began—there next appears a kind of pitting-in or invagination of the blastoderm. According to Heymons, this invagination in *Libellula* (fig. 110) takes the form of a **primitive groove** (*pg*) along the mid-longitudinal line of the ventral plate, running forwards from the point where it first began (known as the *blastopore*). A slight enlargement of the anterior end of this groove marks the position of the future stomodaeal opening (*st*). According to Brandt, in *Calopteryx* the invagination takes place by a sinking-in of the ventral plate at the blastopore in a gradual manner, beginning with the posterior end (fig. 109 B). The result is the same in either case.

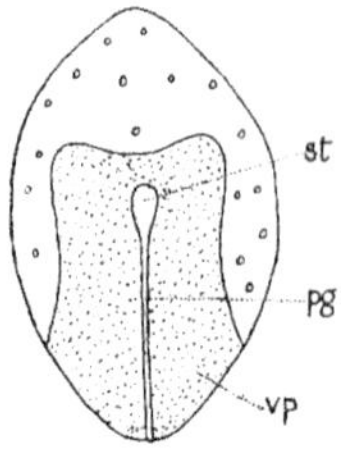

Fig 110. Ventral plate (*vp*) and primitive groove (*pg*) in *Libellula*. *st* stomodaeum. After Heymons.

The cavity formed by the sinking-in of the ventral plate is known as the **amniotic cavity**. As the ventral plate passes in by the *upper lip* of the blastopore (B, C), and as undifferentiated blastoderm is drawn in along the lower lip, the walls of the amniotic cavity are dissimilar. The dorsal wall is formed of a single layer of cells, destined to become the **amnion** (*am*), while the ventral wall is several layers thick, and is formed by the ventral plate. At the sides of the cavity, the amnion merges into the ventral plate. The part of the blastoderm not drawn into the cavity, but remaining surrounding the yolk, may from now on be termed the **serosa** (*se*).

Even before the invagination is completed, the ventral plate begins to change its shape by the upgrowth of its antero-lateral borders, so that the anterior border becomes slightly concave in the middle (fig. 109 C). This marks the beginning of the formation of the head. By the time the invagination is completed, the two large *cephalic lobes* (*cl*) are very distinctly formed, lying just inside the blastopore.

Amnion and Serosa. We may here follow out the further development of these single cell-layers. As the invaginated ventral plate takes on the distinct form of the embryo, the amnion grows out ventrally on either side of it, in two folds. These finally grow completely round the embryo, and unite together to enclose it within the amniotic cavity. The latter is filled with a fluid called the *amniotic fluid.* The cavity is finally completed in the region of the blastopore, where the amnion fuses with the serosa. The function of the amnion is to protect the early embryo from shocks. Except just at the closed-up blastopore, amnion and serosa are definitely separated by a large amount of yolk. Thus all the embryo except the head now lies well within the egg. This condition is known as the "*immersed germ-band.*" Other insects which share with the Odonata the peculiarity of having an immersed germ-band are the Lepidoptera, most Hemiptera, and a number of Orthoptera.

The Germinal Layers.

We must now pause to follow the formation of the three *germinal layers,* or layers of cells, from which all the embryonic structures are derived.

The Ectoderm. This is formed directly from the outermost layer of cells of the invaginated ventral plate. These become differentiated off from the cells underlying them at a very early stage, by the assumption of a compact columnar form, due to mutual lateral pressure. The ectoderm covers the whole of the ventral surface of the embryo. At first it passes insensibly, on either side, into the smaller flattened cells of the serosa. When the amnion is completed, the serosa is cut off from the ectoderm, which then becomes continuous, on either side, with the cells of the amnion. It should be borne in mind that, during this process, vitellophags are at work in the yolk dorsally above the ectoderm, so that a continuous supply of nourishment is furnished to the cells of that region.

The Mesoderm. The main portion of the mesoderm in Odonata seems undoubtedly to be formed by the ingrowth of the primitive groove; though it is very probable that the lateral parts of the mesoderm are formed from isolated cells wandering in from the blastoderm. The cells which form the sides and floor of all the middle portion of the primitive groove become enlarged, so as to differ gradually more and more from the cells continuing the edge of the groove above, which become the ectoderm. Finally, the groove becomes flattened and closed in by overgrowth of the ectoderm above it. The lumen of the groove is lost, and its cells spread out as a slightly convex layer, several cells deep, beneath the ectoderm. This layer is appropriately termed the *lower layer*, since, in many cases, both mesoderm and endoderm are derived from it. If, however, Heymons' observations be correct[1], the endoderm has a different origin in the Odonata; so that we may speak of the lower layer cells in this case as forming the mesoderm.

The mesoderm soon shews signs of splitting into two layers, an upper and a lower[2]. The upper or ventral layer tends to become closely applied to the ectoderm, and is termed the *somatic* or *parietal* layer. From it are developed the segmental muscles. The lower or dorsal layer tends to become closely applied to the yolk-cells lying dorsal to it, and is termed the *splanchnic* or *visceral* layer. From it are developed the muscles of the mid-gut. The

[1] Heymons studied chiefly the Odonata, Orthoptera and Aptera. The more recent work of Hirschler, on the embryology of the Coleoptera (*Zeit. wiss. Zool.*, XCII, 1909), challenges many of Heymons' main conclusions, and appears to offer more reasonable explanations of certain difficult points. But researches made upon the Odonata cannot be *disproved* by work done on the Coleoptera, however excellent it may be, since the embryology of these two Orders differs in many important points. What is required is a thorough re-investigation of the embryology of the Order Odonata itself.

[2] In considering the germinal layers, the ectoderm is taken as the *uppermost*, although actually the *most ventral* layer.

space between the two layers is the true *coelome* or primary body-cavity. It is never very large in the Insecta. Its fate in the Odonata is uncertain, but it is believed to give rise only to the lumina of the gonads and the reduced pericardial space.

It must be borne in mind that the segmentation of the embryo (see below) extends to the mesoderm, subdividing it into blocks forming the **mesodermal somites**, which correspond with the external segmentation. The result of this is that the coelome is constricted off into separate spaces or *coelomic sacs*. These have not been studied in the Odonata. It also follows that the heart and muscles, which are formed from the mesoderm, are formed segmentally.

The Endoderm. The origin of the lowest of the three germinal layers, or *endoderm*, in the Odonata, is still a matter of dispute amongst embryologists. There can be little doubt that Tschuproff is correct in stating that, in the Odonata alone amongst insects, the mid-gut is not formed entirely of endoderm, but consists at first largely of vitellophags. This is clear from fig. 111, where the huge vitellophags (*vit*) are seen to form the greater part of the middle portion of the wall, while small nests of endoderm cells (*en*) lie here and there outside them. After hatching, when the contents of the vitellophags are absorbed, these cells themselves disappear, and the endoderm spreads out to form the deep columnar cells of the mid-gut.

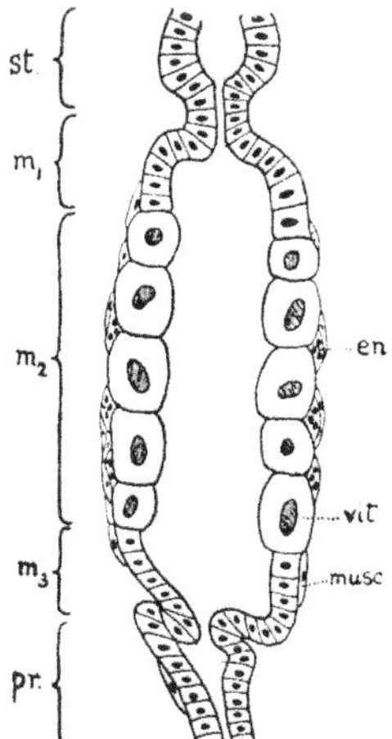

Fig. 111. L.S. through mid-gut of embryo of *Epitheca bimaculata* Charp., late stage. *en* nests of endoderm cells; m_1–m_3 the three portions of the mid-gut; *musc* muscle-layer; *pr* proctodaeum; *st* stomodaeum; *vit* vitellophag. After Tschuproff.

The difficulty, however, lies in determining the origin of these endoderm cells themselves. We can only give our readers the facts so far observed, and leave them to choose between the rival theories. Before, however, we can follow these facts, we must trace the part played by the ectoderm in forming the alimentary canal.

When the primitive groove was formed, its anterior end deepened into a pit, the *stomodaeal invagination*, marking the position of the future mouth of the embryo. This pit gradually deepens and lengthens out posteriorly, so that its blind end at last comes to project close up against the yolk-cells in the region of the future mid-gut. At a later period, when the anal plate (see below) of the embryo is formed, a similar but smaller invagination appears at the extreme posterior end of the embryo, on the eleventh (or possibly the twelfth) segment. This pit is the *proctodaeal invagination*, marking the position of the future anus. It grows longer in an anterior direction, and thus finally pushes its blind end close up to the yolk-cells in the region of the mid-gut in front of it.

When these two pits are small, a cluster of small but very distinct cells appears around the blind end of each, beneath the mesoderm. These clusters soon become crescent-shaped, and then spread out like two pairs of horns advancing towards one another on either side of the mid-gut. Finally the two horns of each side meet and coalesce. These form the endoderm cells (fig. 111, *en*), arranged, as already seen, in a series of nests close outside the vitellophags. But only the middle portion (m_2) of the mid-gut is thus enclosed. According to Tschuproff, a short anterior part (m_1) is formed from the end of the stomodaeum, by constriction off, where its blind end opens to the mid-gut; while a similar short posterior part (m_3) is formed from the proctodaeum.

We thus have this remarkable result in the Odonata, that the whole of the alimentary canal, except the middle portion of the mid-gut, is of ectodermal origin.

Now it is this middle portion which is still in dispute. The two views may be stated thus:

(*a*) Grassi (1884), Nussbaum, Kowalevski, Wheeler and Hirschler, with others, consider that the "nests" of cells which form the mid-gut are undoubtedly derived from the lower layer. Thus they are correctly termed *endoderm cells*, and the middle portion of the mid-gut is of endodermal origin.

(*b*) Heymons (1895), with some more recent supporters, holds that these "nests" are of ectodermal origin, since they first appear close up to the stomodaeum and proctodaeum, which are admittedly ectodermal. If this be so, the alimentary canal is entirely of ectodermal origin, and no true endoderm exists in the Odonata.

Students interested in these rival theories will find them fully discussed in Korschelt and Heider [81] to whom they are referred. Supporters of Gaskell's theory will, no doubt, see in Heymons' statements some further evidence for their views.

Formation of the Segments and Appendages (fig. 109 D–G).

After the first appearance of the cephalic lobes, and the completion of the sinking-in of the ventral plate, the embryo gradually elongates at the end opposite to the future head, the extreme end becoming curved over as in D. This curvature remains for the rest of embryonic life. On the curved end a slight enlargement marks the *anal plate*, on which, soon afterwards, the proctodaeal invagination begins to form. The body of the embryo very soon begins to shew signs of segmentation, by the formation of slight transverse constrictions, indicating the limits of the future body-segments. The constrictions only shew up on the ventral surface of the embryo, which tends to become more convex. On either side of the mid-ventral line, the convexity indicating a future segment very soon begins to take the form of a pair of latero-ventral swellings—the first rudiments of the appendages. The first distinct swellings to be seen are three pairs at about the middle of the ventral surface of the embryo. These represent the rudiments of the future thoracic legs (E, l_1, l_2, l_3). Next to these, nearer the head, three very similar swellings can be made out, destined to become the appendages of the mouth (E, *md*, mx_1, mx_2). At the same time, the cephalic lobes alter in shape. From their more ventral part a pair of swellings arises similar to that mentioned above, but destined to form the antennae. Just in front of this, a median unpaired swelling arises, forming the front of the head. The stomodaeal invagination takes place just posterior to this.

The appendages gradually elongate, but at unequal rates, the most posterior pair (third legs) soon becoming the longest, while the others decrease from behind forwards in order, up to the mandibles (F). The antennae, however, soon become longer than the mandibles and first maxillae. As the appendages grow, they become directed backwards.

Posterior to the third legs, the segmentation of the body continues backwards. Rudimentary pairs of appendages are said to be discernible for a short time on all the abdominal segments; but they soon disappear, except on the eleventh segment, where they persist as the *cerci*.

This stage in the growth of the embryo is completed by its stretching or straightening out, so that the head is drawn inwards towards the posterior pole of the egg, and comes to lie in line with the rest of the body (G). At this stage it will be noticed that the cephalic lobes have extended back more dorsally, while they have also fused ventrally with the median front of the head, which now is seen to form a kind of flattened lobe, destined to become the labrum and clypeus. Thus the head is much more completely formed than it was before the straightening process set in. It should be noticed that the amnion and serosa are still fused at a point near the original position of the blastopore, and that the embryo now lies longitudinally within the egg, with its head downwards and close to the posterior pole.

Rupture of the Amnion, and Reversion of the Embryo (fig. 109 H–K).

The embryo now undergoes a remarkable series of changes in position, during which great alterations take place in the amnion and serosa. The latter appears to thicken and contract, drawing all the yolk left near the posterior pole of the egg forwards towards the anterior pole. The increase of pressure consequent upon this movement, and upon the steady growth of the embryo, causes the amnion to rupture along the portion above the head, where it is fused with the serosa. The amnion here splits into two halves. As the serosa goes on contracting, the pressure causes the embryo to project head first through the split (H). The head of the embryo, once freed from the amniotic envelope, turns round and follows the contracting serosa upwards, towards the anterior pole. The rest of the body of the embryo follows the head in due course, so that the embryo is for a time curved round on itself, while segment after segment passes out through the rent (H, J). Finally, the head comes to lie below the anterior pole of the egg, while the rest of the embryo is in line with it, running back to near the posterior pole, where the bent up end of the abdomen lies. These movements of the embryo are termed *blastokinesis*. All the yolk is now contracted into the space dorsally above the embryo, which lies with all its appendages facing outwards along the ventral surface of the egg.

During the above changes, the remains of the amnion have gone on contracting with the serosa, until finally the two membranes become sufficiently reduced in area to form a tight covering for the dorsal yolk-plug[1]. Thus, at stage K, we may see how the embryo is really only formed laterally and ventrally, while the whole of its dorsal portion consists of this huge yolk-sac. By a gradual extension of the ectoderm and mesoderm laterally, and then latero-dorsally, over the yolk-sac, the latter comes at last to lie completely within the body of the embryo. The final closure takes place along the mid-dorsal line.

Further Changes.

The embryo now grows steadily, so that the head comes closer and closer up to the anterior pole of the egg. The antennae, second maxillae and the three pairs of legs are soon seen to be distinctly jointed. The segmentation of the whole body becomes distinct and annular. Reddish pigment areas on either side of the head indicate the first formation of the compound eyes, which soon appear blackish in the middle. The yolk-plug steadily contracts, and finally becomes all used up, except for the portion enclosed in the mid-gut. The mandibular and maxillary teeth appear, and the bases of

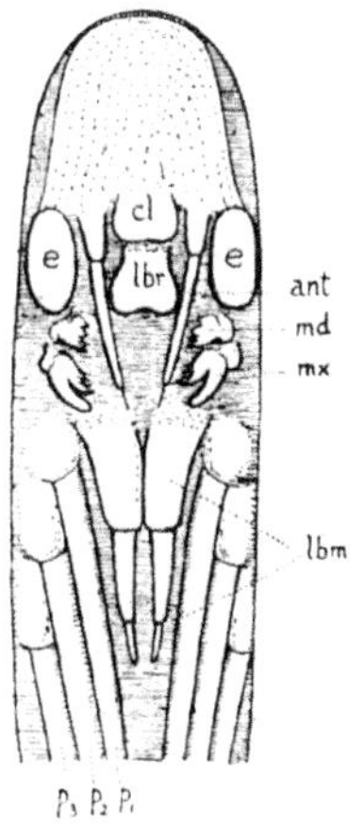

Fig. 112. Anterior half of egg of *Anax papuensis* Burm., just before hatching, ventral view (×60). *ant* antenna; *cl* clypeus; *e* compound eye; *lbm* labium; *lbr* labrum; *md* mandible; *mx* first maxilla; p_1–p_3 legs. Original.

[1] The actual fates of the amnion and serosa are doubtful. It is believed that the serosa gradually breaks up and mingles with the yolk, while the final fate of the amnion is not definitely known.

the three-jointed second maxillae fuse in the middle line, indicating the first formation of the labium. A pale brownish cuticle can be seen enclosing the whole embryo.

Just before hatching, the position of the cephalic and thoracic appendages is as shewn in fig. 112. It will be seen that the original front lobe of the head is now divided into clypeus and labrum. The antennae are two-jointed and slender. The mandibles and first maxillae point rather inwards, and shew their definite shapes. The second maxillae lie parallel and backwards, resembling a pair of legs. From this point the reader may continue the narrative by turning to the account of the hatching of the larva on p. 67.

Origin of the Parts of the Embryo.

A. *Parts of Ectodermal Origin.*

1. *The hypoderm* is derived directly from the ectoderm, which comes finally to enclose the whole body of the embryo as the layer of hypoderm cells.

2. *The cuticle* is secreted by the hypoderm. Two cuticles are formed, one inside the other, during late embryonic life. The outer one forms the *pronymphal sheath* (p. 69).

3. *The tentorium* is developed from two pairs of ectodermal invaginations in the forepart of the head. The details have not been worked out in the Odonata.

4. *The nervous system.* On either side of the mid-ventral line of the body of the embryo, the ectoderm becomes thickened into a longitudinal cord. This pair of cords sinks below the rest of the ectoderm, which grows ventrally over them. The cords then become segmented, and give rise to the ventral chain. The two cords fuse in each segment to form the paired segmental ganglia, from which the longitudinal nerve-cords grow out separately later. The development of the brain and oesophageal commissures has not been worked out in the Odonata.

5. *The compound eyes.* These organs become fairly well developed during embryonic life. Their early formation, however, has not been studied in the Odonata, so far as I am aware. The ocelli do not develop at all in the embryo.

6. *The tracheal system* arises from paired segmental ectodermal invaginations. The original apertures persist, and become the spiracles. The details of the embryonic growth of this system, and of the first formation of the branchial basket of Anisoptera, are unknown. In the embryo, the tracheal system is not filled with air, but with a fluid whose origin is doubtful. The endotrachea and spiral thread are secreted in the embryonic period.

7. *The stomodaeum.* We have already followed the development of this invagination. It gives rise to the whole of the epithelium and intima of the fore-gut, i.e. the buccal cavity and pharynx, with the epipharynx and hypopharynx, the salivary glands, ducts and reservoirs, the oesophagus and crop, and also the gizzard and its armature. Also, we have seen that a small anterior portion of the mid-gut, probably corresponding with the *annular zone* (p. 109), is derived from it.

8. *The proctodaeum.* This has also been dealt with. It gives rise to the epithelium and intima of the anus, the rectum and its pads, the branchial basket of Anisoptera, the small intestine, together with the Malpighian tubules, as blind outgrowths from its extreme anterior end. A small posterior portion of the midgut, probably corresponding with the sphincter region (p. 113), is also derived from it.

B. *Parts of Mesodermal Origin.*

1. *The heart* is formed from two strings of cells called *cardioblasts*, derived from the latero-dorsal walls of the mesodermal somites. These strings become pushed up inwards dorsally, until they meet and fuse in the middle line. They enclose between them a pair of blood lacunae, which later on fuse to form the lumen of the heart. The formation of the ostia, osteoles, alary muscles and other details has not been studied in the Odonata.

2. *The coelome* is the reduced remnant of the cavities of the somites. It forms only the pericardium and the lumina of the gonads.

3. *The blood-corpuscles* are either somatic mesoderm cells which have wandered into the haemocoele, or they may have arisen from the yolk-cells.

4. *The haemocoele* is a secondary enlargement of the blood lacunae, formed in the mesoderm. This increases enormously, and forms the general body-cavity of the insect.

5. *The fat-body* is formed by proliferation of the cells lining the haemocoele.

6. *The muscles and connective tissues* are formed from the somatic layer, except the muscles of the alimentary canal, which are derived from the splanchnic layer.

7. *The gonads and their ducts* are derived from two longitudinal strands of cells formed in the somatic mesoderm. These *genital cells,* from which the gonads arise, are formed very early in embryonic life, as specialized mesoderm cells lying towards the posterior end of the ventral plate. Later, they spread out from the seventh to the second abdominal segments. Undifferentiated mesoderm cells form an epithelium surrounding the true genital cells.

C. *Parts of Endodermal Origin.*

The only part formed in the Dragonfly from the endoderm is the epithelium of the mid-gut, excepting the short anterior and posterior pieces already mentioned as being of ectodermal origin. As already stated, even this is doubted by Heymons.

Length of Embryonic Life.

The time elapsing between the laying of the egg and the hatching of the larva is about three weeks in the case of the Zygoptera, a little less in *Aeschnidae.* But some *Libellulidae,* at any rate, if not all, hatch out much sooner. *Diplacodes haematodes* hatches out five days after the eggs are laid, and some other species almost as quickly.

Plate III

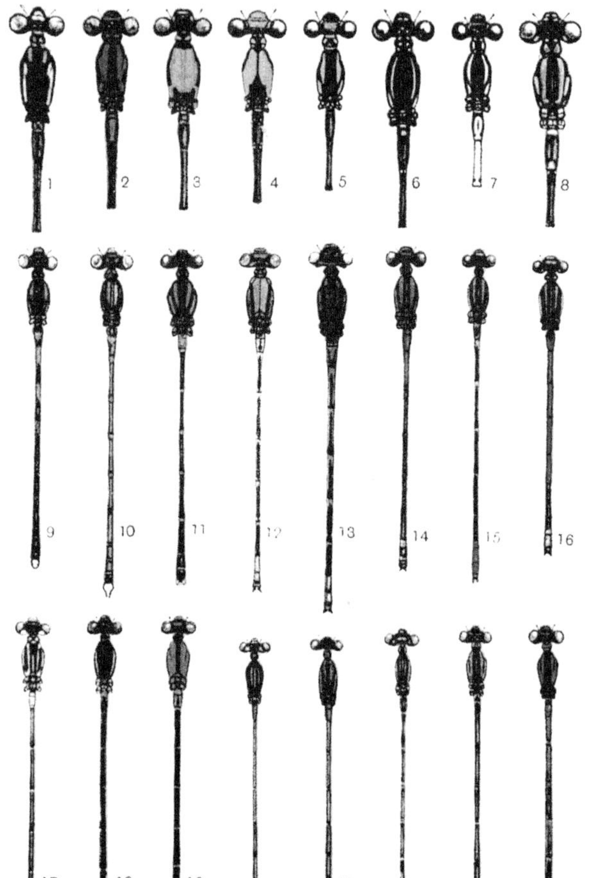

Figs. 1–8, illustrating correlation between coloration and habitat in *Podopteryx* and *Argiolestes*. All the species figured are from New South Wales or Queensland. Fig. 1, *Podopteryx roseonotata* Selys, ♂ (tropical scrub). Fig. 2, *Argiolestes amabilis* Foerster, ♂ (subtropical scrub). Fig. 3, *A. chrysoïdes* Tillyard, ♂ (subtropical scrub). Fig. 4, *A. aureus* Tillyard, ♂ (tropical scrub). Fig. 5, *A. fontanus* Tillyard, ♂ (ferns and waterfalls). Fig. 6, *A. icteromelas* Selys, ♂ (rocky streams). Fig. 7, *A. griseus griseus* Selys, ♂ (mountain bogs, 2000–4000 ft). Fig. 8, *A. alpinus* Tillyard, ♂ (open grasslands, 4000–5000 ft). Fig. 1, $\times 1\frac{1}{5}$; figs. 2–8, $\times 1\frac{4}{5}$. [Segments 4–10 omitted.] Figs. 9–16, some beautiful types of body-coloration in *Lestidae* and *Agrionidae* from New South Wales. Fig. 9, *Austrolestes annulosus* Selys, ♂. Fig. 10, *A. cingulatus* Burm. ♂. Fig. 11, *Nososticta solida* Selys, ♂. Fig. 12, *Pseudagrion aureofrons* Tillyard, ♂. Fig. 13, *Caliagrion billinghursti* Martin, ♂. Fig. 14, *Xanthagrion erythroneurum* Selys, ♂. Fig. 15. *Argiocnemis rubescens* Selys, ♂. Fig. 16, *Ischnura aurora* Br. Figs. 9, 10, 13 $\times 1\frac{1}{5}$; figs. 11, 12, 15 $\times 1\frac{1}{3}$; fig. 14 $\times 1\frac{1}{2}$; fig. 16 $\times 2$. Figs. 17–24, illustrating dimorphism in the females of certain *Agrionidae*. Figs. 17–19, *Ischnura pruinescens* Tillyard ($\times 1\frac{1}{2}$); fig. 17, ♂; fig. 18, normal ♀; fig. 19, heteromorphic (orange), ♀. Figs. 20–21, *Austrocnemis splendida* Martin ($\times 2$); fig. 20, ♂; fig. 21, heteromorphic (rufous), ♀; the normal ♀ has the shape of fig. 21 and colouration of fig. 20. Figs. 22–24, *Agriocnemis hyacinthus* Tillyard ($\times 2$); fig. 22, ♂; fig. 23, normal ♀; fig. 24, heteromorphic (rufous) ♀. (Original water-colour drawings.)

CHAPTER XIII

COLORATION

No Order of Insects can surpass the Odonata in the beauty, variety and brilliancy of its coloration, except it be the Lepidoptera. In intricacy of colour-pattern the latter are undoubtedly supreme. But practically the whole of the colouring of the Lepidoptera is found on the wings; whereas, in the Odonata, not only do we find certain groups with brilliantly-coloured wings, but there is in nearly all cases a wonderful development of body-coloration quite unequalled in the Lepidoptera or in any other Order. The wings of Dragonflies have no efficient unit for the localization of colour-pigment, such as the scales of Lepidoptera. It is not surprising, then, that wing-colours are confined to special groups. The wonder rather is, that those groups should have attained such beauty as they possess, on so insufficient a basis as a wing carrying neither hairs[1] nor scales.

The Nature of Colours.

No investigations have so far been carried out as to the nature of colours in the Odonata. Hagen, however, who knew the Order well, has given an excellent *general* classification of the colours of insects, which seems readily applicable to the Dragonflies, with slight modifications. Simple experiments are easily devised to shew that Hagen's different *types* of colour do all exist in this Order. Of these there are two principal kinds, the second divisible into various separate types:

[1] Recently an undescribed species of *Lestes* from New Guinea has been found with a curious development of hairs at the postero-distal end of the petiole of the wing. This is quite unique in Odonata.

16—2

A. *Optical or Interference Colours.*

These colours do not really exist as pigments, but are produced by the interference of the light reflected at the two surfaces of the double wing-layer, or by combined refraction and reflection of light which falls on to a body-area pigmented in a special manner. If the wing be unpigmented, the interference coloration takes the form of *iridescence*, a kind of delicate rainbow or mother-of-pearl effect varying with the angle of vision. This can be seen in the freshly-formed wings of all Odonata. It seems to be due to the presence of a liquid layer (pale greenish blood) between the two wing-membranes. Thus there are, in all, three superimposed layers with different refractive indices. As soon as the wing is dry, the two wing-membranes fuse, and the *iridescence* gives place to *hyalinity*, in which no interference effects are usually visible. In the *Calopteryginae*, however, the mature wings retain a delicate iridescence of pale green, mauve or purple. This seems to be due to the arrangement of the numerous narrow parallel and unpigmented cells of the wing, which obviously produce an interference effect on the light falling on certain parts of each cell. No cases of permanent iridescence are known in wings in which the cells are large, polygonal, or irregularly arranged.

The phenomenon of *metallic coloration* is the effect produced by superimposing an interference effect upon a dull-coloured underlying pigment. We shall therefore deal with this under the next heading.

B. *Pigmentation or Natural Colouring.*

These colours are produced by the deposition of certain coloured substances either in the wing or the body-integument, in the form of fine dust or particles, called *pigment-granules*. Almost all pigments are elaborated by cells of ectodermal origin, though a few appear to originate in certain mesoderm cells, and to migrate later into the ectoderm. We may classify the pigments found in Dragonflies[1] as follows:

1. **Cuticular or Permanent Pigmentation.** This occurs in all Odonata, but in very varying amount. The pigment is actually

[1] We are here only concerned with *external* coloration. Pigments are also elaborated in many internal organs, e.g. the mid-gut and the tracheae.

embedded in the cuticle in the form of fine granules, called "pigment nuclei," deposited in the outer layers of chitin. Since the pigment is protected from decomposition by the presence of the underlying clear layers of the cuticle, it follows that, at death, when the hypoderm cells decay, the cuticular pigments remain unaffected. Hence they are called *permanent*; though, as a matter of fact, continuous exposure to light, for a long time, causes them to fade. These colours form the *ground-colours* of Dragonfly body-pattern. Blacks and browns of all shades and degrees on the body, the same colours on the wings, with the addition of the yellow stain called *saffroning*, and of the rich reds found in some genera, are all cuticular pigments. Orange and fulvous are formed, in part at least, by the same type of pigment.

Metallic colours, both on wings and body, are to be classed under this heading. They are formed by an interference effect superimposed on a cuticular pigmentation. The effect is produced by a special arrangement of the pigment-granules, whereby the light received becomes split up in a certain manner. All the refracted rays except those of one colour become absorbed by the pigment, while the excepted colour is reflected back, and produces the well-known metallic effect. All these colours are "permanent," in the sense explained above. The commonest metallic colours in Odonata are emerald green, bronze-green, copper, blue, violet and purple.

2. **Hypodermal or Subcuticular Pigmentation.** These are pigments formed either in or above the hypoderm cells themselves. In the Odonata, the pigment is chiefly formed on the outer surface of the cells, and may be extruded to form a definite layer beneath the cuticle. Owing to their position, these pigments are all liable to suffer from decay on the death of the insect. They are, however, always beautifully preserved by immersion in alcohol. Rapid drying may also be employed with more or less success for preserving them.

These colours are nearly all brilliant, viz. sky-blue, bright red, bright orange, lemon yellow, pea-green, French grey, and other rich shades.

3. **Pruinescence or Exuded Pigmentation.** This is a special form of colouring connected, just as in the case of fruits, with

maturation. It is always far more evident in males than in females. The pigment appears to be a product of the process of maturation of the gonads, and is therefore to be considered as of mesodermal origin. It is, however, sooner or later taken up by the hypoderm cells, and excreted through the fine pores of the cuticle, so that it appears as a *supracuticular pigment*, which can often be removed by rubbing. The most extraordinary case known to me in Odonata is that of *Argiolestes griseus albescens*, in which the pruinescence of the thorax resembles an abundant growth of white fungus.

Pruinescence in Odonata is always of a whitish or pale bluish colour. By combination with a subcuticular red pigment it may give a purple appearance to the insect. It is distinguished from all other forms of colouring by its *gradual* growth or onset, beginning long after metamorphosis. The parts first affected appear to be those associated with special sexual organs, e.g. either the second or the ninth abdominal segment. From one or other of these centres it may spread all over the body. In some cases only the thorax becomes pruinescent. In most Odonata the ventral surface of the abdomen becomes pruinescent after a time, in both sexes, even when the rest of the body is not affected. In the females, there seems to be a factor inhibiting pruinescence until late in imaginal life, after most or all of the eggs are laid. A very old female may become rapidly pruinescent, so as to resemble the male in coloration.

Body-Coloration (Plates I, III).

In this section, I propose to give a phylogenetic treatment of the subject of Odonate colour-patterns, based on a careful study of all the forms known to me.

The most obvious fact is that the *bicolorous* type of pattern not only predominates, but forms a basis for the production of those unicolorous patterns known to occur. Many species in which the males are unicolorous have bicolorous females. Moreover, the newly-emerged males also shew signs of the bicolorous pattern for a greater or less period of time. All this points to the fact that the bicolorous pattern is the most archaic form of pattern in the Odonata.

The pattern was evidently originally formed *segmentally*, the pattern of each segment being the same. In the head, owing to fusion of the segments, only certain special features of the pattern can be dealt with satisfactorily. In the thorax, alterations in the original shape and direction of the segments have altered the original plan, which can, however, always be more or less correlated with that of the abdomen. In the latter, certain segments (particularly the second) tend to become specialized beyond the rest. It must be understood, therefore, that the classification of colour-patterns here attempted is based primarily on the study of an unspecialized abdominal segment, such as the fifth or sixth. The colour-schemes of the head and thorax will be correlated with these in subsequent sections.

The origin of the bicolorous pattern is to be explained on the ground of the bilateral symmetry of the disposition of the internal organs beneath the dorsal integument. The ventral parts do not come into the question. Bright colours are called forth by the influence of bright sunlight. An aërial habit is then, above all, favourable to their formation. Thus the Dragonflies share with their Vertebrate analogues, the Birds, the distinction of possessing the brightest body-colourings in the Phylum. As regards the bilateral symmetry of the pattern, the basis for this is the method of deposition of the ground-colour or cuticular pigment. This may be deposited along the sides and middle line of a segment, leaving two clear areas above the pigmented dorsal tracheae, or, apparently, exactly the reverse may be the case. The former I have termed the *central* type, the latter the *marginal*. The terms refer to the position of the chief areas of bright coloration. The marginal type appears to be the older, and probably dates from a period before the pigmentation of the dorsal tracheae set in. We may define the two types as follows:

A. *Marginal Type*. On each tergite, the ground colour forms two longitudinal blocks, one on either side of the middle line. The anterior and posterior borders, the middle line, and a large lateral area on each side, have their cuticle unpigmented, so that the brilliant colours of the hypodermal pigment appear in these areas.

B. *Central Type*. The pigmentation is exactly opposite to that of A. The ground-colour is only absent in the two longitudinal blocks, in which brilliant hypodermal pigment becomes developed. The middle line and all the area surrounding the blocks is darkened by cuticular pigmentation.

We shall now trace the development of these basic patterns in turn:

A. **Marginal Types** (fig. 113).

The most archaic marginal type still extant appears in *Petalura pulcherrima*. Even in this archaic genus itself, the marginal type rapidly tends to become altered, assuming the form which we may call *annulo-marginal*, in which, by further growth, and later by fusion of the two dark blocks of ground-colour, the hypoderm

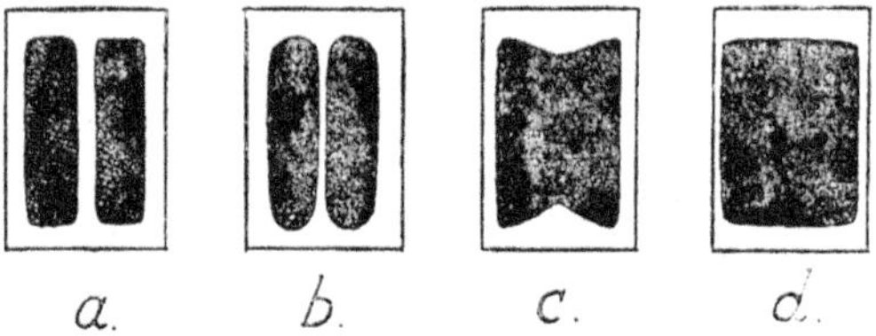

Fig. 113. Marginal types of colour-pattern. *a*, archaic type; *b*, *c*, *d*, stages in the development of the annulo-marginal type (*d*). Original.

pigment becomes at last restricted to narrow transverse zones or annuli at the base and apex of the segment, together with latero-ventral patches not shewn in the diagrams. This *annulo-marginal* type is found, amongst Anisoptera, in the *Gomphinae* and *Petalurinae*, the hypoderm colouring being in all these cases bright yellow, greenish-yellow or brownish-yellow, rarely reddish.

In the Zygoptera, the original marginal type may still be seen in many *Agrionidae*, particularly in the females. Much more commonly, however, it becomes reduced to the annulo-marginal type. The hypodermal pigments in this suborder are generally bright blue or red, sometimes orange, pink, straw, pale green or cream. Dull greens and greenish-blues predominate in the females.

If we now compare the marginal patterns found in the Anisoptera with those of the Zygoptera, we see at once that the former are correlated with a forest-loving habit, the latter with a sun-loving habit. We shall term the former *hylochromes*, the latter *heliochromes*. In the former, the hypoderm pigment is predominantly green or yellow, in the latter, bright blue or bright red.

We cannot here follow out in detail the wonderful specializations of the marginal pattern to be found in the *Agrionidae*. A study of the second segment alone will shew the lengths to which pattern-forming can be carried. Perhaps the highest development of colour-pattern to be found in the Zygoptera is the *tricolorous pattern*, e.g. in *Ischnura aurora* or *Xanthagrion erythroneurum* (Plate III) in which bright blue and red occur in the same insect.

Metallic coloration may be formed on the two blocks of ground-colour, usually giving rise to a dark bronze-black or bronze-green effect. If the annulo-marginal stage is thus affected, the abdomen appears definitely metallic. Finally, the annuli may be suppressed, and a *unicolorous* metallic pattern produced, either bronze, green or blue. These stages can be followed out in the *Lestidae* and *Calopterygidae*.

B. **Central Types** (fig. 114).

The original central type of colour-pattern is preserved for us with very little modification in the long hylochrome series of the *Aeschninae*, in which there is very little tendency towards annulation. The paired longitudinal blocks, which in this type carry the bright hypoderm-colouring, early tend to become oval. By various changes in shape, and by splitting-up, the wonderful patterns in this subfamily have all been formed from the original central type. Fig. 114 shews a series of these. The hypoderm-pigments are usually green or yellow, sometimes bright blue, more rarely brown, grey or cream. The most primitive type is found in the *Petaliini*, where the blocks are still much elongated.

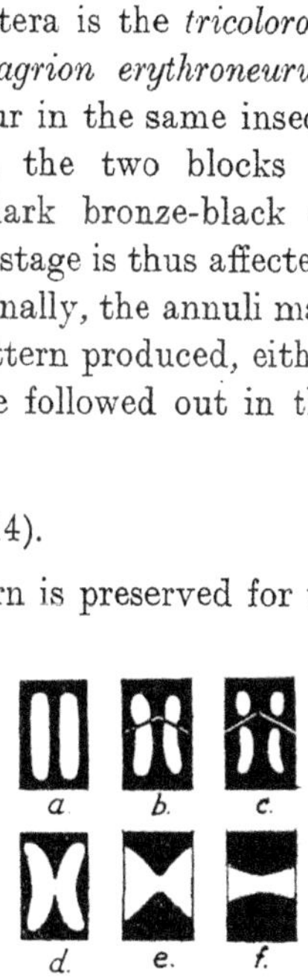

Fig. 114. Central types of colour-pattern. *a*, archaic type; *b*, *c*, stages in the development of the spotted *Aeschnine* type, by interpolation of the supplementary transverse carina. *d*, *e*, *f*, stages leading to the annulo-central type (*f*). Original.

One sub-division of the *Brachytronini* has branched off along a colour-line of its own. The bright hylochrome pattern becomes gradually suppressed, and a uniform dark brown pattern supervenes. Traces of the original pattern are always to be found on the thorax (remnants of bands), on the head, and often on the second abdominal segment. Brown bands of pigment are often developed on the wings. This line culminates in *Telephlebia*, *Austrophlebia* and *Linaeschna*. The effect produced is that of almost complete invisibility during flight, particularly during the late afternoon or early evening, when these insects usually fly. I have termed this form of colouring *aphantochrome* (178). The same tendency is manifested in a less degree in some species of *Aeschna* and *Gynacantha*. In *Anax*, with the adoption of a sun-loving habit, the formation of heliochromes has set in.

Hylochrome patterns of the central type are also retained in the *Synthemini*, and in some *Macromiini*, besides in many other genera scattered throughout the *Libellulidae*. In most *Macromiini* and in the *Cordulegastrinae*, we meet with an advance to the *annulo-central* type (fig. 114, *d–f*). Here the two pale blocks fuse across the middle line, become shortened, and finally form a central annulus across the segment. This type also occurs in the *Tetrathemini* and *Cordulephyini*. We can contrast this with the corresponding annulo-marginal type, where the annuli are formed at the ends of the segment.

In the *Libellulinae*, the main advance has been in the production of *heliochromes*. These insects have forsaken their ancestral forest-loving habits, and now rival the Zygoptera in their love of bright sunshine. Consequently the older bicolorous pattern is changed, either by substitution of bright red or orange for the older yellow or green, or by the production of a metallic colour, or finally by becoming changed to a *unicolorous* heliochrome, generally bright red or pruinescent blue. The stages in this latter process may be followed by studying a series of teneral and mature males and females in such genera as *Libellula*, *Orthetrum*, *Sympetrum* or *Diplacodes*. Taking the three common Australian species of *Orthetrum* as an example, we see that, whereas *O. sabina* has retained the ancient bicolorous pattern in both sexes, in the male of *O. villosovittatum* it becomes overlaid with brilliant red, while in the male of *O. caledonicum* (and later, also, in the female) it is overgrown by blue pruinescence. Metallic developments are the usual rule in the *Corduliinae*, where the various stages in the passage from an original bicolorous pattern to a uniform metallic colouring may be studied in *Somatochlora* or *Hemicordulia*.

C. **The Thoracic Pattern** (fig. 115).

That the basic difference between marginal and central types is not wholly lost on the thorax can be seen by comparing the thoraces of an *Aeschnine* and a *Gomphine* Dragonfly. In both, pale bands alternate with dark areas of ground-colour. But, on the whole, we may say that the pale and dark portions are reversed in the two cases. In the typical *Aeschnine* thorax, the two pairs of broad bands correspond with the two pale central areas of two abdominal segments. In the *Gomphine* thorax, the pale bands correspond with the middle lines and sides of the abdominal segments, so that an opportunity arises for the production of the many-banded pattern seen so often in this subfamily. Likewise, the marginal type is preserved in the formation of an anterior transverse annulus, not complete, it is true, but of great interest, in that, by fusion with the dorsal bands, it produces the peculiar "seven-marks" so distinctive of this subfamily. In connection with this, a study of the thorax of *Petalura* is of interest (fig. 115).

Fig. 115. Dorsal thoracic colour patterns, marginal type. *a*, *Petalura pulcherrima* Tillyard. *b*, *c*, formation of the Gomphine "seven-marks." Original.

In the Zygoptera, the sides of the thorax tend to become uniformly pale, by loss of the metathoracic ground-colour bands. Thus only two isolated pale bands are left, one on either side of the mid-dorsal carina. A very interesting correlation can be established between the thoracic colour-pattern and the nature of the habitat in certain genera, e.g. *Argiolestes* (Pl. III, figs. 2–8), where rock-dwellers tend to become dark, while orange or red are developed in the dense jungle, and blue or grey in the open sunlight.

In many *Aeschninae* the thoracic bands become narrowed, divided up into spots, or even abolished, particularly the dorsal bands. In the *Libellulidae*, when a heliochrome or metallic pattern is produced on the abdomen, the thoracic bands also become masked or abolished. The thorax, however, assumes the heliochrome colouring more slowly than the abdomen, so that many species with bright red or blue abdomens have only brown thoraces.

D. **The Head Pattern** (fig. 116).

We shall only mention here two striking formations which illustrate the effect of the marginal and central type of pattern, respectively, on the head.

1. *The Post-ocular Spots of Agrionidae.* The formation of these by stages from an original *marginal epicranial pattern* is illustrated in fig. 116 A.

2. *The T-mark on the Frons in Aeschninae and Libellulidae.* The formation of this mark by stages from an original *central frontal pattern,* and its subsequent reduction to a "clear" front (usually yellow), is illustrated in fig. 116 B. The jewelled frons appears to have been produced by an exactly opposite process, i.e. *extension* of the ground-colour, instead of reduction.

Indications of the original marginal or central type may also be found by studying the clypeus and labrum.

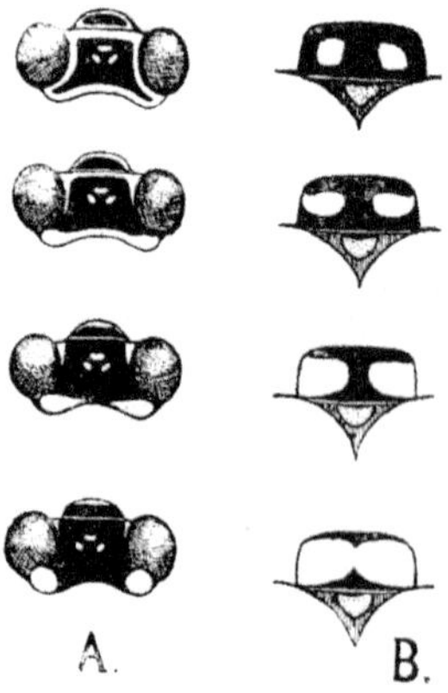

Fig. 116. Colour-patterns of the head. A. Four stages in the development of the brilliant postocular spots of *Agrioninae* from an original marginal type. B. Four stages in the development of the T-mark and of the "clear" frons from an original central type. Original.

We can profitably conclude our account of the Body-Coloration of Odonata by appending a short table of the principal types of colour-pattern.

MARGINAL TYPES

Type		Group
Marginal	Hylochromes	*Petalurinae*
	Heliochromes	*Agrionidae*
Annulo-marginal	Hylochromes	*Gomphinae, Petalurinae*
	Heliochromes	*Agrionidae,* some *Lestidae* and *Calopterygidae*
Tricolorous (Heliochromes)		Some *Agrionidae*
Metallic Unicolorous		*Calopterygidae, Lestidae*

CENTRAL TYPES

Central	Hylochromes	*Aeschninae, Synthemini, Macromiini*
	Aphantochromes	Some *Brachytronini, Aeschna, Gynacantha*
	Bicolorous Heliochromes	*Libellulinae, Eucorduliini*
	Unicolorous Heliochromes	*Libellulinae*
Annulo-central (Hylochromes)		*Tetrathemini, Macromiini, Cordulegastrinae, Cordulephyini*
Metallic Unicolorous		*Eucorduliini,* a few *Libellulinae*

Wing-Coloration (Plate I).

Most Odonata have colourless (hyaline) wings. In a very large number of forms, however, a small deposition of yellowish pigment occurs, known as *saffroning*. This is most commonly found at the base of the hind-wing, less extensively in the fore-wing. *Pantala flavescens* and *Sympetrum flaveolum* are well-known examples of this very common form of colouring. Several species of *Anax* are "saffroned" towards the middle of the wing. In the females of many species, pale yellow or brownish shading occurs at the tips of the wings, or even all over them. In many instances, the tips of the wings are clouded with dark brown or grey pigment.

Red, brown or black pigment often overlies the patch of saffroning at the wing-base, or may entirely replace it. In some forms (e.g. *Tramea*) the basal patch is dark red in the male, brown in the female. In others (e.g. *Libellula, Leucorrhinia*) it may be quite black.

Fugitive pigmentation of the wing is fairly common. In many forms, particularly in females, pigment is deposited at metamorphosis and afterwards lost. In other cases, the pigmentation only sets in with old age.

Apart from the above instances of slight wing-coloration, there are certain groups of Dragonflies in which all, or nearly all, of the species have a definite colour-pattern on the wings. We shall deal with these under four headings:

1. The *Petaliini*. The colour-scheme of the wing in this group (Plate I, fig. 1) is quite unique. It is better developed in females than in males. A series of irregular or rounded spots or patches is arranged along the costal border, each being of a rich transparent brown or ruby red colour, surrounded by black. These spots glow like rich jewels on the wings of the live insect.

2. The aphantochrome *Aeschninae*. The coloration of the wings in this group is purely auxiliary to the body-pattern. In its highest development, it forms a broad brownish band along the entire wing, either along the costa or slightly below it. The huge Dragonfly *Austrophlebia costalis*, with its rich brown body and wing-bands, is quite invisible at a distance of a few feet, when resting on the mid-rib of the frond of a tree-fern. The body looks

like the mid-rib of the frond, the four outspread wings like the mid-ribs of four pinnules of the frond. A transverse dark band crosses the wings in a few forms.

3. The *Libellulinae.* The production of an elaborate colour-pattern on the wing seems to have developed in two groups of this subfamily. In the more archaic forms, it appears first in the *Libellulini,* and culminates in the *Palpopleurini, Perithemis* shewing the highest development in the series. A much more extensive production, however, took place in the more caenogenetic tribes, beginning with the *Sympetrini.* Here we need only mention the lovely *Pseudoleon superbus,* and the remarkable genus *Neurothemis,* where we can trace all stages from the basal black patch in *N. oligoneura* right up to the almost completely brown or black wings of many other species. Correlated with this colour-development there arises a secondary proliferation of cross-veins, apparently to assist the deposition of pigment. Over three thousand separate cells have been counted in a single wing in a male of this genus. The females, on the whole, lag behind the males in colour-production. Thus there arises the peculiar phenomenon of *venationally dimorphic* females, as seen in *N. stigmatizans,* where both deeply coloured and densely-veined (homochrome) and lightly coloured and open-veined (heterochrome) females are known to occur. The latter is clearly the more archaic, and shews the form of the genus before it went in for wing-colouring. In the *Leucorrhiniini,* the genus *Celithemis* offers an interesting study in the formation of beautiful wing-patterns. It is however in the lovely genus *Rhyothemis* (*Trameini*) that the culmination of wing-coloration in the *Libellulinae* occurs. These butterfly-like creatures have sacrificed all the strength of their wonderful wing-formation for the sake of producing a brilliant pattern. They are helpless fluttering creatures which rise up from the ground, as one approaches them, to flutter gaudily in the sunshine, or float gently to and fro upon the breeze. Many of them are easily caught by hand. Some forms have black wings, others brown, or brown and yellow, often with lovely metallic sheen of green or purple or bronze, or intricate pattern like an ancient runic script (*R. graphiptera*). The most beautiful though by no means the most complicated pattern is that of *R. resplendens* (Plate I, fig. 3), where about

two-thirds of the wing is of a glorious metallic blue colour. This latter species has forsaken still waters, and competes with the gaudy *Calopterygidae* upon the broad flowing rivers of Papua and North Queensland.

3. The *Calopterygidae*. This family, so well named the "beautiful-winged" Dragonflies, shews us the development of metallic coloration carried to perfection. The colours are practically confined to the males alone. These are, indeed, the Birds of Paradise amongst Odonata. In a number of forms, e.g. *Anisopleura*, *Diphlebia*, only black pigment is laid down. In the beautiful genus *Hetaerina* and its allies, a lovely transparent red pigment (apparently the same as in the *Petaliini*) more or less overspreads the wing. In a few forms (e.g. *Diphlebia lestoïdes*) a transverse milky-white band appears (cf. *Tholymis* in the *Libellulinae*). In advance of these, we note the glorious effect of the superimposed interference coloration in such cases as *Pseudophaea*, *Matronoïdes* and *Neurobasis* (Pl. I, figs. 6, 7), where either part or all of the pigmented portion of the wing scintillates in radiant blue or green. Finally, in *Rhinocypha* (Pl. I, figs. 4, 5) the play of metallic colours, in a combination of flashing reds, mauves, purples, bronzes and greens, utterly baffles description. The development of small clear areas or "windows" in such a wing as that of *R. quadrimaculata*, or *R. fenestrella*, only serves to heighten the effect of these radiant rainbow hues.

Larval Coloration and Colour Changes.

The coloration of Odonate larvae is very uniform, but may be divided into two classes: (*a*) those shewing no definite pattern, (*b*) those with a pattern. All larvae which live hidden in sand or mud, or under débris, need no colour-pattern, and most of them have none. The few that have owe it apparently to the fact that they have only recently taken to hiding away. Larvae that live either on the river bottom, or amongst water-weed, shew a definite pattern, which helps to render them inconspicuous to enemies and prey alike. In bottom-dwellers, the pattern is a beautiful mottling of pale brown or straw with dark brown or olive (e.g. most *Eucorduliini* and *Libellulinae*) In weed-dwellers, a ground colour of semi-transparent green, yellow, or olive, is overlaid with darker markings in black, dark brown, or deep olive-green (e.g. many *Agrionidae* and *Lestidae*). The most beautiful larvae of all are those of *Anax*[1], dwelling freely in water-weed, with their dark olive-green patterns on a rich green ground-colour. Larvae that dwell on sticks may become dark brown or even jet-black, with either partial or entire secondary loss of their pattern (e.g. many *Brachytronini*). Rock-dwellers are either black, dark grey or brown, according to the colour of the rocks which they frequent.

It is a fact that Dragonfly larvae can change their colours to suit their environment. To shew this, it is only necessary to take larvae of the same species of *Libellulidae* from, say, a pool in a chalk-pit and a neighbouring pond with dark bottom-soil. The latter will be found to be twice as dark as the former. An even more startling colour-change came to my notice one year in the case of *Aeschna brevistyla*. This larva is of an almost uniform brown tint. It happened that a small dam was newly formed. Masses of the reeds, dug out from the banks, floated on the surface of the water. As these decayed, their root-stocks and roots became a bright brick-red. On hauling them ashore I found six *Aeschna* larvae clinging to them. All of these had their ventral surfaces, including the mask and the undersides of the legs, a rich brick-red. Other larvae from the sides of the dam were of the normal brown colour.

These larvae, placed in my aquaria, remained red until the next ecdysis, when they resumed their brown colouring. Hence it is clear that the change is not produced voluntarily, through the eye, as in the case of the Chameleon, but is due solely to what I may term a *chemical plasticity* on the part of the fresh pigments formed at ecdysis. When freshly formed, it seems clear that this pigment can be turned red by red surroundings (as in the case of the red light from the dead reeds acting on the pigment of the *Aeschna* larva), brown by brown surroundings, black by black, or green by green. Whether the change would extend to blue, orange or purple, we do not know. I think that a similar plasticity, *at metamorphosis*, has helped to determine the formation of hylochromes and heliochromes in the imagines.

[1] It is a peculiar fact that these larvae, when young, dwell on twigs or matted roots, and are often found to be whitish, marked with black in broad transverse bands. Later on, they migrate to floating weed-masses and become mottled green.

Colour Dimorphism (Plate III, figs. 17–24).

Apart from what is known as *sexual dimorphism* in colour (i.e. the male and female differing in colour)—a phenomenon of wide occurrence in the Odonata—there is a very interesting group of *Agrionini* in which two forms of females nearly always occur. This group embraces the genera *Ischnura, Ceratura, Anomalagrion, Agriocnemis* and *Austrocnemis*. In all cases, one form of female is commoner than the other. The commoner form may be called the "normal," the rarer one the "heteromorphic" female. Either of these may be similar to the male (*homochrome*) or unlike it (*heterochrome*). Where the heterochrome is normal, it is usually of a dull blackish, olive, or greenish colour. Where the heterochrome is the heteromorph, it usually assumes a rich orange or red form. I offer a short table which embraces all the different types known to me:

Male	Normal Female	Heteromorphic Female	Species
Bronze and blue	Homochrome	Heterochrome (orange and black)	*Ceratura* and many species of *Ischnura*
Bronze and blue	Homochrome	Heterochrome (red and bronze)	*Austrocnemis* (Plate III, figs. 20, 21)
Bronze and blue	Heterochrome (dull blackish)	Homochrome	*Ischnura heterosticta*
Bronze and blue	Heterochrome (dull blackish)	Heterochrome (orange and black)	*I. ramburii, I. perparva*
Bronze and grey	Heterochrome (rich olive-green)	Heterochrome (orange and black)	*I. pruinescens* (Plate III, figs. 17–19)
Bronze and yellow	Heterochrome (bronze and black)	Heterochrome (orange and bronze)	*Anomalagrion*
Red, bronze and blue	Heterochrome (dull blackish)	Homochrome	*I. aurora*
Bronze and red	Heterochrome (dull blackish)	Heterochrome (orange and black)	Most species of *Agriocnemis*
Bronze and red	Heterochrome (dull blackish)	Heterochrome (red and black)	*Agr. hyacinthus*[1] (Plate III, figs. 22–24)
Silver	Heterochrome (dull blackish)	Heterochrome (red and black)	*Agr. argentea*

[1] *Agr. pygmaea* Ramb., of which *Agr. hyacinthus* is probably only a subspecies, has an *orange* female. In series of this insect from the Seychelles, Campion has noted the occurrence of *structural* as well as *colour* dimorphism in the females, the prothorax having its hind-margin distinctly lobed. In view, however, of the facts that the form of the prothorax is closely correlated with that of the appendages of the male, and that no males were taken with this supposed new form of female, it seems more likely that this will really prove to be a new species.

CHAPTER XIV

CLASSIFICATION

The Classification of recent forms adopted in this chapter differs from all hitherto published systems chiefly in attempting a more comprehensive scheme for the Zygoptera. The classification of the Anisoptera has engaged the attention of many authors, prominent amongst whom have been de Selys, Ris, Calvert, Needham, R. Martin, Williamson, Karsch and others. Consequently the points still at issue are only minor ones. We still need a thorough study of the *Gomphinae*, the present division into two tribes being only tentative. The position of the *Petaliini* as a tribe within the *Aeschninae* is open to question on the ground that the differences between them and the rest of the subfamily are perhaps great enough to warrant their elevation to subfamily rank. The true relationships of the *Gomphinae* with the *Aeschninae* on the one hand, and the lower *Libellulidae* on the other, still need careful study. It may be necessary to elevate all the subfamilies of the Anisoptera, except the *Libellulinae* and *Corduliinae*, to the rank of families, the differences between the two latter being of a lower order than those between the others. The division here made of the immense subfamily *Libellulinae* into tribes follows in the main Ris's excellent modern classification, with the exception that Old and New World parallel groups are reduced to the rank of series within a single tribe.

In the Zygoptera, the two chief alterations proposed are the separation of the family *Lestidae* on entirely new characters, and the elevation of the "légions" of de Selys to the rank of subfamilies. Only in the case of the *Platycneminae* is this step open to doubt, since these forms grade very closely into the *Protoneurinae*. The tendency of asthenogenesis has been all along towards the production, by convergence, of closely similar, highly reduced types. It is the duty of the systematist to try to separate the converging threads of descent, to give due weight to the imaginal characters still left over, and, by a judicious study of larval forms, to separate those groups in which the imaginal characters fail or are deceptive. The recent separation of *Synlestes* from the *Megapodagrioninae* is a case in point [176].

The classification here given does not go beyond tribal limits, except in indicating here and there the existence of two or more *series* within a large tribe. For a list of genera, arranged in their correct zoo-geographical regions, the reader is referred to the tables in the section on Entogenic Fauna in chap. XV, and to the section on Palaeogenic Fauna in the same chapter.

Order **ODONATA.**

Suborder ANISOPTERA.

Wings held horizontally or depressed in repose (except *Cordulephya*). Hindwing always more or less considerably broader than forewing near the base. Radius branched, the radial sector crossing two branches of the media. Oblique vein and bridge present. Discoidal cell differentiated into triangle + supertriangle. Eyes large, often meeting mid-dorsally, never separated by a space greater than their own diameter. Labium variable. Male with two superior and one inferior anal appendages, all placed dorsally above anus; the inferior may be bifid or trifid. Penis jointed. Female with superior appendages but no inferior appendage. Ovipositor variable.

Larva proctobranchiate (i.e. with gills in rectum). Anus closed by an anal pyramid formed of three spines, one medio-dorsal (appendix dorsalis) and two lateral (cerci). Gizzard with four to eight dental folds.

Family 1. AESCHNIDAE.

Triangles of fore and hindwings similar or nearly similar in shape, placed equally distad from arculus. Antenodals of first and second series not corresponding, except two conspicuously thickened triangular ones (usually the first or second, and one between the fourth and ninth). Hindwing of male angulated (except in the *Anax*-Series). Auricles nearly always present on seg. 2 in male, sometimes also in female. Labium with well-developed median lobe, not overlapped by lateral lobes; the latter with movable hook and end-hook. Nearly always a bicolorous body-pattern.

Larva with abdomen usually somewhat elongated. Mask flat (except in *Cordulegastrinae*). Gizzard with four to eight fields, radially symmetrical (except *Cordulegastrinae*). Rectal gills simplex or duplex, but never lamellate.

Subfamily 1. **Gomphinae.**

Triangles short or moderately so, that of hindwing slightly more elongated than that of forewing. Subtriangles present in all wings. Median space always free. Anal loop absent or rudimentary. Eyes quite separate; epicranium ending posteriorly in a sharp occipital shelf. Labium with median lobe complete, not incised. Inferior appendage of male strongly bifid. Ovipositor of female reduced to a pair of valvula vulvae at end of eighth sternite.

Larva with variably-shaped abdomen. Legs modified for digging, fore and middle tarsi 2-jointed. Antennae reduced to four joints, and more or less swollen. Mask with flat median lobe, lateral lobes with narrow rounded apex and strong movable hook. Gizzard with four elongated fields, carrying numerous undifferentiated teeth. Gills simplex; undulate or papillate.

Tribe 1. *Ictinini* (fig. 117).

Moderate to large-sized *Gomphinae* with triangles crossed, at least in one pair of wings. Membranule present or absent. Distal side of triangle often

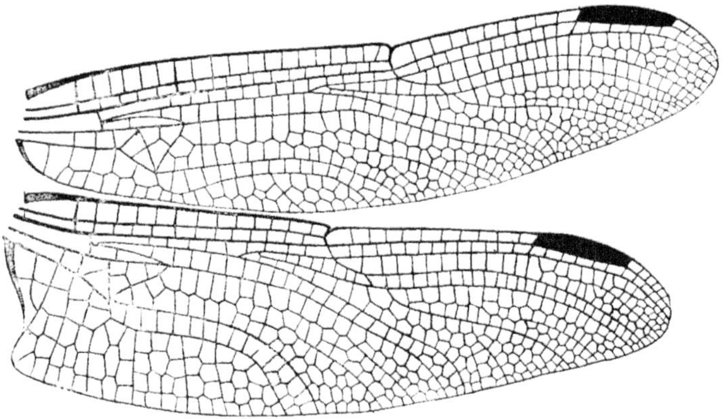

Fig. 117. Wings of *Ictinus australis* Selys, ♂, Queensland. (Hw. 38 mm.) Original.

distinctly broken near distal angle. Larva of variable shape. A heterogeneous tribe including two series:

Series 1. *Gomphoides*-Series. Membranule absent or reduced. Larva with elongated abdomen. New World (except *Davidius* and *Sieboldius*).

Series 2. *Ictinus*-Series. Membranule distinct. Larva with short rounded abdomen (fig. 36 A). Old World (except *Cacus*).

Tribe 2. *Gomphini* (fig. 118).

Small to moderate-sized *Gomphinae* with triangles free. Membranule almost or quite absent. Larva with oval or elongated abdomen (fig. 36 B). Gills simplex papillate (except *Austrogomphus*).

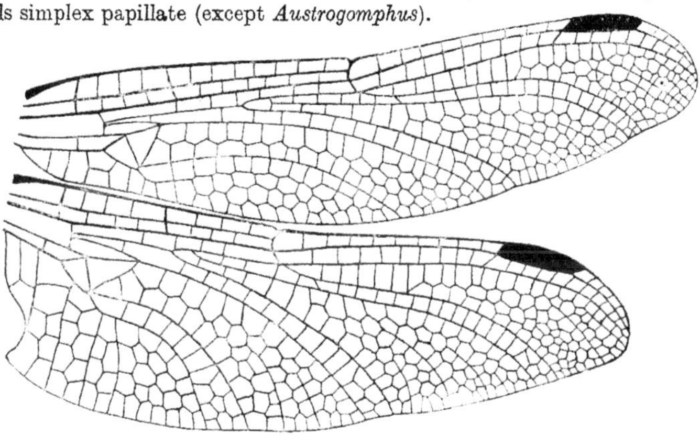

Fig. 118. Wings of *Gomphus pulchellus* Selys, ♂, France. (Hw. 30 mm.) Original.

Subfamily 2. **Chlorogomphinae** (fig. 119).

Triangle of forewing equilateral, that of hindwing transversely elongated. Median space crossed. Hindwing with broad anal field and anal loop. Eyes very slightly separated. Labium with median lobe slightly incised. Larva unknown.

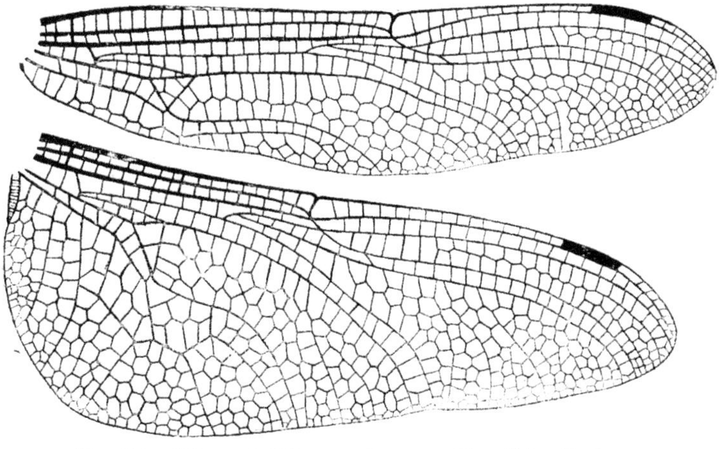

Fig. 119. Wings of *Chlorogomphus magnificus* Selys, ♀, Java. (Hw. 46 mm.) After Needham.

Subfamily 3. **Petalurinae** (fig. 120).

Triangle of forewing slightly elongated transversely, that of hindwing slightly elongated longitudinally. Subtriangle of forewing large, reticulated; that of hindwing simple. Median space free. Anal loop absent or rudimentary. Eyes well separated. Labium with incised median lobe. Female with complete ovipositor. Insects of large size.

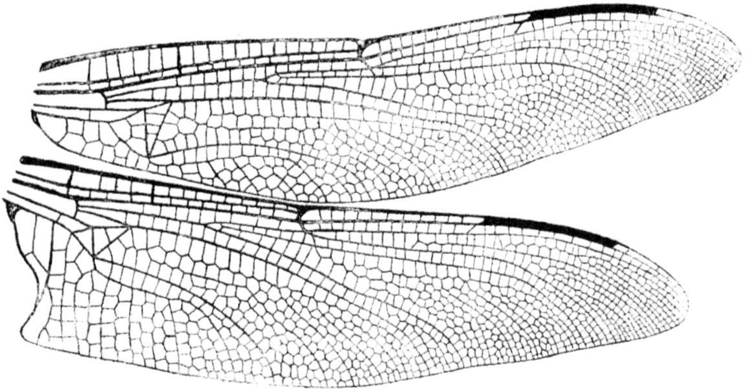

Fig. 120. Wings of *Petalura gigantea* Leach, ♂, New South Wales. (Hw. 57 mm.) Original.

Larva subcylindrical, with short 7-jointed antennae (fig. 43). Mask flat, with thick, slightly hollowed lateral lobes and short movable hook; no setae; median lobe triangular. Gizzard with eight fields, dentition much reduced. Gills simplex undulate.

Subfamily 4. **Cordulegastrinae** (fig. 121).

Triangles equal, slightly elongated; subtriangles incomplete. Median space free. M_2 not waved. Eyes nearly or just touching. Labium with incised median lobe. Female with large projecting ovipositor with valves of seg. 9 absent.

Larva hairy, with oval abdomen, slender 7-jointed antennae, divergent wing-sheaths (cf. fig. 33). Mask spoon-shaped, with mental and lateral setae; lateral lobes with small movable hook and complicated dentition. Gizzard with four bilaterally symmetrical fields. Gills simplex undulate.

Subfamily 5. **Aeschninae.**

Triangles approximately equal, elongated longitudinally, crossed; subtriangles weak or absent. Strong compact anal loop formed between A_2 and A_1. Median space free or crossed. M_2 waved. Eyes meeting mid-dorsally for a long distance (except *Petaliini*). Labium with median lobe slightly incised, or with a longitudinal depression. Female with complete ovipositor.

Larva with elongated abdomen and slender 7-jointed antennae (fig. 29 A–D). Mask flat, without setae (exc. *Gynacantha*); lateral lobes narrow, with apex of variable form; long movable hook. Gizzard with four radially symmetrical fields carrying a few large specialized teeth. Gills duplex, but not lamellate.

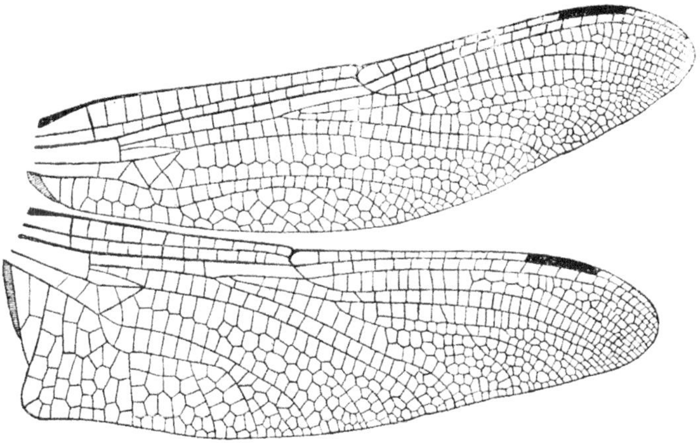

Fig. 121. Wings of *Cordulegaster annulatus* Latr., ♂, Britain. (Hw. 46 mm.) Original.

Tribe 1. *Petaliini* (fig. 122, also Pl. I, fig. 1).

Triangles only moderately elongated, very close to arculus and to base of wing. *Rs* unforked. A series of dark brown or red blotches along costal

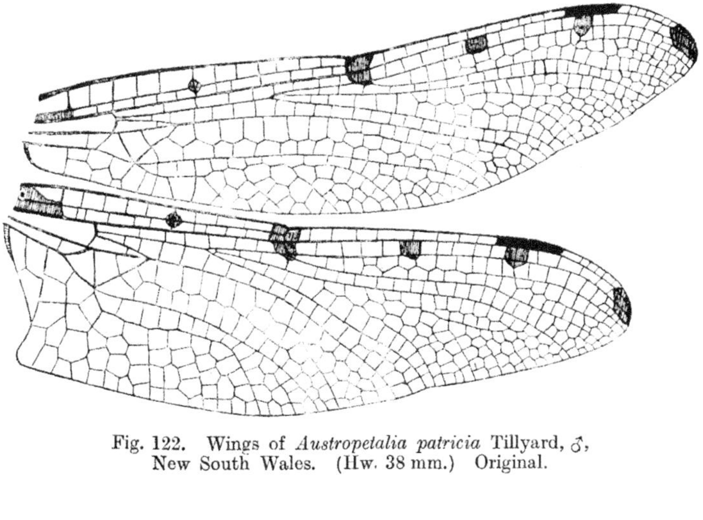

Fig. 122. Wings of *Austropetalia patricia* Tillyard, ♂, New South Wales. (Hw. 38 mm.) Original.

border of wing. Eyes only just touching. Male with inferior appendage trifid, longer than superiors. Larva with broadly upcurved lateral lobes on segments 3–8 (fig. 37).

Tribe 2. *Brachytronini* (fig. 123).

Triangles considerably elongated, reticulated. *Rs* either unforked, or symmetrically forked before level of *pt*. *Rspl* straight, placed only 1–2 cell-rows below the lower branch of *Rs*. *Mspl* straight, similarly placed with respect to M_4.

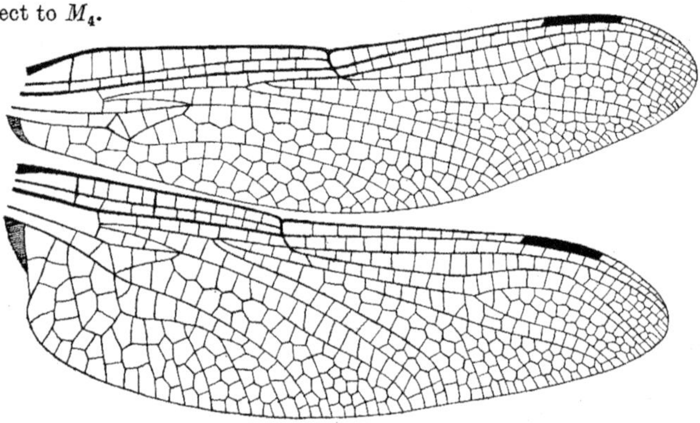

Fig. 123. Wings of *Brachytron hafniense* Müller, ♂, Britain. (Hw. 35 mm.) Original.

Larva with superior appendage pointed at tip, never bifid.

A heterogeneous group including two series:

Series 1. *Boyeria*-Series. *Rs* unforked.

Series 2. *Brachytron*-Series. *Rs* forked.

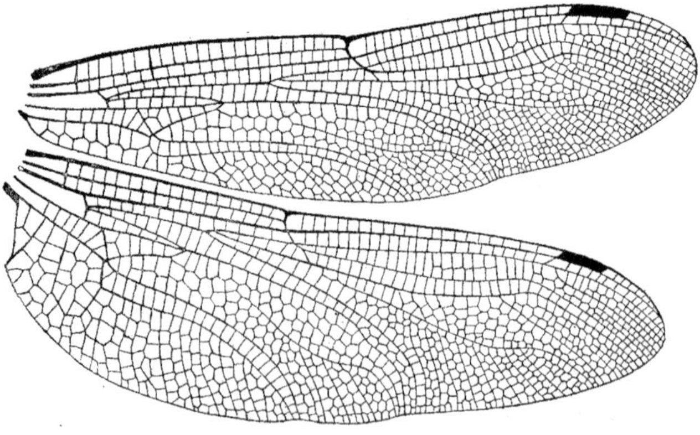

Fig. 124. Wings of *Gynacantha rosenberghi* Selys, ♂, Queensland. (Hw. 49 mm.) Original.

Tribe 3. *Aeschnini* (figs. 124, 172, also Pl. II, fig. *A*).

Triangles much elongated, reticulated. *Rs* either unforked, or asymmetrically forked near or under the level of *pt*. *Rspl* curved concavely to ***Rs***, lying three or more cell-rows below the lower branch of *Rs* at its middle. *Mspl* similarly curved and arranged with respect to M_4.

Larva with superior appendage usually bifid at tip.

This tribe includes three series:

Series 1. *Gynacantha*-Series. Female with a projecting fork under seg. 10; larva with lateral setae on mask.

Series 2. *Aeschna*-Series. Female with dentigerous tubercle under seg. 10.

Series 3. *Anax*-Series. Anal angle of hindwing in male rounded.

Family 2. LIBELLULIDAE.

Triangles of fore and hindwings very different, that of forewing transversely elongated and far removed from arculus, that of hindwing longitudinally elongated and close to or under arculus. Subtriangle of forewing well-developed, often reticulated; that of hindwing often absent. Antenodals of first and second series corresponding, none specially thickened; one or more distal ones may be incomplete. Anal loop present. Eyes united dorsally for a long distance (exc. *Diastatops*). Labium with median lobe reduced to a small triangular piece, completely overlapped and closed in by the large lateral lobes, which have no movable hook. Vertex a vesicle or tubercle with the ocelli grouped round it. Female with ovipositor usually reduced to a pair of simple valvula vulvae or a single vulvar scale (secondarily enlarged in *Uracis* and others). Body-pattern bicolorous or unicolorous.

Larva with broad oval body, and short 7-jointed antennae. Mask spoon-shaped, with wide lateral lobes meeting medially in a series of closely fitting teeth; mental and lateral setae present. Gizzard with four fields, bilaterally symmetrical. Gills duplex, lamellate.

Subfamily 1. **Corduliinae.**

Triangle of forewing not excessively narrowed. Anal border of hindwing angulated in male (exc. *Hemicordulia*). Anal loop very variable. Eyes with a slight sinuous projection near middle of posterior edge. A tibial keel present on forelegs in the male. Anal appendages of male strongly differentiated in the different species, but genitalia of seg. 2 not strongly differentiated. Auricles present in male (exc. *Hemicordulia*). Body-pattern bicolorous, or metallic unicolorous. Abdomen long, corduliform or cylindrical. Larva with deep or moderate dentition of lateral lobes.

Tribe 1. *Synthemini* (fig. 125).

Triangle of hindwing distad from arculus. Median space crossed. Anal loop short, rounded, three or more cell-rows in width. Insects of slender build and graceful flight. Larva resembles that of *Cordulegastrinae*, hairy, with divergent wing-sheaths (fig. 33); lateral lobes of mask deeply dentate; legs short.

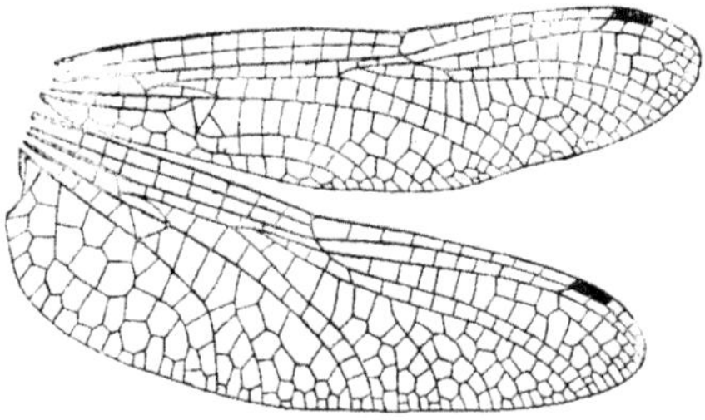

Fig. 125. Wings of *Synthemis claviculata* Tillyard, ♂, Queensland. (Hw. 30 mm.) Original.

Tribe 2. *Macromiini* (fig. 126).

Triangle of hindwing distad from arculus. Median space free. Anal loop as in Tribe 1. Large insects of strong build, *Aeschnine* appearance, and rapid flight. Larva with smooth rounded body, long legs, projecting eyes, and a prominent pyramidal horn on the front of the head, between the antennae (fig. 36 c); lateral lobes of mask deeply indented.

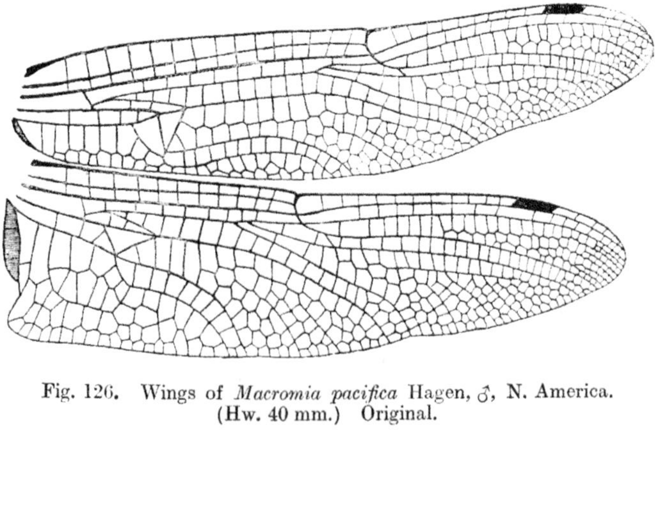

Fig. 126. Wings of *Macromia pacifica* Hagen, ♂, N. America. (Hw. 40 mm.) Original.

Tribe 3. *Idocorduliini* (fig. 127).

Triangle of hindwing distad from arculus. Median space free. Anal loop never more than two cell-rows wide, variable in length, without a widened "toe." Abdomen usually slender, not markedly corduliform. Flight weak, graceful. Larva (*Austrocordulia*) short, with hard spiny body, very flat beneath; short legs; lateral lobes of mask deeply and irregularly broken (fig. 36 D).

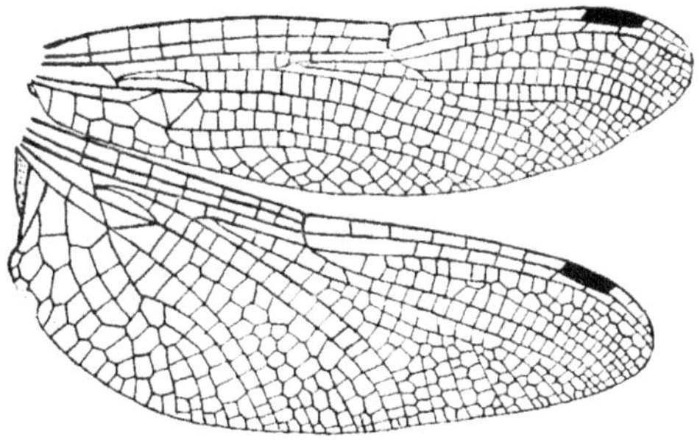

Fig. 127. Wings of *Austrocordulia refracta* Tillyard, ♂, Queensland. (Hw. 29 mm.) Original.

This somewhat heterogeneous tribe includes two series:

Series 1. *Oxygastra*-Series. Forewing with M_4 and Cu_1 parallel (except in *Gomphomacromia*).

Series 2. *Idionyx*-Series. Forewing with M_4 and Cu_1 slightly divergent.

Tribe 4. *Cordulephyini* (fig. 128).

Triangle of hindwing distad from arculus. Costal side of triangles of both fore and hindwing broken. Anal loop reduced to two or three large cells. Hindwing very narrow at base. Reduced forms with zigzag dodging flight; wings folded vertically over abdomen when at rest. Abdomen narrow, cylindrical. Larva of typical *Eucorduline* form (fig. 38), but with peculiar long deep narrow incisions on the lower half of the lateral lobe of the mask.

Tribe 5. *Eucorduliini* (figs. 129, 173).

Triangle of hindwing completely recessed to level of arculus. Anal loop elongate, two cell-rows wide at base, and with a more or less distinctly widened "toe" distally; mid-rib well-developed. Abdomen very distinctly corduliform. Larva with short oval or rounded abdomen, flattened beneath, smooth (cf. fig. 38); long legs; mask with lateral lobes regularly crenated along inner margin.

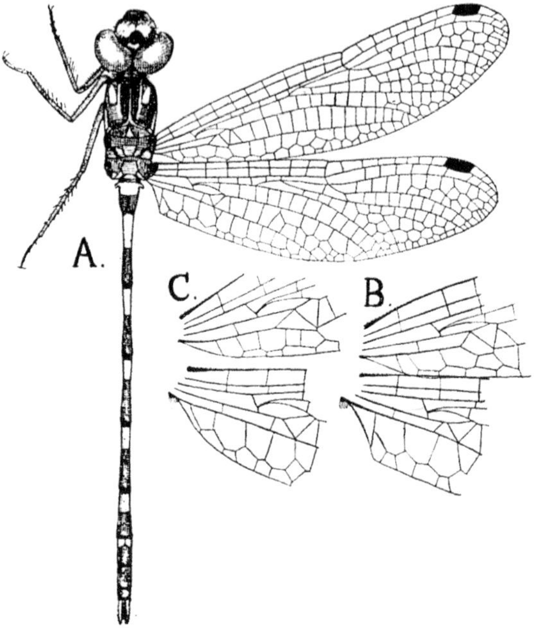

Fig. 128. A. *Cordulephya pygmaea* Selys, ♂, New South Wales. (Hw. 21 mm.) Venation of base of wings enlarged is shewn in B (male) and C (female). Note the hairs which replace the membranule. Original.

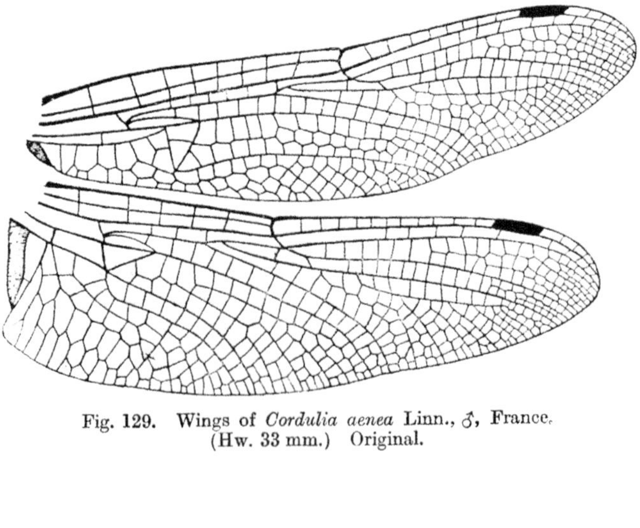

Fig. 129. Wings of *Cordulia aenea* Linn., ♂, France. (Hw. 33 mm.) Original.

This well-marked tribe includes two series:

Series 1. *Neurocordulia*-Series. Forewing with M_4 and Cu_1 slightly divergent.

Series 2. *Cordulia*-Series. Forewing with M_4 and Cu_1 convergent.

Subfamily 2. **Libellulinae.**

Triangle of forewing usually excessively narrowed. Anal border of hindwing rounded in both sexes. Anal loop present, very variable. Eyes globular. No tibial keel on forelegs. Anal appendages of male only slightly differentiated in the different species, but the genitalia of seg. 2 very highly differentiated. Auricles absent. Abdomen mostly relatively short, broad and distended. Body-pattern in males mostly unicolorous, rarely metallic. Larva with the lateral lobes of the mask with very shallow crenulations (exc. Tribe 1 and *Pantala*); legs usually short.

Tribe 1. *Tetrathemini* (fig. 130).

Triangle of forewing nearly equilateral, or with costal side broken. Sectors of arculus united for a long distance, and then separating sharply at an acute angle. *Mspl* absent or rudimentary. *Arc.* usually distad from Ax_2. Base of hindwing narrow; anal loop reduced, small or absent. Small weak forms with a primitive black and yellow body-pattern. Larva (*Nannophlebia*) with lateral lobes of mask moderately crenate (as in *Eucorduliini*).

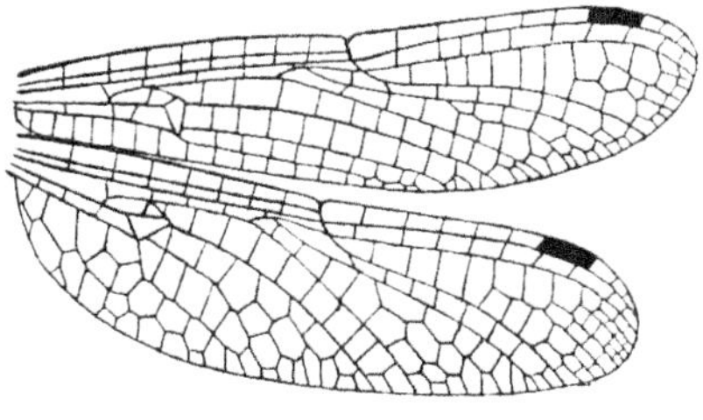

Fig. 130. Wings of *Tetrathemis cladophila* Tillyard, ♂, Queensland. (Hw. 21 mm.) Original.

Tribe 2. *Libellulini* (figs. 131, 155, 175).

Triangle of forewing with costal side considerably shortened, rarely broken; followed by two or more rows of post-trigonal cells. Forewing with *st* well-developed. Anal loop variable, often with outer angle projecting far beyond level of triangle. Sectors of arculus shortly united (exc. *Libellula*). Numerous *Ax* in forewing, last one usually complete. Female mostly with very small valvula vulvae, and lateral edges of seg. 8 frequently widened. Larva with comparatively elongated abdomen, hairy; head squarish in front, with eyes at antero-lateral angles (fig. 36 F).

Tribe 3. *Palpopleurini.*

Wing vari-coloured, relatively broad. Anal loop elongated, with nearly straight mid-rib. Sectors of arculus separate in forewing, slightly fused in hindwing. *Arc.* between Ax_1 and Ax_2. Supplementary *Bx* present. Prothorax with large lobe. Larva with head of somewhat triangular shape, the eyes placed mid-laterally.

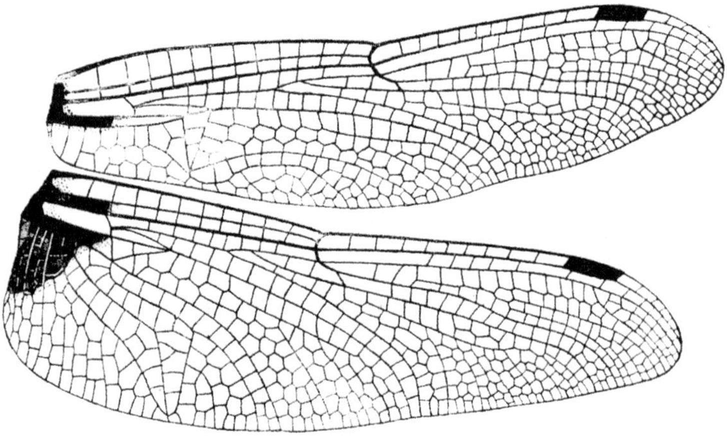

Fig. 131. Wings of *Libellula fulva* Müller, ♂, Britain. (Hw. 33 mm.) Original.

Tribe 4. *Brachydiplacini* (fig. 132, also Pl. I, fig. 2).

Sectors of arculus fused, but less so in fore than in hindwing. *Arc.* mostly between Ax_1 and Ax_2. Anal loop variable, often extending far beyond level of triangle. M_2, *Rs*, and *Rspl* run parallel to one another up to the wing-border; *Rspl* often broken. *Mspl* absent or rudimentary. Triangle of hindwing often distad from *arc.* Supernumerary *Bx* and *Cux* often present. Larva (*Nannothemis*) very hairy, with head rounded in front and behind, eyes well-rounded and placed at antero-lateral angles.

This tribe includes two series:

Series 1. *Brachydiplax*-Series. Old World.

Series 2. *Uracis*-Series. New World.

Tribe 5. *Sympetrini* (fig. 133).

Arculus and its sectors as in Tribe 4. Last *Ax* in forewing usually incomplete. *Rspl* well-formed, concave to *Rs*. *Mspl* usually distinct. No supernumerary *Bx* or *Cux* (exc. *Neurothemis*). Hindwing with broad anal field;

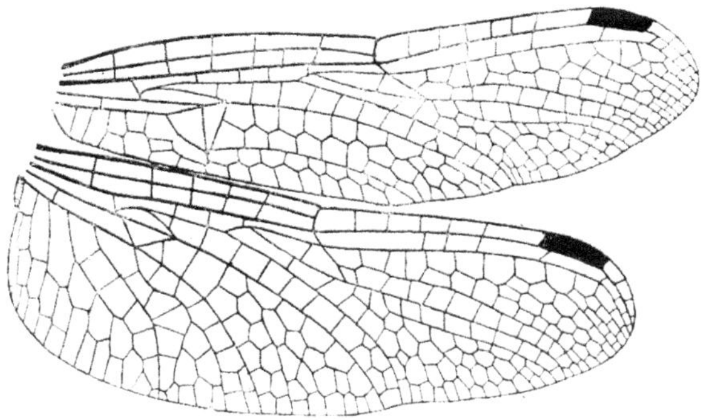

Fig. 132. Wings of *Brachydiplax denticauda* Selys, ♂, Queensland. (Hw. 23 mm.) Original.

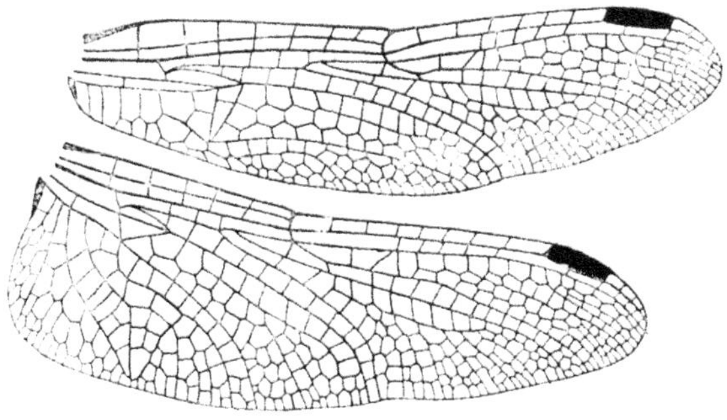

Fig. 133. Wings of *Sympetrum striolatum* Charp., ♂, Britain. (Hw. 28 mm.) Original.

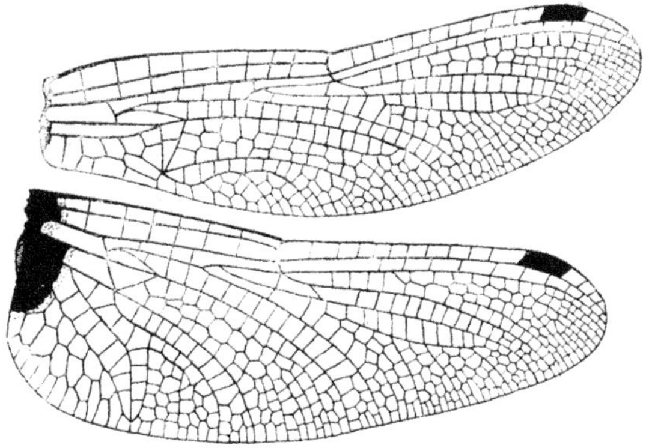

Fig. 134. Wings of *Leucorrhinia dubia* Vand., ♂, Britain. (Hw. 27 mm.) Original.

anal loop well-formed, variable. Genitalia of seg. 2 of male with divided hamuli. Female with prominent valvula vulvae. Larva usually smooth, with large head somewhat triangular in front; eyes large and rounded; body and legs short (fig. 36 G).

Tribe 6. *Leucorrhiniini* (fig. 134).

Sectors of arculus separate or only slightly fused. Lobe of prothorax large, erect, divided and strongly ciliated (as in some genera of Tribe 5 also). Wings at least partly coloured, usually with a dark patch at base of hindwing. Other characters as in Tribe 5.

Tribe 7. *Trithemini* (fig. 135).

Forewing with nodus considerably distad from the middle of the wing; a consequent increase in the number of antenodals; the last one usually incomplete. Sectors of arculus united, less so in fore than in hindwing. *Arc.* usually between Ax_1 and Ax_2. Anal loop long, with outer angle reaching far beyond level of triangle; toe well-developed; midrib complete from base

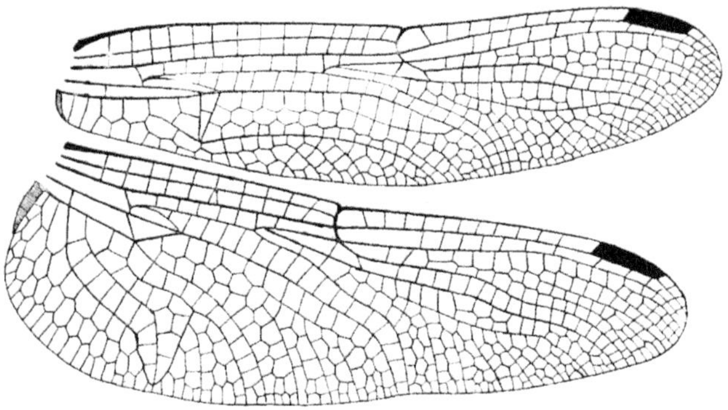

Fig. 135. Wings of *Dythemis velox* Hagen, ♂, Central America. (Hw. 32 mm.) Original.

to toe, and bent at an obtuse angle, sometimes approaching a right angle. Lobe of prothorax usually slender. Male with divided hamuli, but the outer lobe reduced. Female with very small vulvar scale.

Larvae resembling those of Tribe 5.

This tribe includes two series:

Series 1. *Trithemis*-Series. Old World.

Series 2. *Dythemis*-Series. New World.

Tribe 8. *Trameini* (fig. 136, also Pl. I, fig. 3).

Sectors of arculus either separate or only slightly united in forewing. *Arc.* between Ax_1 and Ax_2. Anal field of hindwing exceedingly wide, mostly with numerous cells arranged in numerous rows transversely to the wing-axis. Anal loop long, with well-formed toe, and complete mid-rib bent at an obtuse angle. Often A_2 gives off at its basal third a sector separating off the narrow-

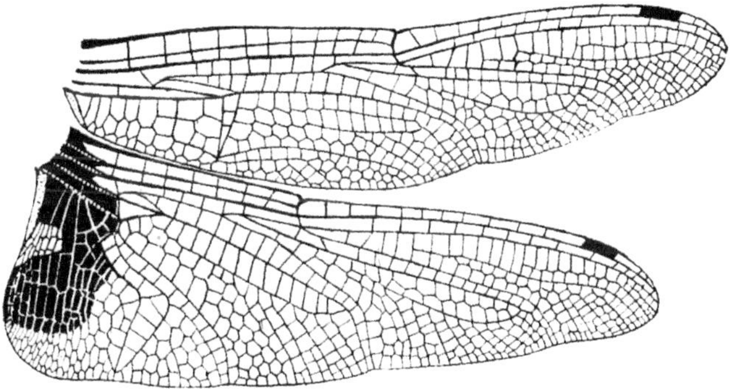

Fig. 136. Wings of *Tramea loëwii* Br., ♂, Queensland. (Hw. 37 mm.) Original.

celled proximal anal field from a portion containing larger cells, between it and the loop. Last *Ax* of forewing incomplete except in the last four genera. Lobe of prothorax small. Larva somewhat *Eucorduline* in appearance, with oval body, longish legs and huge head, the eyes exceedingly prominent, the mask immense (figs. 29 E, F; 36 H).

Suborder ZYGOPTERA.

Fore and hindwings closely similar in form, with narrow or petiolate bases. Radius unbranched. An extra branch of the media (the Zygopterid sector *Ms*) takes the place of *Rs* in the Anisoptera. No bridge or oblique vein (except in *Lestidae*). Discoidal cell in the form of a simple quadrilateral, crossed or free. Wings in repose folded back vertically above abdomen (a few exceptions). Eyes button-shaped, projecting laterally at sides of head, separated by a space greater than their dorsal diameter (except *Epiophlebia*). Labium with middle lobe deeply cleft. Male with four anal appendages, two superior and two inferior. Penis not distinctly jointed. Female with superior appendages only. Ovipositor complete.

Larvae cercobranchiate, with three caudal gills projecting from end of abdomen, one medio-dorsal, the other two latero-ventral; the rectum also used for breathing, but not possessing true tracheal gills as in the Anisoptera. Lateral abdominal gills very rarely present. Abdomen slender, often elongated. Gizzard with eight to sixteen radially symmetrical fields.

Family 1. Calopterygidae.

Wings seldom distinctly petiolate. Nodus not generally close to base of wing. Antenodals generally numerous, always considerably more than two. Venation close and rich, with numerous secondary sectors. *Ms* arises from M_{1-2} far proximad to nodus. M_2 arises from M_1 at subnodus, or a little distad from it. Quadrilateral regular or fairly so, never with a sharply-acute distal angle. Wings iridescent, generally beautifully coloured, often metallic. Body-pattern usually dull unicolorous, or metallic, sometimes bicolorous. Larva very variable; gizzard nearly always with sixteen fields, carrying numerous undifferentiated teeth. Mask of variable form, but without setae.

Subfamily 1. **Epallaginae.**

Sectors of arculus arising close to, or a little above, the middle of the arculus. Quadrilateral regular or nearly so. Pterostigma long, regular. Larva rather broad and thick-set, with rather short abdomen. Mask flat, with median lobe entire or only slightly cleft. Antennae with short pedicel. Caudal gills saccoid. Lateral abdominal gills sometimes present.

Tribe 1. *Epallagini* (fig. 137).

Wings not petiolate. Numerous antenodals, those of first and second series nearly all corresponding. Nodus placed between one-third and middle of the wing-length. Arculus close to base, far removed from nodus. Larva with 6–7 pairs of lateral abdominal gills; caudal gills in the form of ovoid sacs, ending in a point (fig. 87).

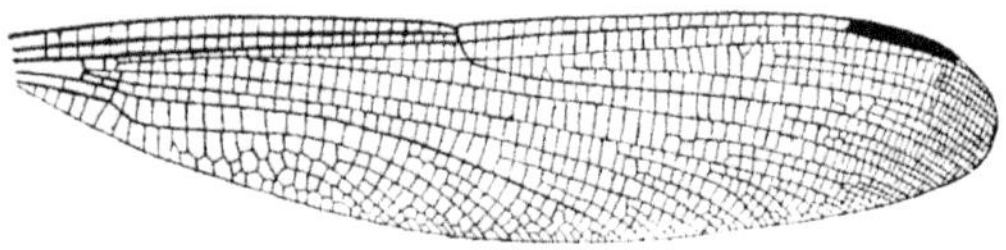

Fig. 137. Forewing (24 mm.) of *Pseudophaea tricolor* Selys, ♂, Borneo. Original.

Tribe 2. *Libellaginini* (figs. 138, 139, also Pl. I, figs. 4, 5).

Wings petiolate. Numerous antenodals, but those of first and second series not corresponding beyond arculus. Nodus placed between one-third and middle of wing-length, usually close to the former position. Arculus

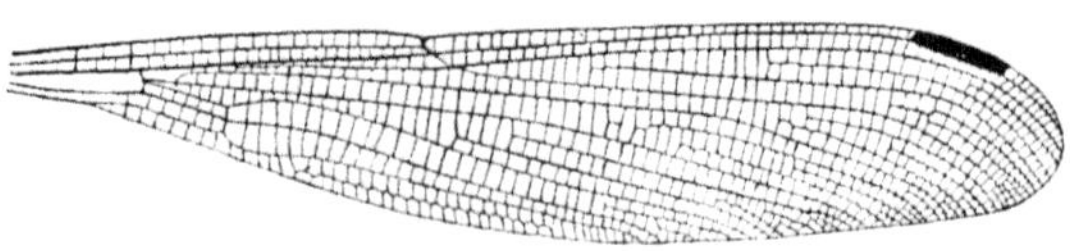

Fig. 138. Forewing (32 mm.) of *Rhinocypha unimaculata* Selys, ♀, India. Original.

about half-way between base and nodus, or nearer to base (much nearer in *Philoganga*). Larva not known for certain[1].

Tribe 3. *Amphipterygini* (fig. 139).

Wings petiolate. Few antenodals of the first series, only 2–3 in second series. Arculus nearer to base than to nodus. Larva without lateral gills; caudal gills of saccoid form, tapering to a point (fig. 35).

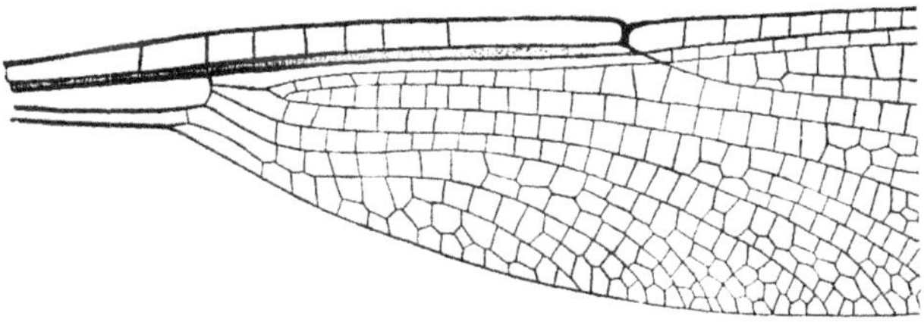

Fig. 139. Basal half of forewing (total length 32 mm.) of *Diphlebia nymphoïdes* Tillyard, ♂, New South Wales. Original.

Subfamily 2. **Thorinae** (fig. 21).

Sectors of arculus arising from top of arculus; hence quadrilateral irregular, with basal side longer than distal side. Pterostigma long, regularly formed.

Larva (*Cora*) similar to that of the *Epallaginae*, with lateral abdominal gills; caudal gills not tapering to a point, but irregularly truncated. Mandibles biramous.

Subfamily 3. **Calopteryginae** (figs. 140, 151, 177, also Pl. I, figs. 6, 7).

Sectors of arculus arising from lower third of arculus. M_{1-2} arching strongly upwards from M_3, so as to approach R very closely, and, in some cases, even fusing with R. Pterostigma much reduced, false, or completely absent.

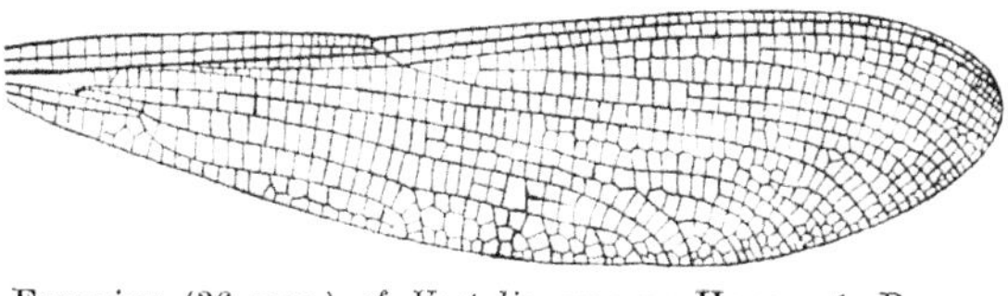

Fig. 140. Forewing (36 mm.) of *Vestalis amoena* Hagen, ♂, Borneo. Original.

Larva of slender build, with moderate or long legs (fig. 40). Antenna with hypertrophied pedicel, longer than the other six segments taken together.

[1] See, however, Karsch., *Berliner Ent. Zeit.*, 1893, XXXVIII, fig. 11. This larva from Togoland resembles that of *Diphlebia*, and is very probably a *Libellago*.

Mask elongated, hollow, with the elongated median lobe deeply incised medially. Caudal gills peculiar, the median one short, lamellar; the laterals long, triquetral. No lateral abdominal gills.

Family 2. Lestidae.

Wings distinctly petiolate. *Ms* fused with M_2 for some distance, leaving it via an oblique vein. A long bridge formed from the latter back to near the bifurcation of M_3 from M_{1-2}. Antenodals variable in number. Very little arrangement of cross-veins into transverse sets. Pterostigma long, regular. Wings rarely coloured. Body-pattern bicolorous, or metallic unicolorous.

Larva of slender build; abdomen long and cylindrical. Mask with highly specialized forms of lateral lobes; gizzard with eight large specialized fields, carrying both large and small teeth. Caudal gills in the form of simple lamellae.

Subfamily 1. **Epiophlebiinae** (fig. 141).

Wings only slightly petioled. Nodus close to middle of wing. Numerous antenodals, those of the first and second series not corresponding, except two (the first, and one between the third and sixth) which are specially thickened (cf. *Aeschnidae*). M_2 arises at subnodus. Bridge connects proximally with

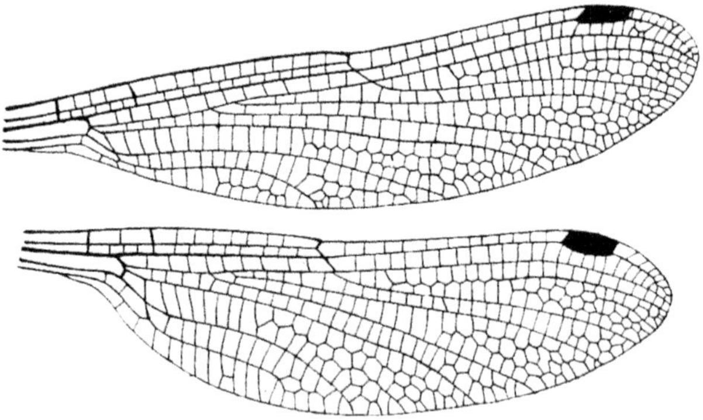

Fig. 141. Wings of *Epiophlebia superstes* Selys, Japan. (Hw. 31 mm.) After Needham.

M_{1-2} a little distad from M_3. Quadrilateral of hindwing broader than that of forewing, with longer costal side. Insect of rather stout build, resembling the *Gomphinae*; eyes close together in male; body-pattern bicolorous. Larva unknown.

Subfamily 2. **Lestinae** (fig. 142, also Pl. II, fig. *E*).

Wings very distinctly petioled. Nodus at about one-third of the wing-length from base, or slightly more distad. Arculus half-way between base and nodus. Sectors of arculus arising high up close to *R*. Bridge connected as in *Epiophlebiinae*. M_2 arising from M_1 distad from subnodus. Antenodals two only (rarely three), the second placed in line with the arculus. Quadrilateral of both wings narrow, with very acute distal angle.

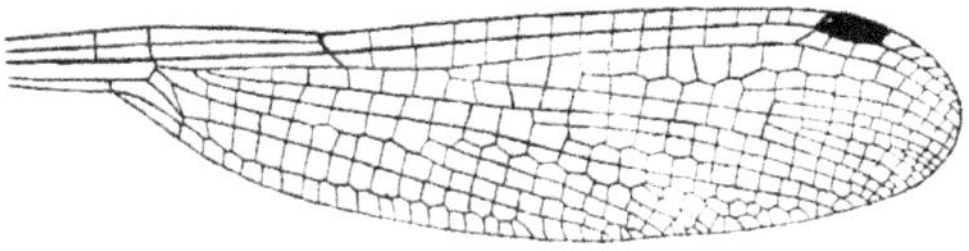

Fig. 142. Forewing (25 mm.) of *Lestes dryas* Kirby, ♂, Britain. Original.

Larva slender, with long legs (fig. 41 B). Mask with incised median lobe; lateral lobes irregularly cleft, with large movable hook bearing two or three setae; mental and lateral setae also present. Caudal gills with secondary tracheae at right angles to gill-axis.

Subfamily 3. **Synlestinae** (fig. 23 A).

Wings very distinctly petioled. Nodus at about one-third of the wing-length from base, or less. Arculus nearer to nodus than to base. Sectors of arculus arising a little above half-way along the arculus. M_2 arising from M_1 far distad from nodus. M_3 arising from M_{1-2} at, or a little proximad to, the subnodus. Bridge connected proximally with M_3 close to its origin. Antenodals two only (rarely three), the second placed in line with the arculus. Quadrilateral of both wings very narrow, with sharply acute distal angle; Cu_1 leaving it with a strong upward arch. General plan of venation resembles that of the *Megapodagrioninae* by convergence. Antenna with very elongated pedicel.

Larva (*Synlestes*) very slender and elongated, with exceedingly long spider-like legs (fig. 41 A). Mask with incised median lobe; lateral lobe narrow, cleft into two sharp teeth, and with a denticulate inner border; movable hook long and slender; no setae present. Antenna with greatly elongated pedicel. Caudal gills with secondary tracheae somewhat oblique to the gill-axis. Gizzard with dentition reduced to a few large teeth on each field.

Family 3. AGRIONIDAE.

Wings distinctly petiolate. Nodus placed at from one-seventh to one-third of the wing-length, rarely a little more distad from base. No bridge or oblique vein. M_2 leaves M_1 far distad from nodus. *Ms* generally arising at subnodus, M_3 a little before (some exceptions). Antenodals two only (exc. *Neurolestes* and *Thaumatoneura*), the second placed either in line with the arculus or a little proximal to it. Quadrilateral variable in shape, but always free. Wings rarely coloured. Cross-veins usually more or less completely arranged in cross-sets. Abdomen slender. Body-pattern variable, often brightly coloured.

Larva variable in form. Caudal gills usually lamellar, constricted, nodate, or subnodate, the division tending to disappear in the higher forms. Gizzard of 8–16 fields, variable in dentition.

Subfamily 1. **Megapodagrioninae** (figs. 23 D, 143).

Nodus placed at from one-third to one-fourth of the wing-length from the base (in *Thaumatoneura*, one-sixth). Quadrilateral somewhat irregular, the distal angle being acute but not sharply so; Cu_1 leaves it without arching strongly upwards. Pterostigma regular, short or long. M_2 leaves M_1 nearer to nodus than to pterostigma, rarely half-way between. Nearly always Ms

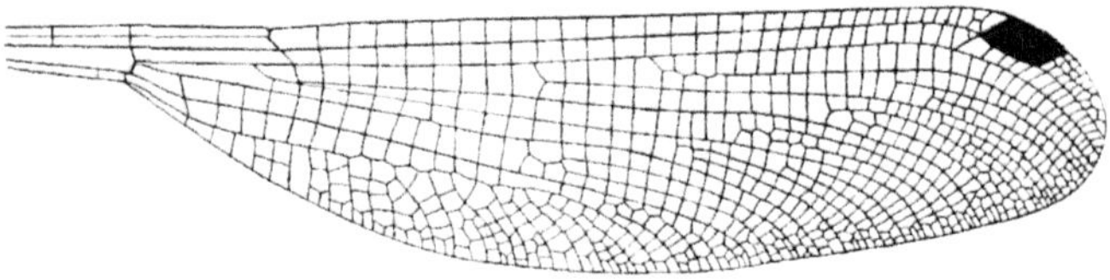

Fig. 143. Hindwing (45 mm.) of *Podopteryx roseonotata* Selys, ♀, Queensland. Original.

arises at subnodus, M_3 somewhat proximad to it. Ax_2 in line with arculus (except *Nesolestes*). Main veins diverging towards distal end of wing, thus allowing development of intercalated sectors.

Larva (*Argiolestes*) short, thick-set (fig. 42 A). Mask short, flat, without setae, resembling that of the *Epallaginae*. Caudal gills in the form of simple lamellae, broadly foliate, horizontally spread out. Gizzard of 14–16 generalized fields.

Subfamily 2. **Pseudostigmatinae** (fig. 144, also Pl. I, fig. 8).

Nodus placed very close to wing-base, at one-sixth to one-seventh of the entire length. M_2 arising very far distad from nodus. Quadrilateral fairly regular, distal angle slightly acute. Postnodals very numerous, continued across the wing as complete transverse veins. Two or more intercalated sectors, with branches in the more richly veined forms. Pterostigma weak, diffuse, abnormal or absent. Abdomen very slender, excessively elongated.

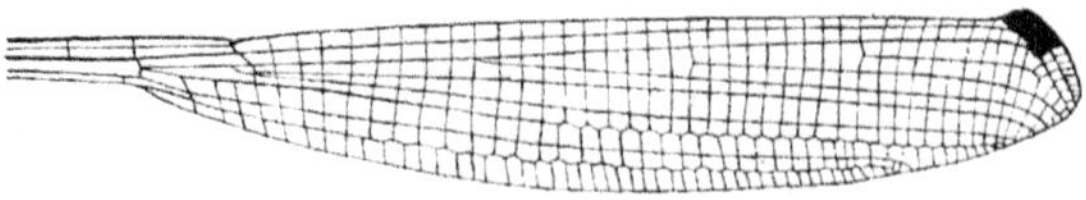

Fig. 144. Hindwing (62 mm.) of *Mecistogaster lucretia* Drury, ♂, Brazil. Original.

Larva (*Mecistogaster*) with comparatively short abdomen, and short legs. Mask with triangular projecting median lobe, not incised; lateral lobes and movable hook small; lateral setae present, but not mental. Gizzard with sixteen generalized folds. Caudal gills with the basal portion forming a stalk, the distal portion foliaceous, pointed.

Subfamily 3. **Platycneminae** (fig. 145).

Nodus placed at from one-fourth to one-third of the wing-length from the base. Quadrilateral regular or slightly irregular (the distal angle acute, but not sharply so). Cu_2 not reduced (except in *Chlorocnemis*). Pterostigma short, regular. No intercalated sectors, except the postnodal one. *Ab* sometimes absent. Tibiae of middle and hindlegs often dilated, especially in the male.

Larva (*Platycnemis*) short, rather broad. Mask with projecting median lobe, not incised; mental and lateral setae present. Gizzard of 16 folds with specialized dentition. Caudal gills in the form of thickened lamellae, with long pointed tips.

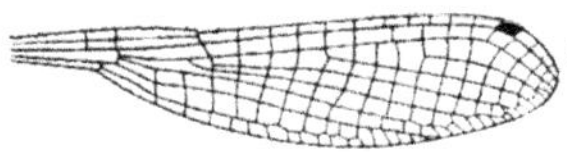

Fig. 145. Forewing (21 mm.) of *Platycnemis pennipes* Pallas, ♂, Britain. Original.

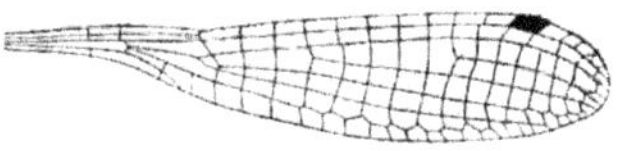

Fig. 146. Forewing (18 mm.) of *Caconeura solitaria* Tillyard, ♂, Queensland. Original.

Subfamily 4. **Protoneurinae** (fig. 146).

Nodus placed at from one-fourth to one-third of the wing-length from the base. Quadrilateral regular. Pterostigma short, trapezoidal (except *Lestoidea*). Cu_2 reduced to a short cross-vein descending to the wing-border; Cu_1 sometimes reduced in length. *Ab* very often absent. No intercalated sectors, except the postnodal one. Small delicate insects with very slender bodies and narrow wings.

Larva moderately short, with short legs (fig. 42 B, C). Mask flat, without setae. Gizzard with sixteen generalized folds. Caudal gills strongly constricted, saccoid or lamellar.

Subfamily 5. **Agrioninae.**

Nodus placed at from one-fourth to somewhat beyond one-third of wing-length. Quadrilateral irregular, with distal angle usually very sharply acute. Pterostigma short, rhomboidal or trapezoidal, rarely abnormal. No intercalated sectors except the postnodal one. *Ab* always present. Cu_2 normal.

Larva of slender build, with short or moderate legs (fig. 42 D, E). Mask with projecting median lobe, not incised; mental and lateral setae present. Caudal gills usually slender lamellae, more or less pointed, and held in the vertical plane; either distinctly nodate, subnodate, or secondarily simple. Secondary tracheae placed obliquely to the gill-axis. Gizzard of 8–16 fields, with specialized dentition.

Tribe 1. *Argiini* (fig. 147).

Wings only petiolate to level of Ax_1. Ac falls on to A' at a point slightly distad from the distal end of the petiole. Moderately robust forms. Legs with long bristles. Larva with caudal gills short, thick, simple.

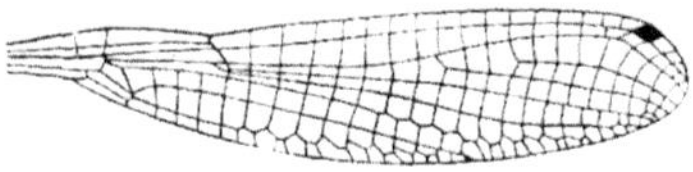

Fig. 147. Hindwing (22 mm.) of *Argia apicalis* Say, ♀, N. America. Original.

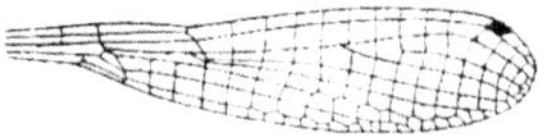

Fig. 148. Forewing (19 mm.) of *Agrion pulchellum* Vand., ♂, Britain. Original.

Tribe 2. *Agrionini* (fig. 148, also Pl. I, fig. 9).

Wings more petiolate than in Tribe 1, to between levels of Ax_1 and Ax_2. Ac falls on to A' at a point slightly distad from the distal end of the petiole. Slenderer forms. Legs with short or moderate bristles. Larva with caudal gills variable, nodate, subnodate or denodate, usually fairly long (rarely reduced and non-functional). Gizzard with sixteen specialized fields.

Tribe 3. *Pseudagrionini* (fig. 149, also Pl. II, fig. *C*).

Petiolation of wings as in Tribe 2. Ac falls on to the angle between the petiole and the curved portion of the posterior border of the wing, or slightly proximad to it; hence A' is completely fused with the wing-border. Slender forms. Larvae as in *Agrionini*, but gizzard sometimes with only eight fields carrying large teeth (*Caliagrion*).

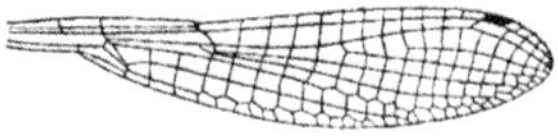

Fig. 149. Forewing (20 mm.) of *Pseudagrion aureofrons* Tillyard, ♂, New South Wales. Original.

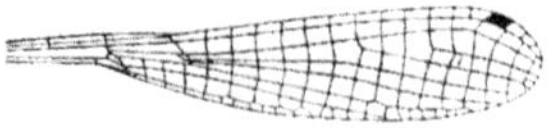

Fig. 150. Forewing (25 mm.) of *Teinobasis rufithorax* Selys, ♂. Torres Straits. Original.

Tribe 4. *Teinobasini* (fig. 150).

Wings very slender, petioled almost to level of Ax_2. Ac falls as in Tribe 3. Ms and M_3 arise exceedingly close together. Extremely slender forms. Tarsal claws reduced, without an inferior tooth. Larva unknown.

CHAPTER XV

ZOO-GEOGRAPHICAL DISTRIBUTION

Space will not allow us to deal with this important question as fully as it deserves. We shall, however, use the somewhat novel method which I have already introduced in this subject [175]. The six main zoo-geographical regions of the earth, as defined by Sclater and Wallace, are accepted as valid, while Polynesia, with its obviously negative or non-continental characters, is treated as an appendix. The fauna of each region is then subdivided into three portions, which may be briefly defined as follows:

1. *Palaeogenic Groups*, or those archaic remnants whose distribution is discontinuous, extending over one or more regions. They are the last remains of groups which were once much more widely spread.

2. *Entogenic Groups*, or those groups which form the autochthonous or peculiar fauna of each region. They are not necessarily confined to a single region, since they frequently develop sufficient energy to spread over two or more regions. A group is, however, only *entogenic* in that region in which it forms a definite *zoocentre* (region of greatest density), placed definitely within the region.

3. *Ectogenic Groups*, or those groups which, being entogenic in a neighbouring region, have invaded the region under discussion, and have gained a footing in some part of it, thus modifying the composition of the fauna of that part. No zoocentre is formed by any group in the region in which it is ectogenic.

It is the *entogenic* groups which give the special character to the regional fauna. The *ectogenic* groups modify the fauna of outlying parts of a region, but are absent or small over the greater part of the region. *Palaeogenic* groups follow no definite rules of distribution. They are always small in number of species, with related forms widely separated, and usually only occupying small isolated areas. Thus the total fauna of a region resembles a picture produced by the "three-colour process," in which the three separate elements, palaeogenic, entogenic, and ectogenic, merge into one complete whole.

A number of genera have become so firmly established in more than one region, that they may be considered to have formed definite zoocentres in two or more of them. These are termed polyentogenic groups (di-, tri-, tetra-, penta- or even hex-entogenic). Special cases of these are known as Holarctic (2), New World (2), Old World (4), Austro-Malayan (2), Bipolar (4), Circumtropical (4), or Cosmopolitan (6) groups. In the table of entogenic genera, these are included under the regions in which they are entogenic, with a number in brackets following each genus, to indicate in how many regions it may be considered to be so; e.g. *Orthetrum* (3), *Pantala* (6).

We shall now proceed to discuss separately the palaeogenic, entogenic and ectogenic Odonate faunas of the World.

The Palaeogenic Fauna.

From the definition of palaeogenic groups, we see that they should not be particularly associated with any one region. Even if to-day a palaeogenic group is only found in one region, we indicate, by including it in this fauna, that we believe it to be, not a true portion of the entogenic fauna of that region, but the remains of a fauna that was once far more widely spread. There are five isolated groups which may be regarded as palaeogenic. Four of these are subfamilies, and one is a tribe:

1. Subfamily *Petalurinae*. This very ancient group has to-day a very typical palaeogenic distribution. In Neotropica, the genus *Phenes* (1 species) is confined to S. Chili. In Nearctica, the genus *Tachopteryx* has one species in Nebraska, and one species in New York and Pennsylvania. In Palaearctica, the same genus has one species in Japan. In Australia, the principal genus *Petalura* is represented by three species; while in New Zealand the genus *Uropetala* is represented by one species. All the species are of large size, occupy only small areas, and bear all the marks

Table of Entogenic Genera of Odonata.

ANISOPTERA

		Subfamily and Tribe	Neotropic	Nearctic	Palaearctic	Ethiopian	Oriental	Australian
AESCHNIDAE	Gomphinae	*Ictinini*	*Gomphoides, Negomphoides, Progomphus, Cyclophylla, Aphylla, Zonophora, Diaphlebia, Cacus*	*Hagenius*	*Sieboldius* (2), *Lindenia*	*Ictinus* (2), *Pseudictinus, Isomma, Diastatomma*	*Ictinus* (2), *Gomphidia, Sieboldius* (2)	
		Gomphini	*Epi-, Neo-, Cyano-, Agrio-gomphus*	*Erpeto-, Dromo-, Octo-gomphus, Lanthus, Gomphus* (2), *Ophiogomphus* (2)	*Gomphus* (2), *Ophiogomphus* (2), *Onychogomphus* (2)	*Onychogomphus* (2), *Meso-, Ocy-, Phyllo-, Podo-, Creni-, Denti-, Corni-, Trago-, Tri-, Neuro-, Noto-, Cerato-, Bursi-, Lestino-gomphus*	*Hetero-, Platy-, Burma-, Anormo-, Aniso-, Cyclo-, Micro-, Lepto-, Mero-gomphus*	*Austro-, Hemi-gomphus*
		Cordulegastrinae	—	*Cordulegaster* (2)	*Cordulegaster* (2), *Anotogaster* (2)	—	*Anotogaster* (2), *Allogaster*	—
	Aeschninae	*Brachytronini*	*Allopetalia, Limnetron*	*Boyeria* (2), *Gomph-, Nasi-, Epi-aeschna*	*Brachytron, Planaeschna, Caliaeschna* (2), *Aeschnophlebia, Boyeria* (2)	—	*Caliaeschna* (2), *Jagoria, Linaeschna*	*Tele-, Austro-phlebia, Dendro-, Noto-, Austro-aeschna*
		Aeschnini	*Sub-, Rhion-, Coryph-, Neuraeschna, Aeschna* (4), *Neureclipa, Staurophlebia, Triacanthagyna, Gynacantha* (4), *Remartinia, Anax* (6)	*Basi-, Oplon-aeschna, Aeschna* (4), *Anax* (6)	*Aeschna* (4), *Anax* (6)	*Aeschna* (4), *Heliaeschna* (2), *Gynacantha* (4), *Hemianax, Anax* (6)	*Gynacantha* (4), *Tetracanthagyna, Platacantha* (2), *Heliaeschna* (2), *Peri-, Amphi-, Anaci-aeschna* (2), *Anax* (6)	*Gynacantha* (4), *Platacantha* (2), *Cora-, Austrogyn-acantha, Anaciaeschna* (2), *Anax* (6)
LIBELLULIDAE	Corduliinae	*Synthemini*	—	—	—	—	—	*Syn-, Eusyn-, Choristhemis*
		Macromiini	—	*Macromia* (4), *Didymops*		*Macromia* (4), *Phyllomacromia*	*Macromia* (4), *Epophthalmia, Azuma*	*Macromia* (4)
		Idocorduliini	*Neocordulia, Gomphomacromia*		*Oxygastra*	*Idomacromia, Nesocordulia, Neophya*	*Idionyx, Macromidia, Metaphya* (2)	*Austro-, Meta-phya* (2), *Syn-, Austro-, Pseudo-, Hespero-, Lathro-cordulia*
		Cordulephyini	—	—	—	—	—	*Cordulephya*
		Eucorduliini	*Aeschnosoma, Paracordulia*	*Somatochlora* (2), *Cordulia* (2), *Tetragoneuria, Neuro-, Platy-, Doro-, Helo-, Epi-cordulia, Williamsonia*	*Somatochlora* (2), *Cordulia* (2), *Epitheca*	*Libellulosoma*		*Pentathemis, Somatochlora, Procordulia, Hemicordulia*
	Libellulinae	*Tetrathemini*		—	—	*Archaeo-, Calo-phlebia, Neody-, Eo-themis, Micromacromia, Allorhizucha*	*Tetra-, Hylaeo-themis, Oda, Nannophlebia* (2)	*Nannophlebia* (2), *Bironides, Microtrigonia, Hypothemis*
		Libellulini	*Or-, Dasy-themis, Cannaphila, Misagria*	*Libellula* (2)	*Libellula* (2), *Orthetrum* (3)	*Orthetrum* (3), *Apatelia, Thermor-, Oxy-, Hadro-themis*	*Orchi-, Amphi, Porno-, Lyrio-themis, Diplacina* (2), *Nesoxenia* (2), *Lathrecista* (2), *Agrionoptera* (2), *Potamarcha, Cratilla, Orthetrum* (3)	*Diplacina* (2), *Nesoxenia* (2), *Lathrecista* (2), *Agrionoptera*(2), *Protorthemis*
		Palpopleurini	*Diastatops, Zenithoptera, Perithemis* (2)	*Perithemis* (2)	—	*Palpopleura*	—	—
		Brachydiplacini	*Argyro-, Poda-themis, Fylgia, Nephepeltia, Elga, Edonis, Oligoclada, Uracis, Anatya, Micrathyria*	*Nannothemis*		*Aethio-, Elea-themis, Chalcostephia, Hemistigma, Thermochoria, Porpax*	*Raphismia, Brachydiplax* (2), *Tyriobapta, Brachygonia*	*Brachydiplax* (2), *Nannophya*
		Sympetrini	*Erythrodiplax, Ery-, Lepthemis, Rhodopygia*	*Pseudoleon, Pachydiplax, Sympetrum*(2)	*Deielia, Crocothemis* (4), *Sympetrum* (2)	*Acisoma, Crocothemis* (4), *Bradinopyga* (2), *Cyano-, Brachythemis* (2)	*Crocothemis* (4), *Neurothemis*(2), *Indo-, Rhodothemis* (2), *Brachythemis* (2)	*Nannodiplax, Diplacodes, Crocothemis* (4), *Neurothemis* (2), *Rhodothemis* (2)
		Leucorrhiniini	*Brachymesia, Planiplax*	*Celithemis, Leucorrhinia* (2)	*Leucorrhinia* (2)	—	—	*Austrothemis*
		Trithemini	*Dy-, Macro-, Gyno-, Paltothemis* (2), *Scapanea, Brechmorrhoga*	*Paltothemis* (2)		*Thalasso-, Helo-, Trithemis, Atoconeura, Pseudomacromia, Olpogastra*	*Pseudagrionoptera, Zygonyx* (2), *Onychothemis*	*Lanthanusa, Huonia, Zygonyx* (2), *Celebothemis*
		Trameini	*Tauriphila, Miathyria, Ephidatia, Antidythemis, Tramea* (5), *Macrodiplax* (3), *Tholymis* (4), *Pantala* (6)	*Tramea* (5), *Pantala* (6)	*Selysiothemis, Pantala* (6)	*Urothemis, Zyxomma* (3), *Tholymis* (4), *Pantala* (6), *Rhyothemis* (3), *Tramea* (5)	*Tramea* (5), *Hydrobasileus* (2), *Aethriamanta, Macrodiplax* (3), *Zyxomma* (3), *Tholymis* (4), *Pantala* (6), *Camacinia* (2), *Rhyothemis* (3)	*Tramea* (5), *Hydrobasileus* (2), *Macrodiplax* (3), *Zyxomma* (3), *Tholymis* (4), *Pantala* (6), *Camacinia* (2), *Rhyothemis* (3)

of being archaic remnants. Ancestors of the group occurred at Solenhofen in Upper Jurassic times.

2. Subfamily *Chlorogomphinae*. This is another archaic group, but not so typically palaeogenic as 1, since it is at present confined to the Oriental region. Its distribution within that region is, however, so discontinuous that it satisfies the definition. The genus *Chlorogomphus* occurs in Sumatra and Java; *Orogomphus* is found in the Himalayas, in Borneo, and also in the Philippines!

3. Tribe *Petaliini*. In the forests of Southern Chili this tribe is represented by *six* species, comprised in the genera *Petalia*, *Phyllopetalia* and *Hypopetalia*. On the Blue Mountains of New South Wales, a single species, *Austropetalia patricia*, occurs. This might be claimed as a case of distribution from an original common Antarctic centre. But the group is so clearly archaic, and had such near relatives at Solenhofen in Upper Jurassic times, that it is much safer to treat it as strictly palaeogenic.

4. Subfamily *Epiophlebiinae*. We have here almost certainly the very last stage in the distribution of a very archaic group. The single genus *Epiophlebia* is represented only by a single species from Japan! The genus seems to be a remnant of an early *Lestid* fauna dating back to Jurassic times, and bears somewhat the same relationship to *Lestes* that the fossil *Heterophlebiinae* bore to our present-day *Synlestinae*.

5. Subfamily *Synlestinae*. This group is represented by two species of *Synlestes* in Eastern Australia, and five species of *Chlorolestes* in South Africa. It is obviously an archaic group, but is still fairly flourishing, and only in the first stages of discontinuity. The commonest species, *Synlestes weyersi*, ranges from Victoria to North Queensland, and is in process of sub-division into a number of regional forms or sub-species.

The Entogenic Fauna.

The entogenic groups, being essentially those which make up the fauna peculiar to each region, must be dealt with separately under the six main headings. I have thought it best to do this by means of a complete table, which will serve the double object of exhibiting at a glance the entogenic faunas of the six main regions (and, in addition, those genera which are entogenic in

Table of Entogenic Genera

ZYGOPTERA

Family	Subfamily and Tribe		Neotropic	Nearctic
CALOPTERYGIDAE	Epallaginae	*Epallagini*	—	—
		Libellagini	*Dicterias, Helio-, Neo-, Cyano-charis*	—
		Amphipterygini	*Amphipteryx*	—
	Thorinae		*Thore, Euthore, Cora, Chalcopteryx*	—
	Calopteryginae		*Lais, Hetaerina* (2)	*Hetaerina* (2), *Calopteryx* (2)
LESTIDAE	Lestinae		*Archilestes* (2), *Ortholestes, Lestes* (5)	*Archilestes* (2), *Lestes* (5)
AGRIONIDAE	Megapodagrioninae		*Thaumatoneura, Paraphlebia, Philogenia, Mega-, Allo-, Hetero-podagrion, Mes-, Dimer-, Neur-, Heter-agrion, Perilestes*	—
	Pseudostigmatinae		*Pseudo-, Micro-stigma, Mecistogaster, Anomisma, Megaloprepus*	—
	Platycneminae		—	—
	Protoneurinae		*Palaemnema, Peristicta, Idio-, Neo-, Micro-, Proto-neura*	—
	Agrioninae	*Argiini*	*Hyponeura, Argia* (2), *Diargia*	*Argia* (2)
		Agrionini	*Ceratura, Ischnura* (6), *Amphi-, Tigri-, Hesper-agrion* (2)	*Chrom-, Anomal-agrion, Agrion* (2), *Ischnura* (6), *Nehallenia* (2), *Hesper-agrion* (2)
		Pseudagrionini	*Oxy-, Acanth-, Anis-, Tel-, Lept-, Hylae-, Anti-agrion, Telebasis, Skiallagma, Argiallagma*	*Enallagma* (3)
		Teinobasini	*Lepto-, Metalepto-basis*	—

of Odonata (continued).

ZYGOPTERA

Palaearctic	Ethiopian	Oriental	Australian
Epallage	—	*Anisopleura, Bayadera, Para-, Dys-, Cali-, Pseudo-phaea* (= *Euphaea*), *Notholestes*	—
—	*Libellago*	*Micromerus, Rhinocypha, Rhinoneura, Philoganga*	—
—	—	*Devadatta*	*Diphlebia*
—	—	—	—
Calopteryx (2), *Mnaïs* (2)	*Pentaphlebia, Phaon, Umma, Sapho*	*Neuro-, Climaco-basis, Matrona, Matronoïdes, Echo, Archineura, Psolodesmus, Mnaïs* (2), *Vestalis*	—
Sympycna, Lestes (5)	*Lestes* (5)	*Mega-, Platy-, Oro-lestes, Lestes* (5)	*Austrolestes*
—	*Neuro-, Allo-, Neso-, Proto-lestes*	*Rhinagrion, Podo-, Rhipido-, Pseudo-, Amphi-lestes, Meso-podagrion*	*Podopteryx, Argiolestes, Wahnesia, Trineuragrion, Metagrion*
—	—	—	—
Platycnemis	*Tato-, Lepto-, Para-, Meta-, Allo-, Meso-, Chloro-cnemis*	*Copera, Coeliccia, Cali-, Priono-, Steno-cnemis*	*Idiocnemis*
—	*Disparoneura* (2)	*Platy-, Proto-sticta, Caco-, Caeno-neura, Disparoneura* (2)	*Noso-, Iso-, Neo-, Austro-, Ori-sticta, Selysioneura, Lestoidea*
—	—	*Onychargia*	*Palaeargia*
Pyrrhosoma, Erythromma, Agrion (2), *Ischnura* (6), *Nehallenia* (2)	*Thermagrion, Ischnura* (6), *Argiocnemis* (2)	*Ischnura* (6), *Argiocnemis* (2), *Agriocnemis* (2)	*Ischnura* (6), *Oreagrion, Agriocnemis* (2), *Austrocnemis, Hemiphlebia*
Enallagma (3)	*Mombagrion, Argiagrion, Ceriagrion* (2), *Pseudagrion* (2), *Enallagma* (3)	*Xiphiagrion, Pseudagrion* (2), *Ceriagrion* (2)	*Cali-, Austro-, Papu-, Xanth-agrion, Xanthocnemis*
—	*Aciagrion* (2)	*Aciagrion* (2), *Archi-, Teino-basis, Amphi-, Peri-cnemis*	*Nesobasis*

more than one region), and of giving, in a fairly compact form, a full list of genera for all the subfamilies and tribes,—except, of course, the five dealt with above in the Palaeogenic Fauna. By combining this table with the classification given in chap. XIV, the student can not only obtain at once a full list of genera for each tribe, but at the same time he will be able to note the *distribution* of those genera. The table is supplemented by a short discussion on the chief characteristics of the autochthonous fauna of each region.

The Neotropical Region.

This region includes the whole of South America and adjoining islands, Central America and the tropical parts of Mexico, and the whole of the West Indies. It is, in the Odonata as well as in other groups of animals, the most self-contained of all the six regions of the earth. Out of 135 genera known to occur, no less than 111, or 83 per cent., are peculiar to or entogenic in the region[1]. Being for the most part a tropical region, it is exceedingly rich in Odonata. The total number of species described from it far exceeds that of any other region, as also does the total number of entogenic genera.

A glance at the table shews us how the various subfamilies and tribes are represented in the region, so that it is not necessary to go into details here on that point. We shall, however, notice the more striking peculiarities of the region. With the exception of two genera (*Hagenius* and *Sieboldius*) the whole of the *Gomphoïdes*-Series of the *Ictinini* is confined to Neotropica. A similar distribution holds for the *Palpopleurini*, the *Uracis*-Series of the *Brachydiplacini*, and the *Dythemis*-Series of the *Trithemini*. The whole of the New World genera of the *Megapodagrioninae*, *Protoneurinae*, *Teinobasini* and all but one of the New World *Pseudagrionini* also belong to this region. The *Thorinae* and *Pseudostigmatinae* (Pl. I, fig. 8) are exclusively confined to Neotropica. As against this wonderful wealth of forms, we notice a great paucity of *Corduliinae* and a complete absence of *Platycneminae*. The *Corduliinae* are essentially a temperate group,

[1] In calculating these "Regional Percentages" for the six regions, genera which are entogenic in more than one region have been excluded.

and, apart from the *Macromiini*, the Oriental region shares with the Neotropical this paucity of forms. The *Platycneminae*, however, are essentially a tropical group. It is peculiar that the Oriental and Neotropical agree in possessing rich developments of *Megapodagrioninae* and *Protoneurinae*, but differ in that the former alone has a rich development of *Platycneminae*.

I am inclined to associate the absence of *Platycneminae* in Neotropica with the presence of the unique *Pseudostigmatinae*. These latter will, I believe, eventually be shewn to be the lost *Platycnemine* stock of the Neotropical Region, which has reached its present form through a hypertrophy of wings and abdomen, brought about in correlation with the adoption of the extraordinary habit of breeding in the epiphytic Bromeliads of the dense tropical forests.

The Nearctic Region.

In this region we include the whole of the North American Continent, together with the temperate parts of Mexico. The region is not nearly so well differentiated, as regards its Odonata, as the Neotropical. Out of 59 genera known to occur, only 24, or 41 per cent., are peculiar or entogenic[1]. The chief reasons for this are (i) the presence of a fairly abundant Holarctic fauna consisting of genera shared in common with Palaearctica, (ii) a considerable southern incursion of ectogenic genera from Neotropica, together with other genera which may be regarded as dientogenic for both Nearctica and Neotropica.

The most striking point about the Nearctic fauna is the rich development of the tribes *Gomphini* and *Eucorduliini*. The seven genera of the latter tribe confined to Nearctica probably represent the highest expression of the ideal Odonate-type yet reached by the Order, and can only be paralleled by the Australian *Hemicordulia*-group.

The Nearctic region is as poor in Zygoptera as the Neotropic is rich. Only 15 genera, belonging to three subfamilies, occur; and of these, only two are peculiar. But for the presence of *Hetaerina* (fig. 151) and *Argia*, which seem undoubtedly to have come originally from Neotropica, there would be left only a

[1] See foot-note on p. 286.

small remnant composed of Holarctic genera and their derivatives. The Tertiary fossil-beds of this region prove, however, that the *Megapodagrioninae* existed fairly abundantly in Miocene times. We can only conclude that this group was unable to resist a subsequent lowering of the temperature over those areas in which it existed, and so became extinct within the region.

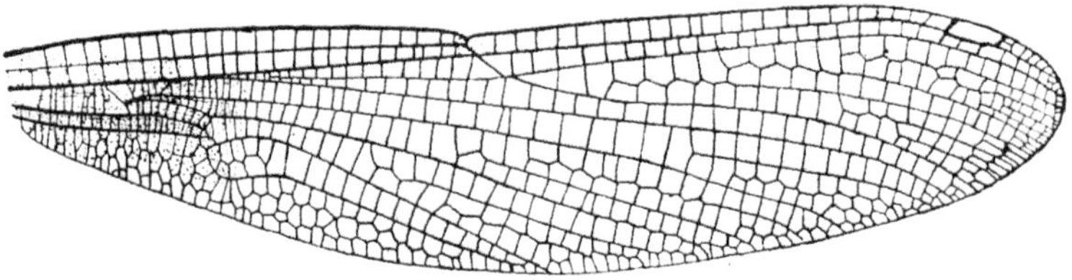

Fig. 151. Fore-wing (30 mm.) of *Hetaerina americana* Fabr., ♀, Mexico. Original.

The Palaearctic Region.

This enormous region includes the whole of Europe, together with the temperate parts of Africa (north of the Tropic of Cancer) and of Asia and the Japanese Archipelago. The boundary is usually drawn along the western watershed of the Indus, thence along the Himalayas into Thibet and China, north of the Yang-tse-kiang watershed. Although the region is of such great extent, it is by far the poorest in Odonata in the whole world. Japan is the only part of it that contains at all an abundant or striking Dragonfly fauna.

Out of 59 genera known to occur, only 14, or 24 per cent., are peculiar to or entogenic in the region[1]. The principal portion of the fauna is supplied by the development of the twelve Holarctic genera shared in common with Nearctica, viz. *Cordulegaster*, *Gomphus*, *Ophiogomphus*, *Boyeria*, *Somatochlora*, *Cordulia*, *Libellula*, *Sympetrum*, *Leucorrhinia*, *Calopteryx*, *Agrion* and *Enallagma*. The development of Anisoptera is on the whole less marked, of Zygoptera more marked, than in Nearctica. There are no peculiar genera of *Gomphini*, and only two of *Corduliinae*, *Oxygastra* (fig. 152) and *Epitheca*. But in the Zygoptera, there are the

[1] See foot-note on p. 286.

five peculiar genera *Epallage*, *Sympycna*, *Platycnemis*, *Erythromma*, and *Pyrrhosoma*.

Japan. This group of islands forms a distinct province of the region. It is marked by two special characteristics: (i) the presence of *Planaeschna*, *Aeschnophlebia* and *Epiophlebia* (peculiar), and *Mnaïs* (Japan and China); (ii) a strong incursion of Oriental

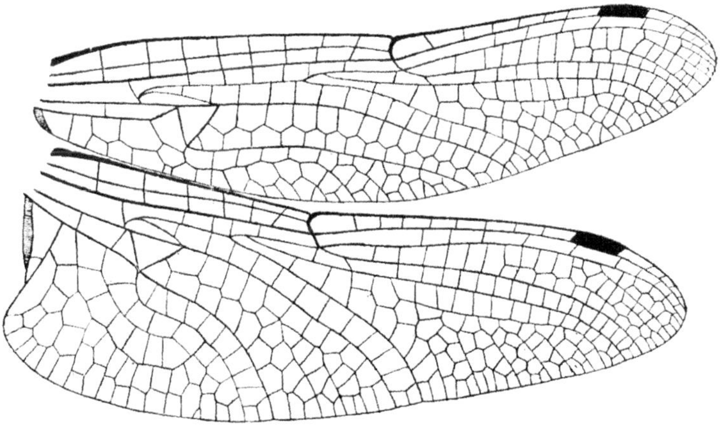

Fig. 152. Wings of *Oxygastra curtisi* Dale, ♂, France. (Hw. 35 mm.) Original.

forms, amongst which the most striking are undoubtedly the magnificent insects of the genera *Epophthalmia* and *Azuma*. Judging from numerous allusions to the Odonata in Japanese literature and poetry, the fauna seems to be one of striking beauty and coloration, and has impressed itself upon the national consciousness.

The Ethiopian Region.

The Ethiopian region includes the whole of Africa and Arabia south of the Tropic of Cancer, Madagascar, Mauritius, the Seychelles and other islands adjacent to them. Though neither so rich nor so varied as those of the Neotropical and Oriental regions, the Ethiopian fauna is a very distinct one, and fairly rich both in genera and species. Out of 105 genera known to occur, 70, or 67 per cent., are peculiar to or entogenic in the region[1]. A great deal of work

[1] See foot-note on p. 286.

still remains to be done on this fauna, and the number of species is sure to be largely increased in the near future. The chief features of the fauna are the numerous genera of *Gomphini* (if indeed most of these can be accepted as anything more than sub-genera!), the rich development of *Tetrathemini*, *Trithemini* and *Platycneminae*, and the peculiarity of the *Corduline* and *Calopterygid* faunas. It is very remarkable that there is an almost total absence of entogenic *Aeschninae*, the only genus that can possibly be considered entogenic being the widely-spread *Hemianax*, possessing only one species. There is a close parallelism between the Ethiopian and Oriental faunas in the following groups: (i) *Idocorduliini*, (ii) *Tetrathemini*, (iii) *Libellagini* (*Libellago* parallels *Rhinocypha* and *Micromerus*), (iv) *Megapodagrioninae* and (v) *Platycneminae*. The fact that *Protoneurinae* are absent (except for *Disparoneura*) coupled with the fact that the Ethiopian *Platycneminae* possess some forms (e.g. *Chlorocnemis*) which are so reduced that they might almost be classed as *Protoneurinae*, points to the origin of the Old World *Protoneurinae* from a *Platycnemine* stock.

Madagascar. As in other groups of animals, so in Odonata, this great island forms a very definite province, marked by many special features. Out of 85 known species, 58 are peculiar to the island, while all but six of the remainder are species occurring only on the mainland of Africa besides. The fauna is distinguished by the presence of seven genera peculiar to it, viz. *Isomma*, *Nesocordulia*, *Calophlebia*, *Neophlebia*, *Archaeophlebia*, *Neodythemis* and *Nesolestes*.

The Oriental Region.

This region includes the tropical parts of Asia south of the Palaearctic region, Ceylon, Sumatra, Java, Borneo, Formosa, the Philippines, and all the other smaller islands to the west of Wallace's Line. Celebes, usually included in the Australian region, may be regarded as Oriental as far as its low-lying coastal regions are concerned, and forms a kind of "no-man's-land" between the two regions. The geographical unconformity of the region makes it difficult to give a satisfactory account of the fauna. Apart from India, Burmah, and Cochin-China, every one of the great islands has a rich and, in some respects, characteristic fauna of Odonata. Borneo, although comparatively little worked, can already boast of more species than can be found in the whole of the great Palaearctic region! The total number of Oriental genera slightly exceeds that for the Neotropical, but the number of known species is considerably less. Also, the presence of a large number of Austro-Malayan genera (i.e. genera entogenic both in the Oriental and Australian regions) reduces the proportion of genera peculiar to the region. Out of 136 genera known to occur, 82, or 60 per cent., are peculiar to or entogenic in the region[1].

The Oriental region is essentially the home of the *Calopterygidae* (exclusive of the Neotropical *Thorinae*). Out of 20 genera and 141 species of *Epallaginae* so far described, 12 genera and 107 species are Oriental. Out of sixteen known genera of *Calopteryginae*, nine occur in this region. The glorious development of

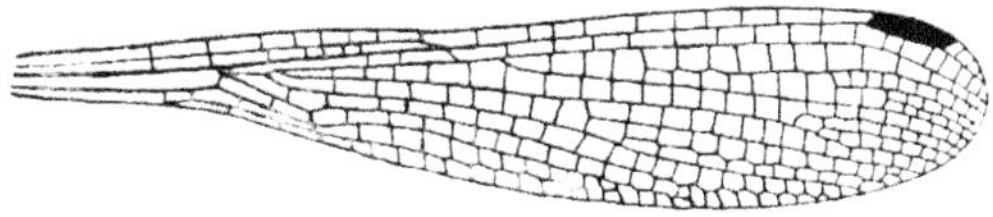

Fig. 153. Hind-wing (17 mm.) of *Micromerus lineatus* Burm., ♂. India. Original.

wing-coloration in these Oriental *Calopterygidae* makes them one of the most striking and gorgeous groups of insects in the world. The tiny *Micromerus* (fig. 153) is the most reduced *Calopterygid* genus known.

The Oriental region parallels the Neotropical in the following points: (i) a great development of *Ictinini*, of the *Ictinus*-Series (paralleling the *Gomphoïdes*-Series); (ii) the large number of species of the *Gynacantha*-Series of the *Aeschnini*; (iii) an exceedingly rich development of the *Libellulinae* (for lists of genera, see the table); (iv) an abundance of *Megapodagrioninae* and *Protoneurinae*, (v) a scarcity of *Corduliinae*. A comparison with the Ethiopian region has already been made. Further striking points about the fauna of this region are (i) the occurrence of the *Chlorogomphinae* (dealt with under the Palaeogenic Fauna), (ii) abundance of genera

[1] See foot-note on p. 286.

of *Gomphini*, (iii) strong development of the *Macromiini*, (iv) the development of the tribe *Teinobasini*, including some rather large species which suggest the beginnings of a hypertrophied or "gerontic" group.

The Australian Region.

Besides the continent of Australia, Tasmania, and the isolated province of New Zealand with its adjacent islands, this region includes the immense Papuan province, which embraces all the islands lying to the east of Wallace's Line. Chief amongst these are Celebes (at least the highlands of this island), Timor, New Guinea, the Solomons, the New Hebrides, New Caledonia, and Fiji. The latter must be included, not only on account of the recent geological demonstration of its continental origin, but also because of the presence of three essentially continental genera, viz. *Synthemis*, *Hypothemis* and *Nesobasis*.

The fauna of the Papuan portion of the region is so different from that of the rest that one may well question the wisdom of including the two within one region. In the Odonata, at any rate, the Papuan fauna, so far as we know it at present, has closer affinities with the Oriental than with the Australian fauna proper. It is advisable, however, not to proceed to any definite decision yet, because the highlands of Papua are not yet explored. It is just at these higher levels that the Australian character of the fauna may be expected to make itself felt, whereas the Oriental portion is flagrantly noticeable along the overheated coast-line, as it is also in North Queensland. Within the last few years such essentially Australian genera as *Synthemis* and *Diphlebia* have been found in Papua, and we may confidently look for more.

Taking the region as a whole, out of 110 genera known, 61, or 55 per cent., are peculiar to or entogenic in it[1]. This low proportion is wholly due to the immense irruption of Oriental forms into the tropical part of the region. This is more fully dealt with in the section on Ectogenic Fauna.

The autochthonous Odonate fauna of the Australian region consists chiefly of a very rich archaic complex of forms not to be paralleled in any other region. Besides the occurrence of *Petalura*,

[1] See foot-note on p. 286.

Uropetala, *Austropetalia* and *Synlestes*, which are included in the section on Palaeogenic Fauna, for the reasons there stated, we note the occurrence of the following groups: (i) two very archaic genera of *Gomphini*, (ii) a very rich *Aeschnine* fauna belonging to the *Brachytronini*, and paralleling the development of *Aeschna* in the Northern Hemisphere, (iii) the very interesting and graceful tribe *Synthemini*, (iv) six genera of *Idocorduliini*, (v) three genera of *Tetrathemini*, (vi) a special development of numerous small bronze and blue *Lestinae* (*Austrolestes*), *bionomically* parallel with *Agrion* and *Enallagma* of the Northern Hemisphere, (vii) a fairly rich fauna of *Megapodagrioninae* and *Protoneurinae*, (viii) six peculiar genera of *Agrioninae*.

The immense development of *Corduliinae*, totalling 16 genera and 52 species, is probably the most striking feature of the Australian fauna. Besides the archaic forms already mentioned, we must also notice the extraordinarily specialized *Cordulephya*, the mysterious *Pentathemis*, and a fairly strong development of *Eucorduliini*, paralleling that of the Northern Hemisphere. The species of the latter have been referred to *Somatochlora*, *Procordulia* and *Hemicordulia*. Of these the last genus is the most specialized, and resembles *Anax* in the adoption of a rounded hind-wing in the male, and also in the startling success of its new career, since it has already overrun the Oriental region and Madagascar.

Besides the very archaic groups which form the basis of the fauna, there are a number of isolated side-branches, all highly specialized, which seem to have found their way into the region long ago, and followed their own lines of development. Of these we may mention *Austrothemis* (paralleling *Leucorrhinia* of the Northern Hemisphere), *Diphlebia* (the most *Agrionid*-like of all *Calopterygidae*), the wonderful little *Hemiphlebia*, and the excessively reduced Australian *Protoneurinae*. These last probably represent the most highly asthenogenetic Dragonflies yet evolved. The very remarkable genus *Lestoidea* (fig. 154) combines an essentially *Protoneurine* venation and facies with certain *Megapodagrionine* characters.

In the Papuan province the most striking facts are the rich developments of the *Gynacantha*-Series and of the *Libellulinae*, together with the total absence of *Gomphinae* (except *Ictinus australis*).

Tasmania. This island is poor in Odonata, though rich in large rivers. Most of the Dragonflies are still-water forms. This suggests that the fauna chiefly crossed from Australia at a time when the land-connection (Bassian Isthmus) was too small to carry many running rivers, and only possessed isolated ponds or shallow lagoons [174].

New Zealand. This isolated province is very poor in all forms of insect life. There are only eight known species of Dragonflies, comprised in the seven genera *Uropetala, Aeschna, Somatochlora, Procordulia, Diplacodes, Austrolestes* and *Xanthocnemis*. Of these, the first and last are peculiar to the island.

For an interesting comparison of the faunas of the three main southern land-masses of Australia, S. Africa, and S. America, the reader is referred to the introduction in Ris [136].

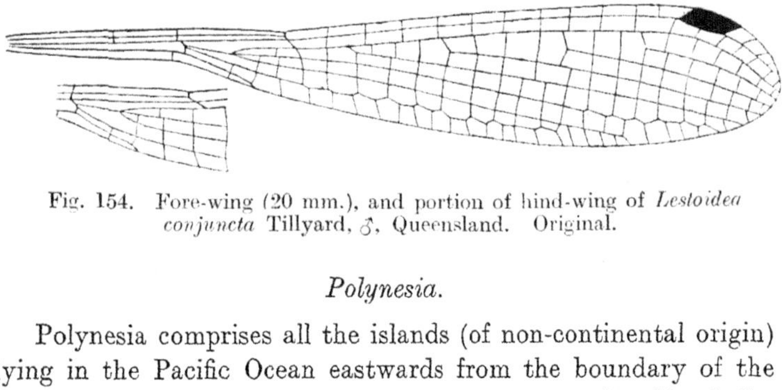

Fig. 154. Fore-wing (20 mm.), and portion of hind-wing of *Lestoidea conjuncta* Tillyard, ♂, Queensland. Original.

Polynesia.

Polynesia comprises all the islands (of non-continental origin) lying in the Pacific Ocean eastwards from the boundary of the Australian region. The principal groups are the Marshalls, Gilberts, Tonga, Samoa, the Marquesas, Tahiti, and the Hawaiian Islands. The interesting fauna of the Hawaiian Islands has been fully dealt with by Perkins [114–116]. There are two peculiar genera, *Nesogonia* (*Libellulinae*) and *Megalagrion*, besides a large number of species referred to the genus *Agrion*, whose larvae live in water collected at the bases of the leaves of plants, and possess reduced caudal gills.

The non-continental character of the fauna of the Polynesian Islands is shewn by the absence of all subfamilies except *Aeschninae*,

Libellulinae and *Agrioninae*. The most noticeable forms are *Anaciaeschna jaspidea* and a few widely spread *Trameini*.

The Ectogenic Fauna.

It is obvious that a group which becomes ectogenic in any given region must be a fairly vigorous group. The more vigorous it is, the more likely it will be to invade other regions and spread far into them. Thus a very large percentage of the ectogenic fauna is made up of *Libellulinae* and *Agrioninae*, which are the dominant and most vigorous groups of to-day. A further point which influences the formation of ectogenic groups is the change of climatic conditions met with in passing the barrier from one region to another. The most obvious case is that of the boundary between a Temperate and a Tropical region, e.g. between Nearctica and Neotropica. In this case, it is clear that change of temperature will be the most important factor. This factor works wholly in favour of the more tropical region. For the groups emigrating from the warmer region to the cooler will hug the coast-line, and will thus have a continuous (or, in the case of a chain of islands, almost continuous) land-area along which they can work their passage onward. On the other hand, the groups emigrating from the colder region to the warmer will tend to seek higher and higher altitudes. As soon as the mean annual temperature of the coastal lands exceeds a certain amount, those lands will be absolutely barred to them. Thence onwards their advance must be confined to the high lands, which usually offer only a discontinuous and precarious foothold.

In the case of the Odonata, it is clear that the nature of the barrier between two regions is of great importance in determining the amount of the ectogenic faunas on either side of it. A desert barrier is impassable to all but the hardiest drought-resisting genera. A high and continuous mountain-range will bar the passage of all but a few alpine forms. On the other hand, a sea-barrier, unless sufficiently wide, will not prove as effective for the strong flying Odonata as it is for most groups of animals. Hence we find that Wallace's Line has not proved a very effective barrier between the Oriental and Australian regions.

The Neotropical Region. This region is very self-contained, being only penetrable along the highlands of the Isthmus of Panama. Hence the ectogenic fauna, received from Nearctica, is exceedingly poor. Several species of *Erpetogomphus*, three species of *Libellula*, one of *Sympetrum*, and one of *Anomalagrion*, appear to form the only undoubted ectogenic fauna. There are, however, a few groups whose zoocentres seem to lie actually in or near to the Isthmus. I have considered these as dientogenic in both Neotropica and Nearctica. *Macromia* seems to penetrate occasionally into the region, and the very beautiful *Pseudoleon* has also a foothold there.

The Nearctic Region. This region is penetrable from Neotropica (*a*) along the coast-line northwards from the Isthmus of Panama, (*b*) along the chain of the Antilles into Florida. It appears also that in fairly recent times Lower California received a considerable influx of Neotropical forms, possibly by a more direct route than any at present existing. Thus we find the fauna of the southern portion of Nearctica, particularly along the coast-line, considerably modified by the influx of a large ectogenic fauna, consisting chiefly of *Libellulinae*. The following Neotropical genera form this ectogenic fauna of Nearctica: *Gomphoïdes*, *Negomphoïdes*, *Cyclophylla*, *Aphylla*, *Gynacantha*, *Orthemis*, *Cannaphila*, *Micrathyria*, *Erythrodiplax*, *Erythemis*, *Brachymesia*, *Dythemis*, *Paltothemis*, *Brechmorrhoga*, *Hyponeura*, *Telebasis*, *Telagrion*. Most of these do not penetrate beyond the Lower Sonoran.

Libellula quadrimaculata and *Enallagma cyathigerum* range through the whole of the Northern Hemisphere, except the tropical parts. There seems to be no evidence to shew whether they originated in Palaearctica or Nearctica. They are best included in the *entogenic* fauna of both regions. The barriers between the two regions are very effective in the case of the Odonata, chiefly because of the lowness of the mean annual temperature at the point where the two approach most closely together (Behring's Straits).

The Palaearctic Region. In spite of its very long frontier to the Ethiopian and Oriental regions, this region seems to be very well contained, the nature of the barriers (deserts and high mountain ranges) making the passage of Odonata very difficult. From Ethiopia, *Hemianax* and *Trithemis* have penetrated into the region. From the Oriental, it is probable that a considerable list of *Libelluline* genera could be drawn up which have reached, in China, beyond the somewhat arbitrary boundary of the Yang-tse-kiang Valley. Southern Japan has also received a fairly rich Oriental fauna, among which we must notice especially the very striking genera *Epophthalmia* and *Azuma* (*Macromiini*), and representatives of the genera *Ictinus*, *Davidius*, *Gynacantha*, *Lyriothemis*, *Rhyothemis*, *Pseudophaea*, *Neurobasis*, *Copera*, *Ceriagrion*, *Pseudagrion*. The widely spread *Diplacodes trivialis* also reaches into Japan. A single species of *Argia* (*A. kurilis*) has colonized the Kurile Islands from Nearctica.

The Ethiopian Region. It is very difficult to determine what is the true ectogenic fauna of this region. Though it is now almost completely shut off by desert from Palaearctica, and by a wide stretch of sea from the Oriental region, yet there is a great deal of evidence to shew that not so very long ago Odonata passed quite easily from both these regions into Ethiopia. Though I have included in the entogenic fauna a large number of *Gomphine* genera which are very closely related to *Onychogomphus* or to *Gomphus*, it might very fairly be argued that these genera actually form a very considerable invasion of essentially Palaearctic forms. Such a conclusion is strengthened by the fact that Ethiopia has no entogenic genera of *Aeschninae* except *Hemianax*. The only difference is that the *Gomphine* settlement is the older of the two. Are these *Gomphine* genera sufficiently differentiated from their parent stock to merit the term "entogenic"? That question cannot be answered except by someone far more familiar with the Ethiopian fauna than myself.

From Palaearctica, the region has received *Platycnemis*, *Agrion*, *Anax imperator*, *Sympetrum fonscolombei* and *Lestes barbarus*. From the Oriental, we notice the following genera: *Heliaeschna*, *Anaciaeschna*, *Hemicordulia* (reaches to Madagascar), *Tetrathemis*, *Agriocnemis*, and the widely spread *Anax guttatus*, *Diplacodes trivialis* and *Ischnura senegalensis*. The close parallelism between certain Ethiopian and Oriental genera, and the large number of dientogenic genera common to the two regions, suggest that a close connection formerly existed between them, via Madagascar, Mauritius and the Seychelles.

The Oriental Region. This is the least isolated of all the regions of the earth, being penetrable from each of the other three regions of the Old World. Palaearctic genera are excluded on account of the great rise in the mean annual temperature south of the barriers. From Ethiopia, the region has received species of *Acisoma*, *Urothemis*, and *Libellago*. There remains Wallace's Line. Here we have a case closely parallel with that of the Isthmus of Panama. A large number of tropical Oriental genera have penetrated eastwards into the Australian region, but very few of the Australian forms have passed westwards into the Oriental. We note especially *Hemicordulia* and *Diplacodes*, which have spread far and wide beyond their Australian zoocentres; the genus *Argiolestes*, which appears to have passed just across the barrier; *Protorthemis*, *Nannophya*, and *Celebothemis*.

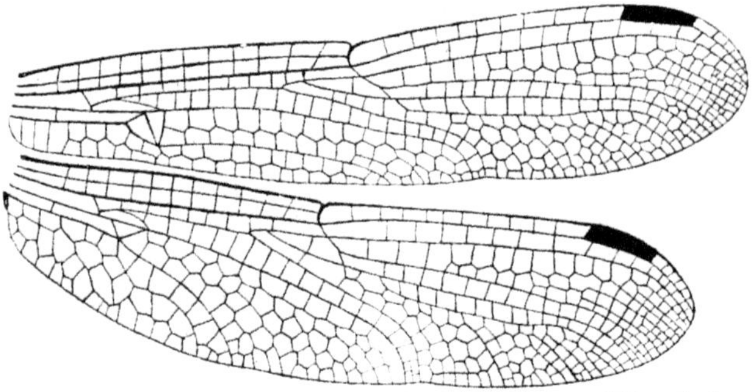

Fig. 155. Wings of *Agrionoptera insignis allogenes* Tillyard, ♂, Queensland. (Hw. 26 mm.) Original.

The Australian Region. This is only now penetrable across Wallace's Line, by strong-flying species. There is evidence, however, of a separate immigration stream into the Australian continent via Timor, at some not very remote period, as well as some indication of the remains of an original Antarctic fauna driven northward through Tasmania. Australia proper, as distinct from the Papuan sub-region, has a rich *Gomphine* fauna, which seems certainly to have entered long ago via Timor, since *Gomphinae* (other than *Ictinus*) do not occur in Papua. The presence of *Diplacodes nebulosa*, *Orthetrum pruinosum*, *Zyxomma petiolatum* and other species on the mainland points in the same direction. The following genera have penetrated into Papua and across Torres Straits into Queensland: *Ictinus, Gynacantha, Heliaeschna, Tetrathemis, Lyriothemis, Potamarcha, Orthetrum, Raphismia, Aethriamanta.* The following genera appear to be well enough established in both the Oriental region and the tropical part of the Australian to be considered as dientogenic: *Anaciaeschna, Macromia, Metaphya, Nannophlebia, Diplacina, Nesoxenia, Lathrecista, Agrionoptera, Brachydiplax, Neurothemis, Zygonyx, Hydrobasileus, Macrodiplax, Zyxomma, Camacinia* and *Rhyothemis.* They are, however, all *ectogenic* for the Australian Continent itself. This shews us how very different the faunas of Papua and Australia proper really are.

Aeschna brevistyla is common in temperate Australia, especially in Tasmania. It also occurs in New Zealand. It seems to be a remnant of an old Antarctic fauna. Species referable to the Holarctic genus *Somatochlora* have been described from New Zealand and Chili; and the presence of the archaic *Austropetalia* on the Blue Mountains may perhaps be a third link in this slender chain of evidence.

Census of the Odonata of the World.

The census here given is based upon Kirby's Catalogue (1890) and upon the Zoological Records up to May, 1913. For the Ethiopian and Australian regions, the records are complete to the end of 1914. For the Nearctic region, Muttkowski's Catalogue (1910) was made the basis. In the case of the *Libellulinae, Corduliinae* and *Aeschninae,* the census has been rearranged on the basis of the Selysian Monographs on these groups by Ris and Martin. By the introduction of the trinomial system in the *Libellulinae* the number of species has been very considerably reduced. A similar result may be expected in the *Agrionidae,* and perhaps also in the *Calopterygidae,* when a similar modern treatment is accorded to those subfamilies.

The numbers in brackets placed after the generic totals indicate the number of genera peculiar to, or entogenic in, the given region, but does not include genera entogenic in more than one region. The "Regional Percentage" gives a fairly reliable measure of the "self-containedness" of each region. It is calculated from the ratio of genera in brackets to the total.

Subfamily	Neotropical Genera	Neotropical Sp.	Nearctic Genera	Nearctic Sp.	Palaearctic Genera	Palaearctic Sp.	Ethiopian Genera	Ethiopian Sp.	Oriental Genera	Oriental Sp.	Australian Genera	Australian Sp.	World[1] Genera	World[1] Sp.
Suborder ANISOPTERA														
Family Aeschnidae														
Gomphinae	13 (12)	79	11 (5)	63	7 (1)	42	19 (17)	43	12 (10)	85	3 (2)	21	52	325
Chlorogomphinae	—	—	—	—	—	—	—	—	2 (2)	7	—	—	2	7
Petalurinae	1 (1)	1	1 (0)	2	1 (0)	1	—	—	—	—	2 (2)	4	4	8
Cordulegastrinae	—	—	1 (0)	6	2 (0)	10	—	—	2 (1)	5	—	—	3	21
Aeschninae	16 (13)	62	10 (5)	22	9 (3)	30	6 (0)	29	12 (5)	52	13 (8)	41	43	208
Family Libellulidae														
Corduliinae	5 (4)	18	11 (8)	47	6 (2)	19	7 (5)	25	7 (4)	24	16 (12)	52	40	185
Libellulinae	42 (31)	202	20 (4)	72	16 (2)	55	41 (28)	148	46 (16)	108	38 (11)	78	118	558
Total Anisoptera	77 (61)	362	44 (22)	212	41 (8)	157	73 (50)	245	81 (38)	281	72 (35)	196	262	1312
Suborder ZYGOPTERA														
Family Calopterygidae														
Epallaginae	5 (5)	7	—	—	1 (1)	2	1 (1)	20	12 (12)	107	2 (1)	6	20	141
Thorinae	4 (4)	40	—	—	—	—	—	—	—	—	—	—	4	40
Calopteryginae	3 (1)	53	2 (0)	17	2 (0)	21	4 (4)	11	9 (8)	19	1 (0)	1	16	120
Family Lestidae														
Epiophlebiinae	—	—	—	—	1 (1)	1	—	—	—	—	—	—	1	1
Lestinae	3 (1)	22	2 (0)	15	2 (1)	11	1 (0)	22	4 (3)	23	2 (1)	13	8	103
Synlestinae	—	—	—	—	—	—	1 (1)	5	—	—	1 (1)	2	2	7
Family Agrionidae														
Megapodagrioninae	11 (11)	42	—	—	1 (0)	2	4 (4)	6	7 (6)	13	5 (5)	21	26	84
Pseudostigmatinae	5 (5)	20	—	—	—	—	—	—	—	—	—	—	5	20
Platycneminae	—	—	—	—	2 (1)	12	8 (7)	20	6 (5)	37	1 (1)	3	15	72
Protoneurinae	6 (6)	36	—	—	—	—	1 (0)	14	5 (4)	53	8 (7)	24	18	120
Agrioninae	21 (17)	165	11 (2)	60	9 (2)	50	12 (3)	52	12 (6)	62	18 (10)	38	52	437
Total Zygoptera	58 (50)	385	15 (2)	92	18 (6)	99	32 (20)	150	55 (44)	314	38 (26)	108	167	1145
Total Odonata	135 (111)	747	59 (24)	304	59 (14)	256	105 (70)	395	136 (82)	595	110 (61)	304	429	2457
"Regional Percentage"	83		41		24		67		60		55			

[1] The Polynesian totals are included in this column.

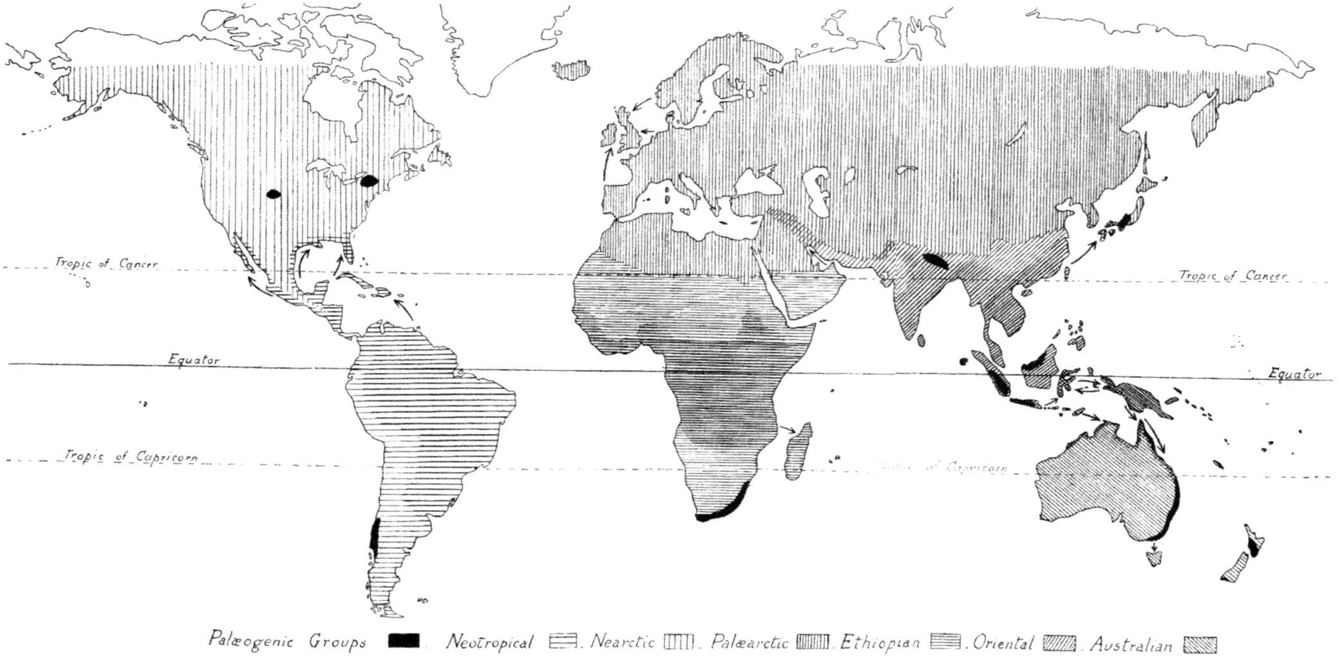
Tropic of Cancer
Equator
Tropic of Capricorn
Tropic of Cancer
Equator
Palæogenic Groups . Neotropical . Nearctic . Palæarctic . Ethiopian . Oriental . Australian

CHAPTER XVI

THE GEOLOGICAL RECORD

It is not to be expected, either from their structure or their habits, that the Insecta could ever have yielded a fossil record commensurate with their great numerical preponderance in the Animal Kingdom. Scarcely any are marine, and comparatively few are aquatic. Consequently, the record is only a partial one. In the case of fresh-water deposits, it embraced only those groups that were wholly or partly aquatic, or others that lived and died amongst dense vegetation, which later on went to form plant-bearing beds, such as our coal-measures. Waifs and strays from land-living groups were also often washed down and embedded in lacustrine deposits, and even in shallow marine beds formed from fine mud derived from adjacent land-areas.

Undoubtedly the most complete record furnished by any group of insects is that of the Cockroaches. No aquatic group furnishes us with a record which can in any way approach that of these dwellers in damp and rotting vegetation. But, poor as it must be judged by comparison, the record of the Dragonflies is more complete than that of any other aquatic group. The actual number of fossil specimens is small; but the ancestry thereby made known to us is a very long one, and very fairly complete.

Chitin, which forms the hard parts of all insects, is very resistant to strong reagents, such as acids and alkalies. But it is nevertheless slowly dissolved by water. Hence it is not found in the fossil state, except in those cases where it has been protected from the action of this liquid, as in the case of the insects enclosed in Baltic amber. Either it becomes impregnated with, and finally replaced by, another substance, or else it simply dissolves away, leaving only an impression of the living organism. Most insect fossils are of this latter character, being usually impressions of wings or parts of wings, very seldom of the whole insect. This fact should make a classification based chiefly on the wing-venation, as in the case of the Odonata, more fully appreciated. In the chain of phylogenetic evidence, the value of a single impression of a complexly-veined Odonate wing may easily outweigh that of a large number of impressions from a group such as the Coleoptera.

Fossil insect remains have been reported to occur as far back as Silurian times. We are, however, on safe ground if we follow Handlirsch in rejecting all those of this Age so far brought to light. The Devonian strata have so far yielded none, nor have the Lower Carboniferous beds. But, in the Upper Carboniferous rocks, we suddenly come upon the remains of a wild riot of teeming insect life. As it first discloses itself to our view, the geological record shews us the Insecta in many ways almost at the maximum of their development, particularly in regard to size. Unless, then, we postulate a rise of startling rapidity, we must admit that our record does not reach back anywhere near the origin of the Class, but rather starts near the middle, at what we may call the Giant Age of Insects. Those who would study interesting theories as to the origin of insects are referred to the hypotheses of Lameere[1] and Handlirsch [67]. In this book, we shall give a review of the various stages in the evolution of the Odonata, as far as our record enables us to do so, together with a short account of the conditions which probably prevailed in the four great Odonate-bearing beds from which our information is chiefly derived, viz. Commentry, Solenhofen, Florissant and Oeningen.

The Palaeozoic Record.

Our record opens, then, in the rich coal-measures of Commentry, near St Etienne in France, where the labours of Brongniart [17] and Meunier [93] have exposed for us a wonderful collection of fossil insects of bewildering variety. These are characterized by two points—the large average size of the insects, and the density of their wing-reticulation. Even the Cockroaches, which are the first to be recognized as approximating to a present-day type,

[1] *Manuel de la faune de Belgique*, Tome II, A. Lameere, Brussels, 1900.

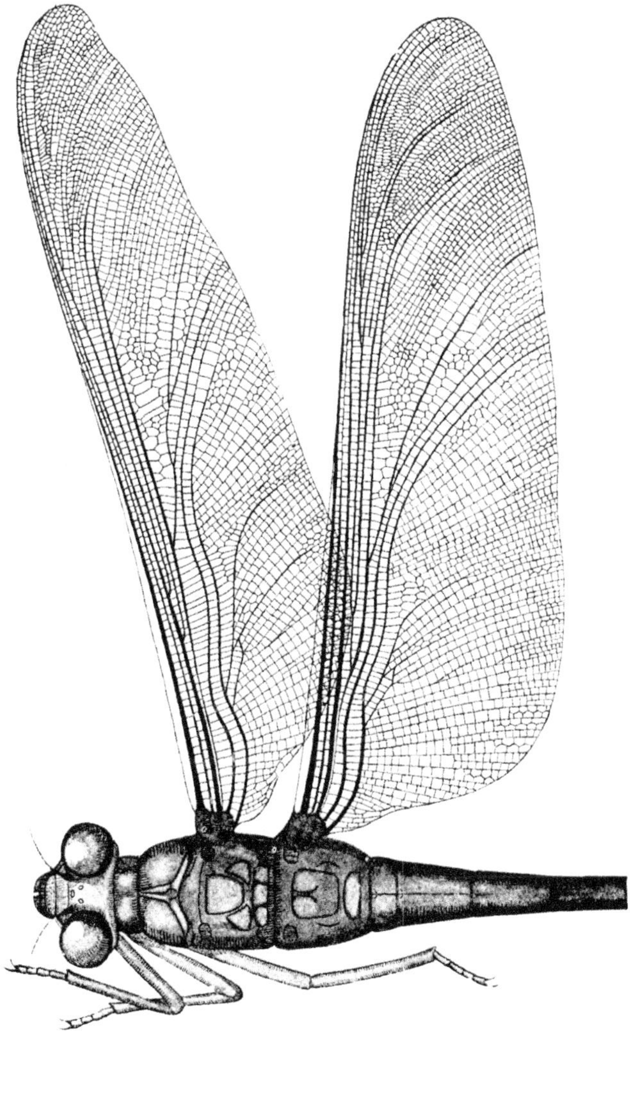

Fig. 156. *Meganeura monyi* Brongniart. Upper Carboniferous, Commentry. ($\times \frac{2}{5}$.) Restoration, altered from Brongniart and Handlirsch.

had a very rich venation, and a size that would be considered embarrassingly large in our modern world. Of the others, it may be briefly said that a general survey shews that none of them can be strictly placed in existing Orders, although the Orthoptera, Plectoptera, Hemiptera and Odonata are almost certainly direct descendants of forms similar to those found at Commentry. Most of these insects, however, are placed in a separate Order, the Palaeodictyoptera. From these, probably, side-branches broke away, to give origin sooner or later to our existing Orders, while the main mass died out.

One of these side-branches, undoubtedly allied to the Palaeodictyoptera, but just as undoubtedly allied to the Odonata of to-day, was the Protodonata. Only some seven or eight species are known, comprised in five genera and three families. But the fossils are so well preserved, and have become such a centre of interest, that we cannot underestimate their value towards a correct view of the origin of the Odonata. Probably the best known, and certainly the most wonderful, of all the Commentry fossils is *Meganeura monyi* (fig. 156), the gigantic Dragonfly whose restoration was first attempted by Brongniart. This magnificent insect had a wing-expanse of 27 inches, thus far exceeding the largest insects at present existing. It was undoubtedly a Dragonfly, and would fall within the Order Odonata in its general form of body and wing, and in the definition of its mouth-parts. But it had four-jointed tarsi, and its wing-venation lacked those well-known specializations characteristic of our present-day Odonata, the pterostigma, nodus, and the quadrilateral or triangle. Unfortunately, in Brongniart's time little was known of the true significance of much of the details of venation as we view them to-day. Handlirsch [67] has already corrected Brongniart's restoration, as far as it fails to agree with the obvious design of the preserved parts of the fossil wing. In fig. 156, I offer a further restoration, including the probable design of the head and thorax, based on a study of the allied *Meganeurula selysi*. This latter species differs from *Meganeura monyi*, and agrees with our recent Zygoptera, in possessing an unbranched radius in all four wings. In *M. monyi*, the radius is unbranched in the hind-wing, but in the fore-wing it has two large branches arising rather close together. This condition is unparalleled amongst recent forms. The other interesting points about the wings of this insect are:

(1) The presence of an unveined membranous precostal area, not found in any recent Dragonfly, (2) the passage of *Sc* almost to the extreme tip of the wing, (3) the slight separation of *M* basally as a fine vein just below *R*, foreshadowing the complete fusion seen in recent forms, (4) an undulation of *Cu* and *A* near the base, (5) the regular arrangement of the interpolated sectors and cross-veins.

In *Meganeurula selysi*, there is a further marked advance towards the true Odonate type by the fusion of *Sc* with the anterior border of the wing, at a point less than half the total distance from base to tip. This foreshadows the formation of a true nodus. The radius, too, runs into the same margin at about the region of the more modern pterostigma, thus suggesting the origin of that organ as a very long and narrow membranous thickening between *C* and *R* at this position. One of the most archaic of surviving Anisopterid subfamilies, the *Petalurinae* (fig. 18 B), has a pterostigma not far removed from this form.

The recent discovery by Bolton [11] of portion of a *Meganeurid* wing (*M. radstockensis*) in the Forest of Dean coal-measures in England, throws further light on the genus. Only parts of the basal half are preserved, but we can note two new peculiarities. The precostal area is absent, but the costa itself is much thickened basally, and has a coriaceous border. A remnant of this condition still obtains in *Petalura*. Further distad, the costa is ribbed in the regular manner seen in recent Dragonflies. More interesting still, the slanting cross-vein *Ac*, recently shewn by me [176] to indicate the point where the true anal vein diverges from *Cu*, is clearly to be seen. Hence it would appear that *Cu* and *A* were fused basally, as in all recent forms. Thus the gap between Protodonata and Odonata is being gradually lessened, until to-day we may almost certainly see in the *Meganeuridae* the giant relatives of the direct ancestors of some at least of our recent families.

Only two other families of Protodonata are known, each represented by a single genus and species. The *Paralogidae* (*Paralogus aeschnoïdes* Scudder) had a rather short and broad wing, with *Sc* ending about half-way, and $R + M$ completely fused basally. The *Protagrionidae* (*Protagrion audouini* Brongniart) had *R* and *M* separate, with a kind of arculus joining them some way from the base. Such an arrangement places them as a side-branch of the Order, which left no descendants even amongst Mesozoic forms.

An interesting fact in connection with Commentry is the non-occurrence of larval forms. Without unduly stressing this fact, we may couple it with the immense abundance of imaginal forms which seem to have been suited to a life amongst damp vegetation. Then we may well ask ourselves whether the larvae of the Protodonata did not dwell in damp earth rather than in water. The formation of the larval tracheal system (chap. IX) undoubtedly proves that this was at one time the case. It may well have been so in Carboniferous times.

The insect-beds of Commentry occupy a comparatively small portion of the total strata. They belong to the highest division (Upper Zone) of the highest or Stephanian stage of the Upper Carboniferous rocks of that region. Thus they really form a passage-bed between the Carboniferous proper and the Permian. The principal plant remains of this zone are the Giant Mare's-tail (*Calamites*), characteristic of the coal-measures, and the large Cycad *Zamites*, most abundant in Jurassic strata. Had the insect-beds been isolated, instead of occurring at the top of a long series of undoubted Carboniferous rocks, such a flora would have placed them in the Permian. Emphasis is laid on this fact, because the only known Permian Dragonflies are *Meganeurids*. One of these, *Ephemerites rückerti* Geinitz, is from the Lower Permian of Germany, where a similar flora prevailed. The other, *Tupus permianus* Sellards [150], is from the Permian of Kansas. It is of interest in possessing a branched radius in all four wings, with signs of a weak bridge and an oblique vein.

We may picture to ourselves the giant insects of Commentry as inhabiting the shores of a large, shallow, nearly stagnant lake. In the muddy ooze around its borders grew forests of the Giant Mare's-tail; while, further back, on the sandy slopes, the graceful Cycads and other extraordinary plants formed a more diversified medley. There, amidst rotting vegetation, these insects lived and bred. In such almost amphibious conditions, it may well be that the larvae of Protephemeroidea and Protodonata first began that series of adaptive changes which finally led them to adopt a purely aquatic mode of life.

The Mesozoic Record.

A. *The Liassic Dragonflies.*

With the close of the Palaeozoic Age, the Protodonata pass from our ken, and the Giant Age of Insects gives place to a period of great activity, in which reduction in size begins to work hand in hand with specialization in structure and function. The record of the Trias proper is very poor in insects, and no Odonata are known from it. But, in the Lias, the curtain is once again raised. We see before us a Transition Period, in which the giants, already considerably reduced in size, are either dying out, or evolving to higher forms.

Three Liassic beds yield fossil Dragonflies. These are the beds of Cheltenham and Dumbleton in Britain, and Dobbertin in Mecklenburg. The first belongs to the Lower, the other two to the Upper Lias. The fauna of all three is closely similar, and is remarkable for the occurrence of an undoubted *Gomphine*, *Gomphoïdes brodiei* Buckmann. This establishes the first connection with present-day Anisoptera. For the rest, putting aside a number of incomplete wings which were very probably also Anisopterous, but in which the base of the wing is missing, there remains that remarkable Dragonfly fauna called by Handlirsch the Anisozygoptera. These consist of three separate subfamilies, the *Archithemimae*, *Tarsophlebiinae* and *Heterophlebiinae*. Judged by the condition of the radius, which is unbranched, and by the absence of any true triangle-formation, these all seem to be undoubtedly Zygoptera of an early unreduced form. As such I propose to treat them, placing them as subfamilies of the *Calopterygidae* and *Lestidae*.

The *Architheminae*[1] (fig. 157) had a complete arculus, but suffered from weakness in the region of the quadrilateral, which seems to have tended to become merely an area of undifferentiated

[1] =*Diastatommidae* Handlirsch. *Diastatomma* Giebel is preoccupied by *Diastatomma* Brauer, a genus of recent *Gomphinae*.

polygonal cells. Hence they probably soon died out, from weakness of flight, and left no direct descendants.

The *Tarsophlebiinae* (fig. 158) had an incomplete arculus. Consequently the quadrilateral was open basally, as is the case in the hind-wing of the living *Hemiphlebia* (fig. 22). They were undoubtedly *Calopterygidae*, and were chiefly remarkable for their very long slender legs, with elongated four-jointed tarsi. Probably they used their legs for both walking and climbing.

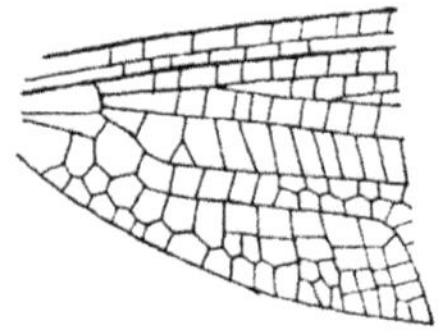

Fig. 157. Basal portion of wing of *Archithemis brodiei* Geinitz (× 2·6), Lias, Dobbertin. After Handlirsch.

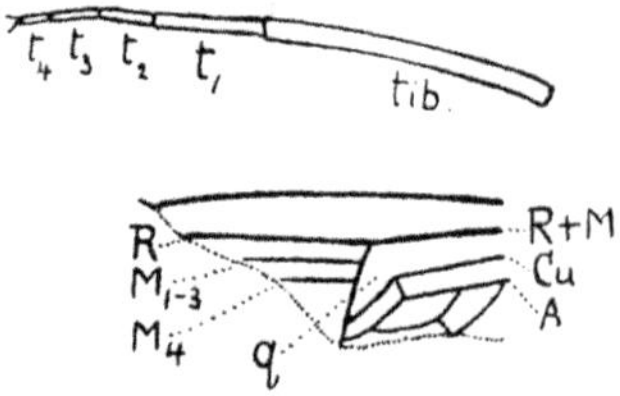

Fig. 158. Portions of leg and wing of *Tarsophlebia westwoodi* Giebel (× 1), Lias, Cheltenham. *q* quadrilateral; t_1–t_4, the four joints of the tarsus; *tib* tibia. After Hagen.

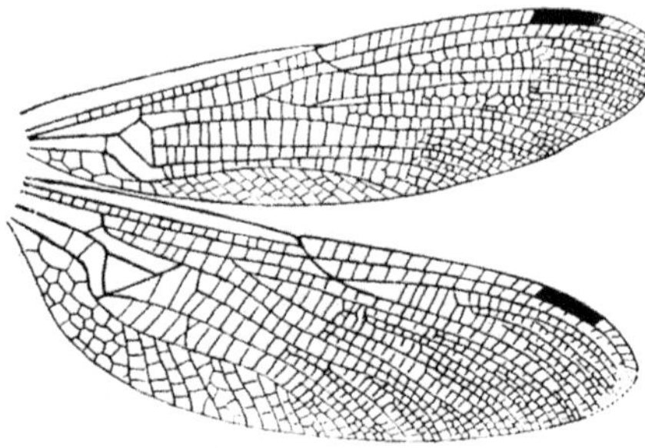

Fig. 159. Wings of *Heterophlebia dislocata* Brodie and Westwood (× 2), Lias, Cheltenham. After Westwood.

The *Heterophlebiinae* (fig. 159) were fairly closely related to the living *Epiophlebia* of Japan. The fore-wing quadrilateral resembled that of the latter genus, but was considerably wider. The hind-wing had a peculiarly stretched and distorted quadrilateral, with a horizontal cross-vein separating off a lower triangular portion. This formation was essentially unstable, since the upper side of the "triangle" was not continued by any main vein, as it is in the true Anisoptera. The most important point to notice in this genus is the undoubted presence of the oblique vein and elongated bridge of the *Lestidae*. Further, since this bridge-vein is continued back to join M_3 instead of M_{1-2}, we must place the genus in close relationship with the *Synlestinae*. Its relationship to this latter subfamily seems to be closely similar to that which *Epiophlebia* bears to the *Lestinae*.

The only other Liassic genus, *Protomyrmeleon*, is an extraordinary aberrant Zygopterid, with *Sc* reduced to a mere remnant. It seems to have had no near allies, and certainly left no descendants. Probably it was the last of some Protodonate side-branch.

It is perhaps fitting to record here the recent discovery of a beautiful Dragonfly wing from the Ipswich coal-measures of Queensland (Trias-Jura). I have named it *Mesophlebia antinodalis* (fig. 160), placing it in a new subfamily *Mesophlebiinae*, probably belonging to the Anisoptera. Only the distal two-thirds

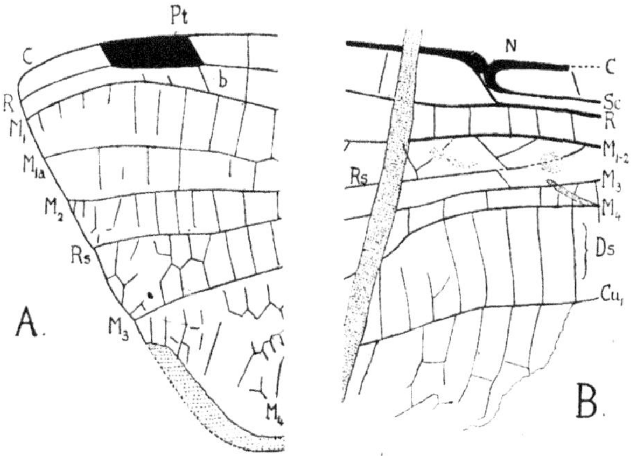

Fig. 160. Portion of hind-wing of *Mesophlebia antinodalis* Tillyard (× 3·3), Trias-Jura, Ipswich, Queensland. A. Tip of wing. B. Region of nodus. Original.

of the wing is preserved, but it shews a beautiful pterostigma, and a very distinct nodus of remarkable form. There is also a great reduction in the length of the wing compared with its breadth. The row of elongated post-trigonal cells, *Ds*, resembles that of *Epiophlebia*.

B. *The Jurassic Dragonflies.*

The record of the Upper Jurassic beds is not only very complete, but is the most valuable chapter of Odonate phylogeny that has yet been revealed to us. From the Dogger of Ust Balei, Siberia, we have a series of records of Dragonfly larvae, the *Samarurinae*. These had no lateral abdominal gills, but three short oval caudal gills. They may be provisionally classed as *Calopterygidae*. In Britain, the Purbeck beds, and the Stonesfield Slates of Oxford, supply a few interesting fossils. But it is to the splendid mass of material derived from the lithographic slates of Solenhofen and Eichstatt in Bavaria that we owe our chief knowledge of the Dragonflies of this horizon [60].

These beds are probably best known because of the discovery there of the wonderful fossil *Archaeopteryx*, which so beautifully links the Birds with the Reptiles. Thus, as students of the Odonata, we gaze for the last time upon the remnants of the former giant race, the Kings of the Air, whose domain is now about to be wrested from them by the Birds. From henceforth they are condemned to follow the difficult path of high specialization either for rapid flight or for concealment, if they are to escape destruction at the hands of their more powerful enemies.

A remnant of the Palaeozoic giants is found in the wonderful subfamily *Isophlebiinae*, of which the best known species is *Isophlebia aspasia* Hagen (fig. 161). This beautiful insect had a slender body six inches long, while the expanse of its wings was about eight inches. The wings were long and slender, and evidently of *Calopterygid* design, except for the peculiar structure of the anal area, particularly noticeable in the hind-wing. Cu_2 descends sharply from the quadrilateral, and gives off a proximal branch almost meeting the strongly curved anal vein at its ending on

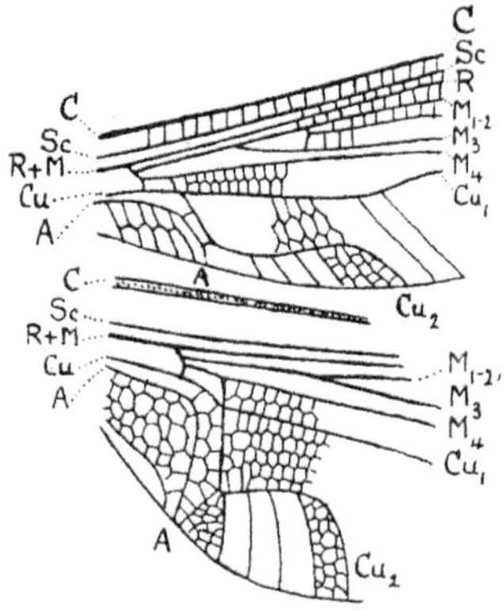

Fig. 161. Portion of bases of wings of *Isophlebia aspasia* Hagen (×1), Upper Jurassic, Solenhofen. After Hagen.

the wing-margin. The result is to broaden the hind-wing basally. Even so, the wing is only of moderate width when its great length is taken into account.

In the *Stenophlebiinae* we meet with another group of large Dragonflies, ranging from five to seven inches in expanse, and having excessively slender wings. The quadrilateral in this group had a sharply acute distal angle. A short cross-vein divided it into a weak triangle and super-triangle. The radius, however, was unbranched, and in other respects these insects exhibited distinct *Calopterygid* structure.

Three species of *Tarsophlebia* give us a close connection with the Lias.

Zygoptera belonging to recent subfamilies are for the first time represented, by two genera of *Epallaginae*, *Euphaeopsis* and *Pseudoeuphaea*. The former (fig. 162) had the costal margin

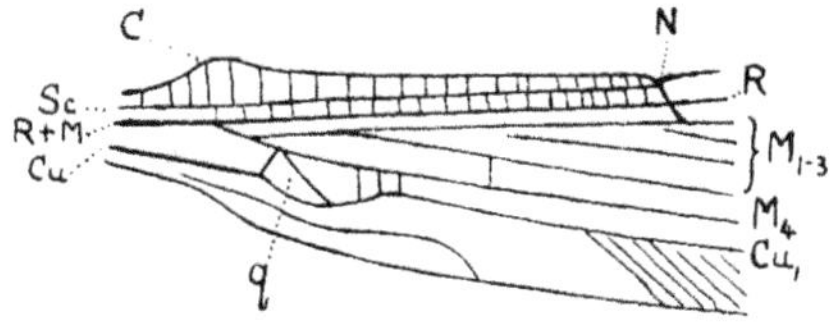

Fig. 162. Part of hind-wing of *Euphaeopsis multinervis* Hagen (×2), Upper Jurassic, Solenhofen. *q* quadrilateral. After Hagen.

raised near the base into an obtuse-angled swelling, recalling the similar formation in *Cora*. The genus *Steleopteron* appears to have had *Lestine* affinities; but the nodus was weakly formed, and there were five antenodals.

In the Anisoptera, we find a considerable assemblage of forms which can be placed definitely in the *Gomphinae*, *Cordulegastrinae* and *Petalurinae*. The triangular region of *Mesuropetala koehleri* Hagen closely resembles that of *Petalura*. Except for greater density of venation, *Cymatophlebia* comes very close to *Cordulegaster*. The *Gomphinae* are represented by *Nannogomphus*, while *Protolindenia* seems only doubtfully referable to the same subfamily. The most interesting of all the Anisoptera represented at Solenhofen are undoubtedly the *Aeschnidiinae*, characterized by the density of their wing-reticulation, and by having the triangles elongated transversely to the wing-axis in both wings. Of these, the best preserved is *Aeschnidium densum* (fig. 163), remarkable for the width of the wings, and the immense number of cross-veins and tiny areolets filling up the spaces between the main veins and their branches. Between *C* and *Sc* prenodally, and again between *C* and *R* postnodally, these areolets form two or more rows. The boundary between each pair of rows tends to become a straight line, thus forming a slender longitudinal vein parallel to the main veins. By this means there arose an apparent prolongation of *Sc* beyond the nodus, comparable with a similar formation in the recent genera *Telephlebia* (fig. 19 D) and *Staurophlebia*, but more extensive. The *Aeschnidiinae* seem to have died out in Cretaceous times: unless, indeed, the recent *Chlorogomphus* (fig. 119) be a close ally or a descendant of them.

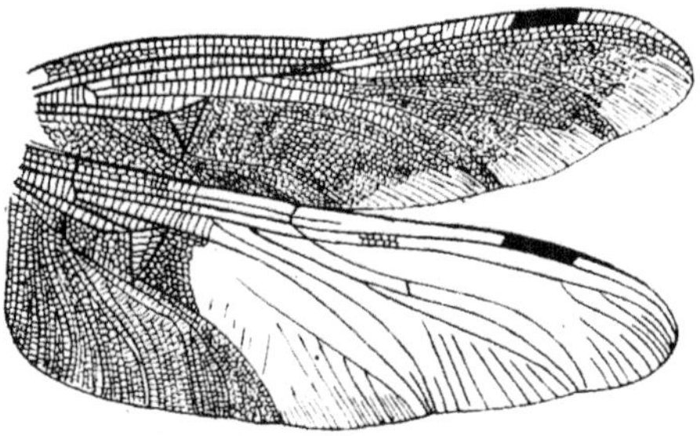

Fig. 163. Wings of *Aeschnidium densum* Hagen (×1·3), Upper Jurassic, Solenhofen. From Handlirsch, after Deichmüller.

There are two outstanding features of the Solenhofen beds. These are, firstly, the extreme fineness of the grain of the stone, and, secondly, the great scarcity of fossils. With regard to the first, the Solenhofen slate was undoubtedly formed from a very fine sediment deposited in shallow *marine* bays and inlets, and derived from the adjacent land. In such places, Crustacea, Fishes and Reptiles lived together, and their remains are beautifully preserved for us in the fine-grained slates. Insects do not strictly represent a part of the true fauna of these bays. But a number of insects, including Dragonflies, were carried thither, probably by floods, from the neighbouring streams. Hence their remains are mingled with those of the true fauna. With regard to the second feature,—the paucity of fossils—it must be clearly grasped that these beds are in reality exceedingly poor in specimens. It would be impossible for anyone, visiting the quarries, to pick up more than one or two fossils, even though he searched assiduously for days. It is only because of the commercial value of the slate, which is in great demand by lithographers, and the consequent opening up of immense quarries extending for many miles, from which millions of tons of stone have been stripped, that we can reckon the *total* of Solenhofen fossils as a great one, and, in the case of the Odonata, exceeding in number and variety the total from any other beds known to us. Every fossil has been carefully preserved by the workmen, who obtain high prices for them, chiefly because of the exquisite outline of the impressions in the fine-grained slate.

The colour of the stone is for the most part pale ochreous, and so finely grained as to appear artificially smoothed. At first sight, one of the Solenhofen Dragonflies might well be taken for a "fake," so exquisitely is the outline of body and wing preserved. It is only when one comes to study the details of venation that the disadvantages of this fine grain become obvious, many of the finer cross-veins being found to be almost or quite obliterated. This detracts much from the value of the fossils for detailed study. Their charm lies chiefly in the perfection of outline and of the main venational plan.

In concluding this account of Solenhofen, it may be remarked that we cannot expect these beds to yield much further information in the future. They have been very extensively worked, and it seems most unlikely that any more new forms will be found. We may, however, still hope for the discovery of better preserved specimens of some of the more interesting genera, from which to gain further information of their finer venational characters.

C. *The Cretaceous Dragonflies.*

The only Cretaceous fossil Dragonfly known is *Aeschnidium flindersiense* Woodward. It was found, in company with the fresh-water bivalve *Aucella*, in the chocolate-coloured limestone of the Flinders River beds in Queensland. It shews us that the *Aeschnidiinae* were once widely spread, and perhaps lingered on in Australia after their extinction elsewhere.

The Tertiary Record.

This, as might be expected, is fairly extensive. Fossil Dragonflies are known from the following beds:

In the Eocene, the Monte Bolca beds of Italy, and the Green River beds of Wyoming, U.S.A. In the Oligocene, the beds of Roan Mountain, Colorado, U.S.A. (where the first fossil *Corduline*, *Stenogomphus carletoni* Scudder, was found), from Sieblos in Bavaria, Schlossnitz in Schleswig, Covent and Aix in France, and in Baltic amber. In the Miocene, the Radoboj beds, the beds of Gurnet Bay, Isle of Wight, of Falkenau in Bavaria, of Florissant in Colorado, and of Oeningen in Switzerland. It is in the two latter beds only that Odonata occur at all abundantly. Hence we may profitably confine our attention to them alone, in this chapter.

A. *The Florissant Beds.*

The chief importance of Florissant lies in the interesting evidence which it affords us of the true descent of the *Agrionidae*. The deposit is only a small one, of Miocene Age and lacustrine origin. It is situated a few miles west of Pike's Peak, in Central Colorado. Its present altitude is about 8000 feet. The lake was about fourteen miles long, but narrow. Numerous volcanoes existed in the neighbourhood. The deposit was formed of fine volcanic dust, which was either washed down or fell into the lake, until it was finally silted up. Insect remains are numerous, in conjunction with the leaves of many interesting plants, and the skeletons of fresh-water fishes. Probably most of the insects were killed during eruptions, and were then washed down into the lake by storms and floods. Thus they became buried in the silt, which has now become a fine-grained soft shale of a greyish or buff colour.

The flora of Florissant was of a warm temperate type. Palmetta and Sequoia grew side by side with poplars and pines, while many well-known Nearctic shrubs and plants seem to have been common. Cockerell [45] claims that the Proteaceae were represented there. But his figures of *Lomatia* leaves are quite unconvincing to anyone who knows the Australian flora. As regards the Insecta, the most abundant Order was the Hymenoptera, particularly bees and ants.

Many beetles and flies occur; also the larvae of Plectoptera, *Chironomus* and other aquatic insects. Dragonflies are not common, and not all of them well-preserved.

The question of altitude is a very interesting one. At the present day, no *Libellulinae*, *Gomphinae* or *Calopterygidae* reach a high elevation in a climate like that which Florissant possessed. A typical high-altitude Odonate fauna would consist of *Corduliinae*, *Aeschninae* and *Agrionidae.* The two latter groups only have so far been found at Florissant. We may therefore conclude that the lake lay at a considerable altitude in Miocene times, even as its remains do to-day.

The Odonata of Florissant have been chiefly collected and described by Scudder [148, 149] and Cockerell [43–47]. The *Aeschninae* all appear to belong to the older genera of the tribe *Aeschnini*, such as *Oplonaeschna*, having an unbranched *Rs* and a long pterostigma.

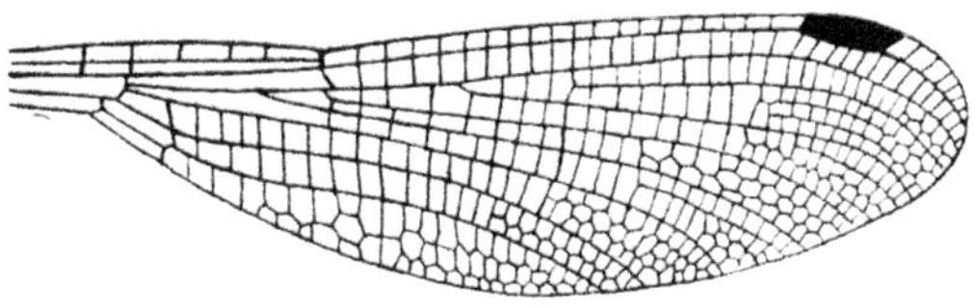

Fig. 164. Fore-wing of *Phenacolestes mirandus* Scudder (× 4), Miocene, Florissant. Original restoration from nearly complete wings figured by Calvert.

The *Agrionidae* are of greater interest, since it is here that we find positive evidence of the former existence of a stage in the evolution of that family, in which the reduction of the prenodal area was still incomplete, and the number of antenodals still greater than two. Here we refer particularly to the beautiful and interesting genus *Phenacolestes* (fig. 164), of which two species, *P. mirandus* Scudder, and *P. parallelus* Cockerell, have been found, both in a very good state of preservation. One specimen of the latter is almost complete, and lies with wings outspread. This suggests that *Phenacolestes* used to rest with outspread wings—a habit still preserved in the allied Australian genera *Argiolestes* and *Podopteryx*, but long since lost in the great majority of *Agrionidae.* The venation is typically *Megapodagrionine*, except for the extra antenodals, and the greater distance of the nodus from the wing-base. The Neotropical genus *Thaumatoneura* is another *Megapodagrionine* remnant which has stopped at about the stage exhibited by *Phenacolestes* in the Miocene; but it is not so closely related to the latter as are some existing genera with only two antenodals. Calvert [33] has recently discussed *Phenacolestes* very fully in an excellent paper, to which the reader is referred for further information.

Cockerell [44] placed *Dysagrion*, from the Oligocene of the Green River beds, together with *Phenacolestes*, in a new subfamily *Dysagrioninae*. It is clear that this subfamily cannot stand, but must be merged in the *Megapodagrioninae*. The restored wing of *Phenacolestes* makes this quite evident.

Another interesting *Agrionid*, with two antenodals, and a general close resemblance to *Argia*, was described by Scudder as *Trichocnemis aliena*. The generic placing is here evidently wrong. An *Agrionid* larva with caudal gills complete has also been found. Scudder described it as *Agrion telluris*. But the remarkable banded pattern of the legs is very marked in the *Megapodagrioninae*, to which subfamily this larva probably belonged.

B. *The Oeningen Beds.*

The Oeningen beds are situated in the Rhine Valley, between the towns of Constance and Schaffhausen in Switzerland. They consist of a series of marls, sandstones and limestones, some being marine and some fresh-water. Beds of brown coal are interpolated here and there. It is the uppermost strata which contain the insects, in company with the remains of fresh-water fishes, a few Crustacea, reptiles, birds and mammals, a number of Arachnida, the very common bivalve *Anodonta*, and an abundant flora shewing distinct Nearctic affinities. The complete formation covers the whole of the flat lands of Switzerland, some 150 square miles, and is of considerable thickness. The uppermost strata are known as the Upper Fresh-water Molasse, and are of late or Upper Miocene age.

More than three hundred species of insects are known from these beds, which have evidently been formed by slow accumulation of sediments in a lake fed by springs of water strongly impregnated with carbonate of lime. The region was volcanic; hot springs and geysers probably occurred. Thus we have at once a comparison and a contrast with Florissant. For, on the one hand, the general conditions were somewhat similar; but, on the other hand, the altitude was very much less. In a low-level lake at the present day, we should expect to find an Odonate fauna chiefly consisting of *Libellulinae* and *Agrionidae*, with some *Aeschninae*, while a few *Calopterygidae* might visit the lake, though breeding in neighbouring rivers. This is just the kind of fauna revealed to us by the Oeningen beds.

A very extraordinary circumstance is the occurrence at Oeningen of a very large number of *Libellulid* larvae, while the imagines are much less common. In a thin stratum from one of the quarries, larvae in all stages of growth were found jumbled together in perfect swarms. Some had the labial mask projecting as if in the act of striking their prey. The only possible explanation of this phenomenon seems to be that, in the spring of one year, before many of the imagines were on the wing, a sudden catastrophe overtook the lake-bed. Through volcanic agencies, either by an eruption, or a sudden uprush of hot water, or a burst of poisonous gases, these larvae met a sudden and violent end, and died in their natural positions[1].

The Oeningen specimens are not well preserved, the rock in which they are enclosed being rather hard and brittle, and cleaving very unevenly. In such a case, specific and even generic determinations are not to be absolutely relied upon. The fauna seems, however, to have been undoubtedly closely allied to that of Holarctica as we know it at the present day.

General Conclusions.

The Cockroaches are the only group of insects in which the geological record is so complete that a fairly accurate phylogenetic tree can be constructed. The record of the Odonata, as insect records go, must be reckoned a rich and valuable one. Moreover, there are hopes that more careful working of some of the lesser known beds, and perhaps also the discovery of new ones, may yet fill up the wide and lamentable gaps to which we have at

[1] *Die Urwelt der Schweiz*, O. Heer; Zürich, 1879.

present to confess. I would here refer to two disturbing questions that inevitably face us. Firstly, what was the true relationship between the Protodonata, as we take farewell of them in the Permian, and the abundant and varied Mesozoic Odonate fauna which we meet with in the Lias? Somewhere between these two points, in late Permian times may-be, or in the Triassic, the "old order" changed, "yielding place to new." To-day the veil is not yet fully lifted, and we can only partly grasp the nature of the profound changes by which the beautiful Odonate wing, with its nodus, its pterostigma, and, in the Anisoptera, its intricate triangle and radial-sector formations, came into being. For European workers the problem is not to be solved, since the Permian and Trias in that region offer little scope for such discoveries. We may turn with some hope to the rich coal-measures of New South Wales, and to the Mesozoic shales and later coal-measures overlying them in that State and in Queensland. There, bearing in mind that, all along, Australia has probably lagged somewhat behind Palaearctic in the types of her insect fauna, we may still hope to find the solution of this interesting problem.

Secondly, what is the history of the *Libellulinae*, that fascinating dominant group of to-day? Its origin is lost in obscurity. Neither palaeontology, ontogeny, nor comparative morphology has so far been able to enlighten us. The absence of *Libellulinae* from Solenhofen, and again from Florissant, may possibly be only due to the nature of these deposits. We need to discover *low-level lacustrine deposits* for the solution of this problem. We know only of a single generalized species of *Corduline* from the Oligocene. On the whole, it seems likely that the *Libellulinae* are a later Tertiary offshoot of an older *Corduline* stem. They may have arisen in Oligocene or early Miocene times, spread and increased with great rapidity, and thus quickly become dominant over the older groups of Anisoptera.

On the other hand, let us look at the positive results of our record. It tells us for certain that the *Gomphinae*, *Petalurinae*, *Cordulegastrinae* and *Epallaginae* were well established in Jurassic times, and differed little from these groups as we know them to-day, except in a tendency to the possession of useless excess cross-veins. Both ontogeny and comparative morphology agree with the

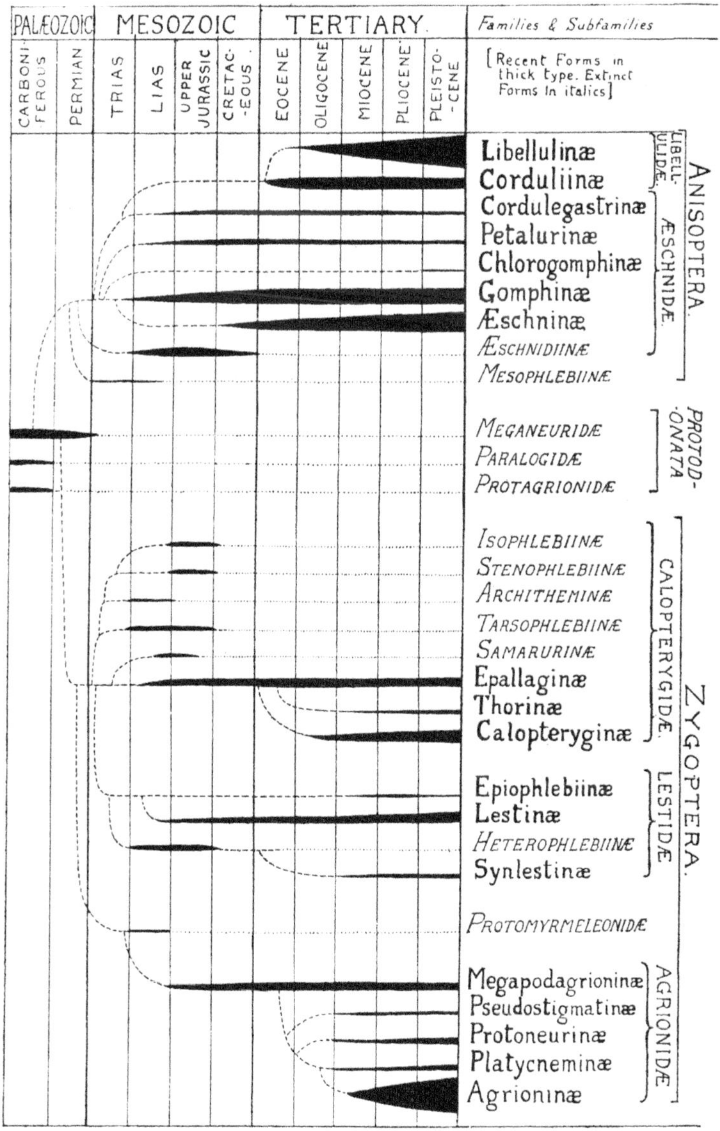

Fig. 165. Phylogenetic diagram of the Odonata, from Carboniferous to recent times. Original.

geological record in placing these groups as the most archaic of present-day Odonata. So that here, at any rate, we are on safe ground.

Again, the value of the Florissant beds for us lies in their undoubted demonstration of the descent of the *Agrionidae* from more complexly veined forms, with a larger prenodal area and less petiolation of the wing-base. As our studies in the ontogeny and morphology of this family point with ever-increasing certainty in the same direction, we may dismiss for ever the old idea that a simple venation was necessarily archaic. We shall thus learn to see, in the *Agrionidae*, the highly specialized and excessively reduced descendants of *Calopterygid*-like ancestors, which—though by the very opposite means, it is true—have attained to a success only comparable with that of the *Libellulinae* of to-day, and which far outnumber all other existing families.

The attached phylogenetic diagram (fig. 165) is an attempt to reconstruct the phylogeny of the Odonata from the available evidence. A *dotted* line indicates that there is no *direct* evidence for the supposed line of descent. Where definite evidence is to hand, the line of descent is represented by a *continuous black line*, whose thickness is made to vary somewhat in proportion to the relative abundance of species in each group. Thus, at the present day, the *Libellulinae* and *Agrioninae* are the two dominant groups, and lie at the two opposite extremes of menogenetic and asthenogenetic specialization respectively. Hence, in the diagram, they not only have the thickest lines of descent, but are placed farthest apart.

Plate IV

Fig. 1.

Fig. 2.

Fig. 1. A mountain swamp near Moss Vale, N.S.W ; elevation 2000 ft. Typical dragonfly fauna: *Petalura, Synthemis, Nannophya, Argiolestes.* The white-trunked trees in the background are tea-trees (*Melaleuca*) growing on the edge of the swamp.

Fig. 2. A rapid mountain creek at Wentworth Falls, Blue Mts, N.S.W.; elevation 2800 ft. Typical dragonfly fauna: *Austrogomphus, Austropetalia, Austroaeschna, Telephlebia, Eusynthemis, Cordulephya, Diphlebia, Synlestes, Argiolestes.* The tree-ferns belong to the genera *Dicksonia* and *Todea.*

(From photographs taken by Mr F. W. Carpenter, M.A.)

CHAPTER XVII

BIONOMICS, ETC.

In this chapter I have collected together not only those facts which are strictly classed as bionomical, but also others which, as they fail to fit in anywhere else, may perhaps be permitted to be included here. The result forms a heterogeneous mass of facts which will, I trust, be found of considerable interest to the student of Odonata.

Habits and Flight.

Although Dragonflies pass their early stages in fresh-water, yet the imagines are by no means confined to the same neighbourhood. It may be said of the females of almost all groups that they quickly remove themselves to cover, and seldom appear flying upon the water except when pairing or ovipositing. A collector who confines himself solely to catching the Dragonflies which are to be seen upon the ponds and rivers will most certainly return home with a very large majority of males. The only exception to this rule is the *Aeschninae*, in which both sexes hawk about with equal vigour. Owing to the superior speed of the males, it is quite usual for more females to be captured. He who would add good series of females of the other groups to his collection must search carefully in out-of-the-way places, clearings in the forest, paths through long grass, low bushes and hedges, and even railway cuttings.

Many groups have their home in the forest. This applies to the great majority of the *Gomphinae*, some of which disappear into the forest immediately after emergence, and are seldom or never seen on the wing. The *Calopterygidae* which have iridescent but uncoloured wings, in the Oriental region, live in the forest and jungle, while those with brilliantly coloured wings fly up and down the rivers.

Most of the *Lestidae* and *Agrionidae* congregate close to the banks of rivers, lakes or ponds. They tend to form "colonies"; i.e. a species may be very abundant on one pond, or along a certain stretch of river, but absent from another near by. Many of these colonies may be due to the phenomenal success of the progeny of a single pair, or of a few pairs, which visited the locality for ovipositing in a previous year. Besides these brightly coloured little gems, the most sun-loving of all the Dragonflies are the *Libellulidae* and the genus *Anax*.

The differences in flight are very great, both in style and in speed. The Zygoptera on the whole use their wings in the manner known as "sculling," their passage over the water close to the surface no doubt suggesting this metaphor. Amongst the *Aeschnidae* in general, a "darting" method of flight is very evident, the insect progressing by a series of jerks. This is varied, in the larger *Aeschninae*, by "hawking," a term which indicates a strong steady flight up and down some particular "beat," with a view to driving off all other intruders, and securing all the food to be found. In the *Libellulidae*, we miss the jerking movement as well as the power of the *Aeschnine*-flight, which is, in most cases, replaced by a "skimming" or "soaring" mode of flight. The *Trameini*, which have the broadest hind-wings of all, float airily about with very little effort. Some of the smaller *Libellulidae* develop a great power of "dodging." The *Sympetrini* foil attempts at capture by short darts forward, sideways, or even backwards behind one's net. *Cordulephya* and the *Tetrathemini* perform zigzag spirals up into the air in the most tantalizing manner. For perfection of evolutions during flight, the *Eucorduliini* must be given first place. Many of these can fly backwards for short distances!

As regards speed, many Dragonflies are notoriously swift. Estimates of the speed of a small object are very apt to go astray. Of the genera known to me, *Macromia* and *Austrophlebia* are by far the swiftest on the wing. As the latter flies regularly up and down small mountain streams, I had an opportunity once of timing it over a measured stretch of between eighty and ninety yards. This distance was covered in three seconds; so that this Dragonfly can fly at the rate of nearly sixty miles per hour! I doubt if any greater speed than this occurs amongst Odonata.

Resting-Positions (fig. 166).

The resting-positions of Dragonflies are well worth studying and recording. Ability to escape from an enemy depends not only on the speed of flight, but also on the method and quickness of the "take-off" from rest. Most Zygoptera choose their resting-position for concealment. When danger threatens, they are as likely to "orientate" themselves, by moving round the

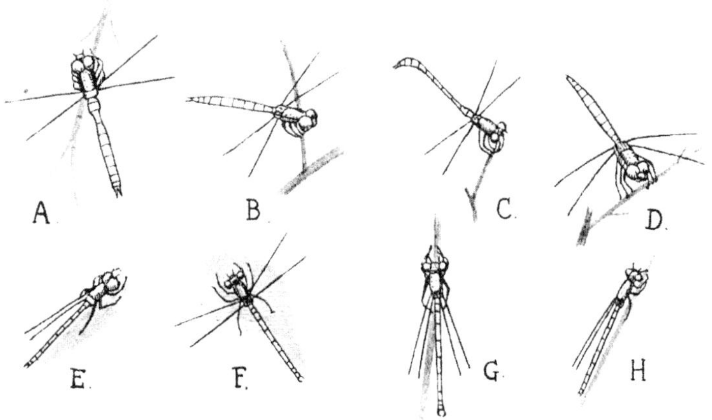

Fig. 166. Typical resting positions of Dragonflies. A. *Aeschninae*. B, D. *Libellulidae*. C. *Gomphinae*. E. *Cordulephya*. F. *Diphlebia* and *Argiolestes*. G. *Lestes*. H. *Agrionidae*. Original.

reed-stem to the side away from the approaching enemy, as they are to fly off. *Diphlebia* and *Argiolestes*, however, resemble the *Gomphinae* in their method of rest. They sit on rocks, sand or twigs, with wings horizontally poised for flight. From such a position they move off very rapidly. The *Aeschninae* alone, of all the groups, regularly adopt a "hanging" position, which is advantageous for concealment, but undoubtedly unsuited for a quick take-off. In the *Libellulinae*, there is a similar habit of rest, on the whole, to that of the *Gomphinae*, but with a tendency

21—2

to depress the wings. Many of the smaller forms are very quick "off the mark," but the larger *Trameini* are very clumsy and easily captured. The most difficult genus to capture, known to me, is *Aethriamanta*, which is always a few feet up in the air (when one strikes at it) above the position one believed it to occupy! *Cordulephya* rests on rocks or tree-trunks with wings closed; but it is very quick at the take-off, and not easy to catch.

The question of protective colouring need not be gone into here; the reader is referred to chapter XIII, from which he can draw his own conclusions. The bright red tip developed on the abdomen of certain very small *Libellulinae* (Plate I, fig. 2) gives these insects a vicious and wasp-like appearance, but is not a case of true mimicry, since no true wasps occurring with them have a similar coloration.

Size (Plate I).

No Dragonfly at present existing can compare with the immense *Meganeura monyi* of the Upper Carboniferous, whose expanse of wing was somewhere about twenty-seven inches. The Liassic and Jurassic Dragonflies were not very markedly larger than those of to-day, except perhaps in the average size of the Zygoptera. This latter suborder has undergone, on the whole, a progressive reduction in size and wing-venation, culminating in the smallest known forms of to-day. On the other hand, the *Pseudostigmatinae* have all the marks of a hypertrophied group, and may be rightly considered to be now much larger than the more *Platycnemine*-like ancestors from which they probably sprang.

As size cannot be measured by any single character, I have selected three points on which we may judge the size of a Dragonfly. These are (i) length of abdomen, (ii) expanse of wings, and (iii) robustness.

(i) *Length of abdomen.* There is nothing to approach the *Pseudostigmatinae* (Pl. I, fig. 8) in this, but the abdomen in this group is excessively slender. The longest abdomens in robust species are found in *Petalura* and *Tetracanthagyna*[1]. The shortest

[1] Dr F. F. Laidlaw (in litt.) has recently informed me of the discovery in Borneo of a species of this genus exceeding in size any Dragonfly yet described. I have, however, no details concerning this.

abdomens in slender species occur in the genus *Agriocnemis*; in robust species, *Nannophya* (Pl. I, fig. 2).

(ii) *Expanse of wing.* We may judge of this by measuring the hind-wing. Here again the long slender wings of the *Pseudostigmatinae* are easily first (Pl. I, fig. 8). In the robust species the same two genera as above lead the way. For the small species, the rather long-bodied *Hemiphlebia* vies with *Agriocnemis* (Pl. I, fig. 9) in shortness of wing.

(iii) *Robustness.* For general robustness of build, but especially for the great thickness and strength of its wing-veins, *Petalura* stands unrivalled. For excessive slenderness of build, one might single out perhaps *Aciagrion*, *Protoneura* and *Caconeura* as the slenderest amongst a large number of delicate species.

These results are shewn in the following table:

Species	Length of abdomen in mm.	Length of hind-wing in mm.	Breadth of abdomen in middle of body	Remarks
Mecistogaster lucretia Drury	♂ 120–130, ♀ 100–110	♂ 59–65, ♀ 59–85	—	The longest abdomen known
Megaloprepus coerulatus Drury	♂ 96–102, ♀ 76–85	♂ 56–64, ♀ 70–94	—	The female has the longest wing known
Petalura ingentissima Tillyard	♂ 92, ♀ 94	♂ 71, ♀ 76	—	The most robust species known
Tetracanthagyna vittata MacLachlan	♂ 80, ♀ 70	♂ 75, ♀ 80		
Tetracanthagyna plagiata Waterhouse	♂ 69–76, ♀ 74–75	♂ 73–75, ♀ 84		
Aciagrion fragilis Tillyard	—	—	♂ 0·15 mm.	The slenderest species
Hemiphlebia mirabilis Selys	♂ 19–23	♂ 10·5–12		
Agriocnemis hyacinthus Tillyard	♂ 16–18	♂ 10–10·5		
Agriocnemis rubricauda Tillyard	♂ 16–17	♂ 9–9·5		The smallest of all

Courtship.

A few interesting observations on this subject are available. In the case of *Rhinocypha fenestrella*, the mature male has the surface of the tibiae whitened. These it displays to the female, while dancing in the air before her [82]. The male of *Calopteryx maculata* possesses a shining white ventral spot at the tip of its abdomen, which it displays by curving the abdomen upward and forward, the fore-wings being held motionless, the hind-wings fluttering rapidly [198]. A more prolonged courtship takes place in the case of the little *Hemiphlebia mirabilis*. This metallic green Dragonfly is almost invisible on green reed-stems, except for its long ribbon-like white appendages (fig. 14 L). These are displayed as a sign to the female, by raising the abdomen and bending it slightly sideways, while walking up the reed-stem. The female replies by moving the whitened end of her abdomen from side to side in a peculiar manner. Finally, the little creatures fly out from the reeds, and engage in a miniature "pas de deux" before pairing [173].

Comparative Abundance of the Sexes.

The causes of the capture of more males than females by most collectors have been already indicated (p. 321). It is quite possible to reverse the proportion by choosing the right locality. Thus, an afternoon's collecting along a deep railway-cutting in the tropics (including the mouth of a tunnel, in which I secured several females of *Gynacantha*) yielded me over sixty females of various species, but only two or three males. An even more extraordinary result than this was obtained by working three or four small streams in dense palm jungle in Queensland. *Argiolestes amabilis* was to be seen on every rock. In a week I had taken 195 females, but only a single male! Though I sought for the males high and low, they were not to be found. Probably they lived in the sunlight, on the tops of the giant trees, more than a hundred feet above my head.

As the result of breeding out many species during the last ten years, I find that two conclusions may safely be drawn.

(i) The number of males and females is *approximately equal*. My records shew about 5 per cent. excess of females. The most vigorous larvae were nearly always females.

(ii) The females emerge, on the average, a few days before the males. They should therefore be sought for early in the season, before pairing has begun.

Habits of the Larvae.

The habits of the larvae are, if anything, more varied than those of the imagines. A general résumé of this question has already been given, as far as the correlation between habitat and colour-pattern goes, on p. 255. We may offer on p. 328 an attempt at a classification of the different habitats of Dragonfly larvae. In this table, the letter A indicates the commonest habitat for the group, while B, C, D indicate less usual habitats, in descending order.

As the table is based chiefly on observations made on Australian species, it is to be hoped that observers in other regions will compare it with the case of their local fauna.

For the question of the mode of progression and method of stalking the prey, the reader is referred to the account of the rectal and caudal gills in chap. IX, and to the account of the labial mask in chap. IV.

The larvae of Dragonflies are not without *means of defence* against their enemies, though they rely chiefly on concealment, both for avoiding their enemies and for the capture of their prey. If a leg or a caudal gill be seized, the breaking-joint at the base of the organ enables it to be cast off at once. The sharp dorsal and lateral spines on some larvae, and the acutely pointed anal pyramid in the *Brachytronini*, may be used as a means of defence if the larva is seized. These same spines are probably of great use to the larvae in the case of a sudden rush of water during flood-time.

Food.

Little definite is known as to the food of Dragonflies, beyond the fact that it is captured on the wing, and consists exclusively of live insects. Poulton [125] and Campion [35] have collected together all the definite records, which include representatives of the Plectoptera, Trichoptera, Lepidoptera, Hymenoptera, Coleoptera and Diptera, besides, of course, smaller Odonata. As definite observations concerning gnats and mosquitoes being captured for food seem to be few or lacking, I may add here that I have repeatedly seen *Aeschninae* feeding on these pestilent insects in the late afternoon. A specimen of *Telephlebia godeffroyi* was once observed flying round and round a small bush about 7 p.m., when the mosquitoes were particularly troublesome. After ten minutes, it was captured. I found its mouth so full of mosquitoes that it was unable to shut it. There must have been over a hundred,

		Running Water								Still Water								
		Fast				Slow				Backwaters		Lakes and Lagoons		Ponds and Water-holes				
	Subfamily	Clinging to rocks	Buried in sand, etc.	In débris	Clinging to banks, roots, weeds, etc.	Buried in mud or sand	On bottom	In débris	On weed	Bottom	Weed	Bottom	Weed	Bottom	Weed	Swamps	Mountain bogs	Water collected at bases of leaves of plants
ANISOPTERA	*Gomphinae*	—	A	A	—	B	—	B	—	C	—	—	—	C				
	Cordulegastrinae	—	—	A														
	Petalurinae	—	—	—	—	—	—	—	—	—	—	—	—	—	—	—	A	
	Aeschninae	B	—	—	A	—	—	—	B	—	B	—	B	—	B	C		
	Corduliinae	C	—	A	—	—	A	B	—	A	—	A	—	A	—	B		
	Libellulinae	—	—	D	—	—	C	C	—	A	—	A	—	A	—	A	B	
ZYGOPTERA	*Epallaginae*	A	—	A														
	Thorinae	—	—	A														
	Calopteryginae	B	—	—	B	—	A											
	Synlestinae	—	—	—	A													
	Lestinae	—	—	—	C	—	—	—	A	—	A	—	A	—	A	B		
	Pseudostigmatinae	—	—	—	—	—	—	—	—	—	—	—	—	—	—	—	—	A
	Megapodagrioninae	A	—	A	—	—	—	A										
	Protoneurinae	A	—	B	—	—	—	B	C									
	Platycneminae	—	—	—	—	—	A											
	Agrioninae	—	—	—	C	—	—	—	A	—	A	—	A	—	A	B	—	C

all tightly packed into a black mass. I have frequently seen *Aeschna brevistyla* take gnats and mosquitoes in dozens, while on the wing. There can be no doubt that those Dragonflies which fly late in the day are of great value in checking the spread of the various objectionable *Culicidae* that are on the wing from just before sundown.

The largest Dragonfly that I have ever seen captured was *Synlestes weyersi*, seized by *Aeschna brevistyla* while on the wing. The wings were quickly bitten off, and fluttered to the ground. The whole of the long body was chewed up in a few minutes, but the thorax was mostly discarded.

As regards Dragonfly larvae, they seem to relish almost anything they can catch, but more particularly smaller specimens of their own or other species of Odonata. When young, they feed readily on Protozoa. A little later, they are expert at catching water-fleas and other small Crustacea. Later on, I find they prefer the larvae of May-flies or mosquitoes to anything else. As an experiment, I once kept a larva of *Synthemis macrostigma* without food for a week. I then offered it mosquito larvae in large numbers. It used its mask with such vigour that I counted no less than *sixty* larvae swallowed within ten minutes. After that, nothing would tempt it. As regards the large *Aeschnine* larvae, I find that they will stalk *Agrionid* or *Lestid* larvae with persistency and cunning. They will also stalk one another, and woe be to the one that is in front! They also attack tadpoles and even small fish, but always refuse to eat more than a small portion of them.

Enemies.

In the imaginal state, the Dragonfly has a number of enemies, most of which can, however, only compass his destruction if they come upon him just after metamorphosis, before his wings are ready for flight. Amongst the plants, the giant sun-dew (*Drosera binata*) takes heavy toll of those Australian Zygoptera which frequent swamps and marshes (fig. 167). Spiders prey upon Dragonflies, both by capturing them in their webs, and by pouncing upon them at metamorphosis [196]. *Gomphinae* appear particularly liable to attack at this latter time, since they do not usually leave the ground. Lizards and snakes are, curiously enough, expert at catching Dragonflies. I once saw *Diphlebia lestoïdes* alight upon the tail of a lizard which was sunning itself on a rock. The lizard caught it, by a stroke as quick as lightning! Birds frequently attack Dragonflies: but, as far as my observations go, seldom succeed in catching them. The kingfishers are an exception, as they are wonderfully expert at catching Dragonflies skimming close to the water. The most deadly enemy of all is the trout. In Tasmania, the introduction of the English trout has reduced the Odonate fauna of the rivers to a minimum. A 2-lb. trout which I caught on the Macquarie River in Tasmania had in its stomach the undigested heads of thirty-five Dragonflies, twenty-eight belonging to the rather rare species *Procordulia jacksoniensis*. There were also the remains of one beetle.

Fig. 167. *Austrolestes analis* Ramb., ♂, and a number of smaller insects, captured by *Drosera binata* Labill. (×⅔). Original drawing by P. Tillyard.

The trout is also an enemy of the larvae, especially of those *Libellulidae* which live on the bottom of slowly moving rivers. The voracious larva of the beetle *Dytiscus* frequently attacks Dragonfly larvae. But the latter has no enemy to compare with his own near relatives, if they happen to be of a somewhat larger size.

Parasites (figs. 168, 169).

These may be divided into *true* parasites and *false* parasites. Of the former, the best known are the tiny Hymenoptera of the families *Tetragrammidae* and *Mymaridae*, or "fairy flies." Three species (*Anagrus incarnatus*, *Polynema ovulorum* and *P. natans*) have been described as laying their eggs within the eggs of *Calopteryx*. The grub hatches out and feeds on the contents of the Dragonfly's egg, which it devours completely in a few days. It then pupates, and emerges as an imago ten or twelve days later. The female uses her wings for swimming, and by this means seeks out the eggs of *Calopteryx*, which are laid in the leaves of water-lilies [94].

I have found the epithelium of the mid-gut in the larva of *Ischnura heterosticta* almost completely destroyed by a large undescribed species of Gregarine Sporozoan (fig. 169), four or five of which completely filled the lumen of the gut. The larva can live for days with this parasite inside it, but gradually becomes weak and sickly-looking, owing to its inability to digest any food.

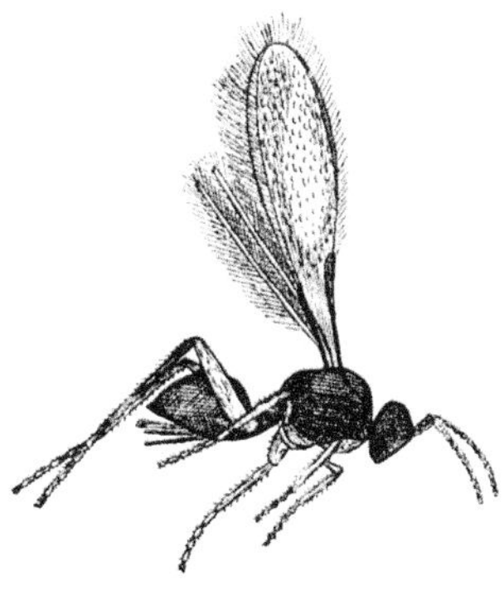

Fig. 168. *Polynema natans* Lubbock, ♀, Britain (× 20). After Miall, from Lubbock.

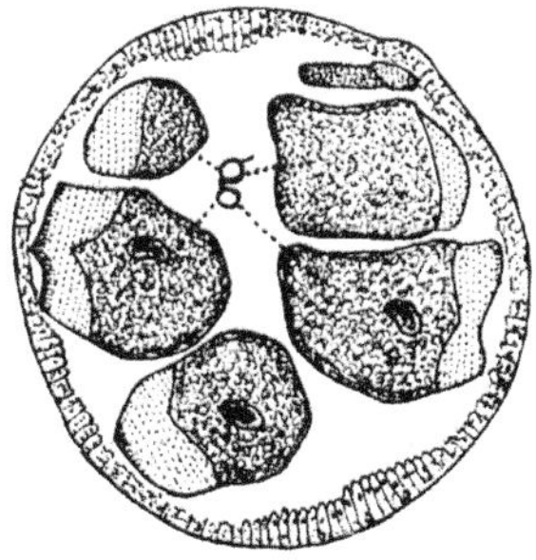

Fig. 169. T.S. through mid-gut of larva of *Ischnura heterosticta* Burm., containing several large Gregarinida (*g*) affixed to its wall (× 75). Original.

De Selys found a large *Filaria* in the abdomen of *Sympetrum flaveolum*.

The imagines do not, however, appear to suffer much from true parasites. Dragonflies whose larvae live in still water (*Libellulinae*, *Agrioninae*) are frequently found covered with a species of small red mite (family *Hydrachnidae*). These are false parasites, usually found clinging to the underside of the thorax, the bases of the wings, abdomen and legs of males and females alike. All the mites on the Dragonfly are young ones. I have recently succeeded in finding the mature female of this species, which is a rather large animal (3–4 mm. diameter) of a brilliant carmine colour.

It seems probable that it attacks the Dragonfly at metamorphosis, placing either eggs or viviparous young in the positions already described. The young mites cling to the Dragonfly without doing it any harm, and are carried by it to other ponds or lagoons, where some of them at any rate drop off. Thus the Dragonfly is used as a means of dispersal by this peculiar Arachnid [34].

The Confervoid Alga *Oedogonium* is very abundant in stagnant water or slowly-running streams. I have found it growing, not only on rocks, wood, and even stones and soil, but also on various animals which live in such localities, such as aquatic Hemiptera, and also the larva of *Aeschna brevistyla*. As the latter takes two years to mature, it usually retires for the winter, and becomes very inactive. In the early spring (Aug.–Sept.) this larva can be frequently taken around Sydney with a thick growth of *Oedogonium* upon it. There is no question of parasitism here. I found, by means of sections, that the *Oedogonium* does not penetrate the cuticle of the larva, but simply grows on it as it grows on everything else in such places. On one larva of *Aeschna* there were no less than three species of *Oedogonium*, fifteen species of Diatoms, and a large number of *Vorticella*. The exercise of a little imagination can, of course, elaborate a theory of symbiosis in a case like this. But, as this method of erecting a scientific mountain out of a small natural mole-hill has already been used to perfection by Kammerer [78], I shall content myself with stating my opinion that such methods are not to be considered scientific. Kammerer's *Aeschna* larvae could not avoid becoming overgrown with *Oedogonium* in the filthy surroundings in which he found them. But to give a list of "advantages," both on the side of the alga and on the side of the larva, is quite another question, and one that can be safely left for common-sense to decide.

Larvae in Brackish Water.

Dragonflies are known to breed in coastal lagoons, where the water is brackish. Osburn [110], by a series of experiments, shewed that the larvae could withstand a salinity up to a density of 1·01 (sea-water 1·026). At that point the larvae could live, but could not undergo metamorphosis. At a density of 1·015 the larvae suffered greatly, and soon died. The Dragonflies known to inhabit brackish water are (as might be expected) certain species of *Libellulinae*, *Anax*, and some *Agrioninae*; to which we may add, in Australia, a few species of *Austrolestes*.

Hibernation.

In Europe, only one small Dragonfly (*Sympycna fusca*) passes the winter in the imaginal state. On warm days it may come out and fly about, but it spends most of the time hiding away under dead leaves. In Sydney, where the winter is mild and short, both *Diplacodes bipunctata* and *Austrolestes leda* hibernate regularly through June, July and the early part of August. On sunny days in those months, I have occasionally found them flying.

Migration.

As Dragonflies are frequently seen and even captured far from land, it has been assumed that the Odonata as a whole are readily dispersed, and overcome natural barriers with great ease. The evidence does not, however, warrant this. Dragonflies quickly take cover when bad weather is at hand, and are seldom likely to be caught in a storm or driven out to sea. The only species that are at all liable to spread are those with strong migratory instincts. Of these, *Libellula quadrimaculata* is the most conspicuous example. Immense swarms of this species have been recorded in different parts of Europe on many occasions. These swarms sometimes pass out to sea, and may travel hundreds of miles. A few large *Aeschninae* sometimes accompany them. In Australia, *Hemicordulia tau* occasionally swarms in like manner, and appears to travel a long distance, though not yet recorded at sea. It has, however, recently colonized Tasmania, across a strait two hundred miles wide! On the other hand, the strong-flying and equally common *Anax papuensis* has so far failed to do this. *Pantala flavescens* and some species of *Tramea* travel far and wide, and have overspread the whole of the Tropics. Apart from these few examples, the migratory instinct is not developed in Odonata. It may be safely stated that a very great majority of the Order never travel far from their favourite haunts, and are as strongly "held" by natural barriers as are many insects of much weaker flight.

Drought.

It is a widespread belief that the appearance of a swarm of Dragonflies heralds a drought. There is some truth in this. Observations which I have made on the drying-up of water-holes in the Sydney district lead me to conclude that, of all the larger aquatic life in the pond, the Dragonfly larvae will last out longest. As the volume of water decreases, the amount of food to be obtained with ease increases, so that the larvae take toll of everything that comes to hand. Thus, if they mature before the final drying-up of the pond, a huge swarm of Dragonflies will emerge, and precede the worst part of the drought. The other side to the picture is not so likely to be noticed, but it is this:—In the spring of 1908, the pond I was observing dried up early in September, whereas it usually lasted until December or later. On the hot caking mud of the pond, I counted over three hundred nearly full-fed larvae of *Anax papuensis.* Every one of these perished. In a week or less, had the water held out, they would have appeared as a huge swarm of Dragonflies, "heralding" a drought that lasted for several months!

The larvae of certain Dragonflies can resist prolonged drought. In the case of *Synthemis* [170] I have kept the larva in dry sand for three months. At the end of this period they were apparently dried up, and so light that, when placed in water, they floated! But they soon revived, and, after some struggling, forced their way to the bottom, where they at once burrowed in the usual way. Several of these were bred out later on. The larvae of *Austrocordulia* can also resist drought, as probably also can those of many of the *Sympetrini.* But lack of water is fatal to most Zygoptera and *Aeschninae.* The peculiar larva of *Telephlebia* has a slimy integument, and prefers to live out of the water rather than in it [178].

Economic Value of Dragonflies.

The Odonata appear to touch the life and activities of man only very lightly. Were they to disappear from the earth, most of us would probably not feel their loss in the slightest. On the other hand, there are so many injurious and obnoxious insects about us at every turn, that we may take a pleasure in contemplating the fact that the Dragonflies are a beneficent and not a maleficent race. Apart from the single record of *Anax* larvae destroying the young fry in fish hatcheries, all the evidence is in favour of them. We do not yet know the value of their services to us in the destruction of millions of other insects, both larvae and imagines, in the areas which they frequent. I feel persuaded that this service is greater than it appears. Not only are the Dragonflies the most powerful determining factor in preserving the balance of insect life in ponds, rivers, lakes and their surroundings, but they do most certainly make war upon the flies, mosquitoes and gnats which we all desire to see exterminated. I believe that a successful checking of the mosquito pest in the ornamental waters of parks and gardens could be readily obtained by the introduction of species whose larvae, as well as the imagines, would prey upon the nuisance. If a successful planting of a colony of Dragonflies in such a position were to be tried, the species selected might also be chosen for its colouring, and thus add a new note of interest to the locality. The glorious red *Orthetrum villosovittatum* has now become well established in the Botanical Gardens at Brisbane, and certainly adds a vivid touch of colour to its lovely surroundings.

It is extraordinary how the idea that Dragonflies are harmful, and possess a powerful sting, continues to hold sway even amongst people who claim to be educated. The old proverb "Give a dog a bad name, and hang him" is particularly applicable to this case. If an insect is known to all the country-side as a "horse-stinger," how can a mere scientist hope to convince them that it does not sting horses. In the course of my travels in Australia, I have collected evidence, on the solemn word of various settlers in Queensland, that the gigantic *Petalura* and *Anax* have not only done to death various horses and cows, but in one instance even attacked the farmer's wife! Fortunately the Dragonfly now forms one of the stock object-lessons in the Nature Study curriculum of the Queensland schools. So we may hope that this ignorance will soon be a thing of the past.

There is a record of Dragonflies being used as food by man. It is said that the inhabitants of certain Malay Islands, where these insects are very numerous, smear long sticks with a kind of bird-lime. By deftly wielding these sticks, they capture a large number of Dragonflies, which are made into a kind of paste for food. Neither this nor the locust is ever likely to figure in the white man's menu.

The Dragonfly has unconsciously rendered one very great service to mankind, by providing in itself the natural model for the modern aeroplane. As long ago as 1883, Amans [2] suggested that the Dragonfly would serve as a suitable model for a flying machine, to be propelled by electricity. The idea took root in the fertile minds of his countrymen, and France has ever since led the way in this new branch of science. One of the first French monoplanes to be manufactured on a large scale was called the "Demoiselle," thus commemorating Amans' original idea. Later models, if anything, approach the natural model even more completely than the earlier ones. A study of the different effects on flight of angulated and rounded hind-wings, as well as of the arrangements of braces and cross-pieces suggested by various parts of the Dragonfly's wing, might well lead to further improvements in our models, and might even suggest a solution for "hovering" on simpler lines than anything yet attained.

Folk-Lore.

I have been able to gather very little on this interesting topic. In North Queensland there is a very remarkable Dragonfly, *Podopteryx roseonotata* (Plate IV, fig. 1) of a rich black and pink colour, which sits about with outspread wings in the dense tropical jungle. The aboriginals know this Dragonfly well, and call it "Wongera-wongera." They seem to regard it with some reverence, but I have not definitely ascertained that it figures in their system of totemism.

Folk-lore has gathered around the Dragonfly in Japan, where the people all know these lovely insects and are proud of them. It is said that this love of the Japanese for their Dragonfly fauna arose from the remark made by an Emperor who ascended Fuji-yama. Viewing his kingdom spread out before him, with its widely indented coast-line, he is said to have compared it with a gigantic Dragonfly with outspread wings. The remark was, of course, taken up by the courtiers, and the phrase "The Land of the Dragonfly" passed into current use. Representations of Dragonflies in art, and allusions to them in literature, are, I am told, very numerous in Japan.

Local Names.

The commonest epithets for Dragonflies in English-speaking countries are "Horse-stingers," "Mosquito-hawks," "Devil's Darning-needles" and "Snake-doctors." The small *Agrionidae* are known as "Damsel-flies." In Victoria, the larvae of *Hemicordulia* are called "Mud-eyes" (an excellent description!) and are much sought after as bait for trout and perch.

Foreign Names.

Country	Name	Country	Name
France	Demoiselle (maiden)	Russia	Strekoza
,,	Libellule	Denmark	Guldsmed
Germany	Wassernympfe (water-nymph)	Holland	Juffer
,,	Wasserjungfer (water-maiden)	Sweden	Trollslanda
,,	Libelle	Japan	Tonbo
Italy	Cevettone		
,,	Saëtta (arrow)		

The Name *Libellula*.

The origin of this name, selected by Linnaeus as the generic title for what is now the Order Odonata, appears to be somewhat obscure. We find in Littré the following guess: "Etym. On croit que *Libellule* est un diminutif de *libellus* (petit livre), et que cette dénomination vient de ce que ces insectes tiennent leurs ailes étendues comme les feuillets d'un *livre*." Littré might have gone farther, and given the etymology as *libellulus* instead of *libellus*. For we find the double diminutive *libellulus* used in Martianus Capella (about 425 A.D.). Hence we have the following derivation[1]:

Lat. *Līber*, properly the *bark* of a tree (and hence "a book").

Diminutive *Lĭbellus*, the *inner bark* of a tree.

Double diminutive *Lĭbellulus* (with same meaning as *libellus*).

Generic derivative *Lĭbellula* Linnaeus, a genus of Dragonflies.

Professor MacCallum has, however, suggested that Linnaeus used the feminine termination as a diminutive from *lībella* (a balance) on account of the way that these insects poise their wings in flight or at rest.

[1] I have to thank Mr R. H. Bode, M.A., of Sydney Grammar School, for tracking down the etymology of this word.

CHAPTER XVIII

BRITISH SPECIES

The table of British Species given in this chapter is based mainly on a study of a large collection of British and Continental specimens received by exchange during the past ten years from many kind correspondents, and embracing specimens from almost every country in Europe. It is many years since I collected British Dragonflies in their native land. Some of the rarer species I have never seen alive. Information as to coloration, habits, larvae, etc., has been obtained from the excellent works of Lucas[88], Ris[133] and from numerous smaller publications. Lack of space forbids the complete description of any species here. Only those characters which I have found most reliable are included. To use the table, the student must first make himself thoroughly conversant with the classification in chap. XIV down to Tribes, as far as this comes within the purview of the British List; always remembering that the most important characters, without which the specific diagnosis may well go all awry, are therein to be found. The measurements which are given for every species seek to include the extremes of possible variation in size. In the case of closely related species, figures illustrate the most reliable points of difference. In *Aeschna* and *Somatochlora*, the form of the male appendages is a sure guide (figs. 170, 174); in *Sympetrum* the hamuli of the male (fig. 176) and the valvula vulvae of the female, though the latter cannot always be seen in dried specimens, unless a chitin preparation be made. Undoubtedly the most difficult genus is *Agrion*. Here the males may be reliably determined, preferably from newly caught specimens, by the colour-patterns of segments 1–3 (fig. 179). These are, however, subject to individual variation within limits, so that a certain allowance should be made for possible changes in details. The females of *Agrion* are very difficult to separate. The best rule is always to try and capture males from the same spot; if possible, in coitu. Where two or more species of *Agrion* occur together (as on the Norfolk Broads) pairs in coitu should be taken. The student may then study them carefully until he is familiar with their facies. A reliable key can be constructed from a comparative study of the outlines of the prothorax in both sexes.

The classification of the larvae is a difficult matter, partly owing to the great similarity of the larvae in some genera (e.g. *Sympetrum*, where even the expert is at a loss to find specific differences) and partly because many species are still not known for certain in their larval forms. The student is referred to figures of larval forms typical of the different genera. Having mastered these, he may then use the short descriptions given, which summarize the principal differences for the known larvae.

(N.B. All measurements are given in millimetres.)

Suborder ANISOPTERA (Characters, p. 259).

Family AESCHNIDAE (Characters, p. 259).

Subfamily **Gomphinae** (Characters, p. 260).

Tribe *Gomphini* (Characters, p. 261).

Genus I. *Gomphus* Leach (fig. 118). Coloration black and yellow; thorax with dorsal, ante-humeral and humeral black bands.

1. *G. vulgatissimus* Linn. ♂ *abd* 32—34, *hw* 29—31, *pt* 2·5, dark brown. *C* black. Occiput straight, yellow, with black hairs. *Abd.* with a mid-dorsal yellow line on 1—7; 8—9 considerably dilated. *App. sup.* short, black, suddenly pointed; *inf.* with divergent cornute branches. ♀ *abd* 34—35, *hw* 31—32, *pt* 3. Stouter; *abd.* thicker, more cylindrical; yellow markings in general more evident. Very rare. June.

Subfamily **Cordulegastrinae** (Characters, p. 262).

Genus II. *Cordulegaster* Leach (fig. 121). Robust *Aeschna*-like species. Coloration black and yellow; ♀ with hard black ovipositor reaching beyond end of 10 (fig. 101).

2. *C. annulatus* Latreille. ♂ *abd* 55—58, *hw* 41—48, *pt* 4, black; *memb.* whitish; *C* yellowish in front. *Thorax* with three pairs of wide and one pair of narrow yellow bands. *Abd.* black with median narrow yellow bands

interrupted dorsally. *App. sup.* short, black, divergent, pointed; *inf.* truncated. ♀ *abd* 58—62, *hw* 46—50, *pt* 5. Closely resembles ♂.

Fairly common. June—Aug.

Subfamily **Aeschninae** (Characters, p. 262).

Tribe *Brachytronini* (Characters, p. 264).

Genus III. *Brachytron* Evans (fig. 123). *Rs* symmetrically forked proximad to *pt*. *Rspl* and *Mspl* only one cell-row below *Rs* and M_4 respectively. *Pt* very narrow. *Aa* of *hw* angulated in ♂. *Eyes* meeting only for a moderate distance. Auricles present in ♂.

3. *B. hafniense* Müller (= *B. pratense* Müller). ♂ *abd* 40—41, *hw* 35—36, *pt* 4·5, yellowish brown; *memb.* small, white. *Thorax* and *abd.* hairy, dark brown with bands and spots of yellow, green or blue. *App. sup.* long, curved; *inf.* short. ♀ *abd* 38, *hw* 35. Dorsal thoracic bands reduced to spots; other markings yellower. Local. July.

Tribe *Aeschnini* (Characters, p. 265).

Genus IV. *Aeschna* Fabr. (Plate II, fig. *A*). Sectors of *arc.* arising near middle of *arc.*; *pt* long, of moderate width. *Rs* unsymmetrically forked close to level of *pt*. *Rspl* and *Mspl* curved, 3—4 cell-rows below *Rs* and M_4 respectively at middle. *Aa* of *hw* angulated in ♂. Auricles present in ♂.

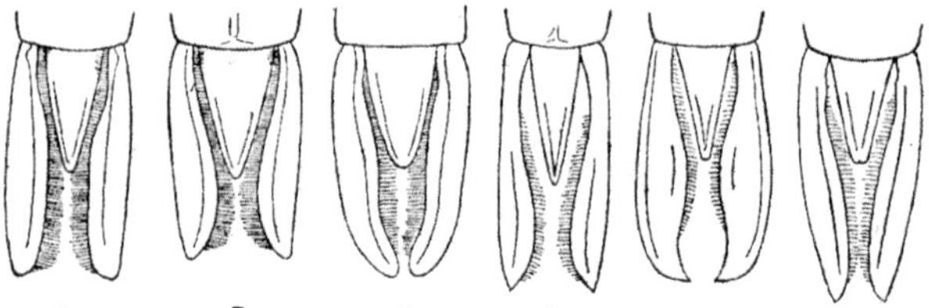

Fig. 170. Anal appendages of males of *Aeschna*, dorsal view. A. *Ae. grandis* Linn. B. *Ae. juncea* Linn. C. *Ae. coerulea* Ström. D. *Ae. mixta* Latr. E. *Ae. cyanea* Müller. F. *Ae. isosceles* Müller. Original.

4. *Ae. grandis* Linn. ♂ *abd* 55—57, *hw* 48—50, *pt* 3. *Wings* yellowish-brown all over. *Aa* of *hw* as in No. 5. *Eyes* closely contiguous. *Frons* without a T-mark. Coloration brown, *thorax* with two pairs of lateral yellow stripes. *Abd.* constricted at 3, with very small blue or yellowish spots. *App. sup.* 4·5, *inf.* 2; shaped as in fig. 170 A. ♀ *abd* 51—53, *hw* 47—51, *pt* 3·5. *Abd.* not constricted at 3, almost uniformly brown; *apps.* 3, leaf-like. Fairly common; Southern. July—Sept.

5. *Ae. juncea* Linn. ♂ *abd* 53—59, *hw* 43—48, *pt* 4. *Wings* hyaline, *C* yellowish. *Aa* of *hw* 2-celled (fig. 171 A). *Eyes* closely contiguous. *Frons* with a T-mark. *Thorax* with narrow dorsal and broader lateral yellow bands.

Abd. much constricted at 3; large blue and small yellow spots on all segments. *App. sup.* 4, *inf.* 3; shaped as in fig. 170 B. ♀ *abd* 53—55, *hw* 43—44, *pt* 4·5. *Abd.* not constricted; blue replaced by green; *apps.* 4·5, slender, leaf-like. Fairly common. June—Sept.

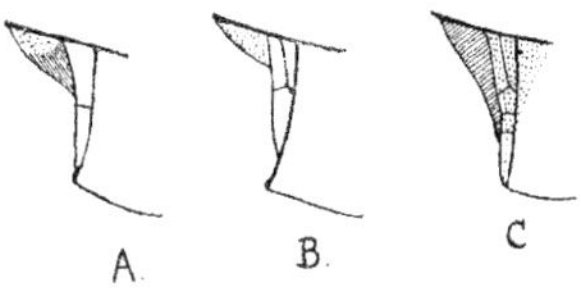

Fig. 171. Anal angle of hind-wing of male in *Aeschna*. A. *Ae. juncea* Linn. B. *Ae. cyanea* Müller. C. *Ae. isosceles* Müller. Original.

6. *Ae. coerulea* Ström. (= *Ae. borealis*, Zetterstedt). ♂ *abd* 45—48, *hw* 39—40, *pt* 3—4. *Wings* hyaline, *C* yellow. *Aa* of *hw* as in No. 5. *Eyes* only meeting for a short distance. *Frons* with a T-mark. *Thorax* with narrow dorsal and lateral blue stripes. *Abd.* much constricted at 3; 1—2 mostly blue, rest with numerous large blue spots. *App. sup.* 4, *inf.* 2, shaped as in fig. 170 C. ♀ *abd* 43—45, *hw* 38—40, *pt* 4. *Abd.* not constricted, spots smaller, some yellowish; *apps.* 4, leaf-like. Rare; Northern. June.

7. *Ae. mixta* Latreille. ♂ *abd* 44—45, *hw* 37—39, *pt* 3. *Wings* hyaline, *C* brown. *Aa* of *hw* as in No. 8. *Eyes* closely contiguous. *Frons* with T-mark. *Thorax* with small dorsal yellow specks and broad lateral yellow bands. *Abd.* slightly constricted at 3; fairly large blue spots on each seg. *App. sup.* 4·2, *inf.* 2·5, shaped as in fig. 170 D. ♀ *abd* 44—46, *hw* 38—40, *pt* 3·5. No dorsal specks on thorax. *Abd.* not constricted, spots smaller, blue-green and yellow; *apps.* 5, slender, leaf-like. Rare; South-eastern. Aug.—Oct.

8. *Ae. cyanea* Müller. ♂ *abd* 54—58, *hw* 45—50, *pt* 2·5. *Wings* hyaline. *Aa* of *hw* 3-celled (fig. 171 B), rarely 4—5-celled. *Eyes* moderately contiguous. *Frons* with T-mark. *Thorax* with large oval yellowish-green dorsal marks and broad yellow lateral bands. *Abd.* constricted at 3, with fairly large blue and smaller yellow spots. *App. sup.* 5, *inf.* 2, shaped as in fig. 170 E. ♀ *abd* 55—58, *hw* 48—52, *pt* 3. Dorsal marks of thorax yellow. *Abd.* not constricted, colours as in ♂; *apps.* 4·5, leaf-like. Fairly common; Southern. July—Oct.

9. *Ae. isosceles* Müller (= *Ae. rufescens* Vanderlinden). ♂ *abd* 47—49, *hw* 41—42, *pt* 3. Base of *hw* deeply saffroned. *Aa* of *hw* 3—6 cells with large dark *memb.* (fig. 171 C). *Eyes* closely contiguous. *Frons* without a T-mark. *Thorax* brown with yellow lateral lines, no dorsal markings. *Abd.* brown, scarcely constricted at 3, a large yellow triangle on 2. *App. sup.* 5, *inf.* 2, shaped as in fig. 170 F. ♀ *abd* 50—52, *hw* 41—42, *pt* 3·5. Slightly stouter than ♂, closely similar; *apps.* 3·5, leaf-like. Rare; Fens and Norfolk Broads. May—June.

Genus V. *Anax* Leach (fig. 172). Upper sector of *arc.* arising close up to *R*, lower one at about middle of *arc.* *Rspl* and *Mspl* much curved, 5—7 cell-rows below *Rs* and M_4 respectively at middle. *Aa* of *hw* rounded in both sexes. Auricles absent.

10. *A. imperator* Leach (= *A. formosus* Vanderlinden). ♂ *abd* 55—56, *hw* 46—47, *pt* 4·5. *Wings* hyaline; *hw* with long narrow *memb.*, whitish basally. *Eyes* closely contiguous, blue. *Thorax* greenish, with dark lateral

lines. *Abd.* much swollen at 1—2, slightly constricted at 3, bright blue with an irregular mid-dorsal black band. *App. sup.* 5, thickened in middle, inner border curved, tips rounded; *inf.* 1·5, broad, truncated. ♀ very similar to ♂, but *eyes* and *abd.* greenish; *apps.* 5·5, leaf-like. Fairly common; Southern. May—Aug.

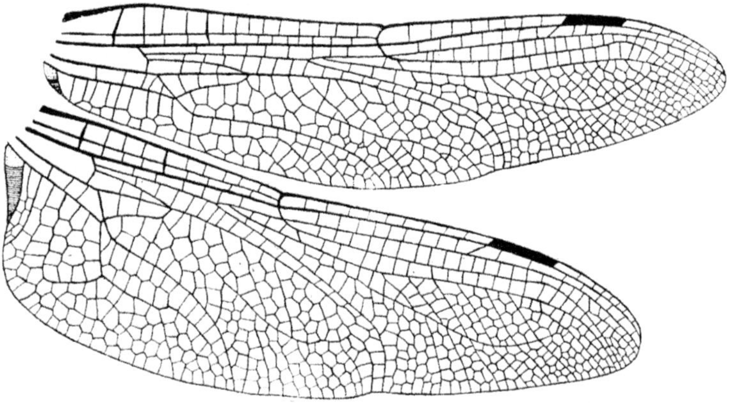

Fig. 172. Wings of *Anax imperator* Leach, ♂ (*Hw* 46 mm.). Original.

Family Libellulidae (Characters, p. 265).

Subfamily **Corduliinae** (Characters, p. 265).

Tribe *Idocorduliini* (Characters, p. 267).

Genus VI. *Oxygastra* Selys (fig. 152). *Fw* with 2 rows of post-trigonal cells. *Al* without a toe. *Hw* with a weak *st*; *t* free in all wings.

11. *O. curtisi* Dale. ♂ *abd* 36—38, *hw* 33—36, *pt* 2. Bases of wings slightly saffroned; *C* yellowish brown; *memb.* whitish. *Eyes* only slightly touching. Colour metallic green; *abd.* slightly clubbed, with long lanceolate mid-dorsal yellow markings on 1—7; 10 with a compressed yellow piece projecting between *apps.* *App. sup.* short, subcylindrical, hairy, divergent; *inf.* blunt, slightly bifid. ♀ *abd* 33—36, *hw* 32—34, *pt* 2·5. Wings clouded with yellow. Coloration as in ♂; no projecting piece on 10; *apps.* short, conical. Very rare; Southern. June—July.

Tribe *Eucorduliini* (Characters, p. 267).

Genus VII. *Cordulia* Leach (fig. 129). *Fw* with 2 rows of post-trigonal cells, followed later by 3. No *st* in *hw*; *t* crossed in *fw*, free in *hw*; *memb.* large. *Al* with a distinct toe and a straight sole. *App. inf.* of ♂ forked.

12. *C. aenea* Linn. ♂ *abd* 33—36, *hw* 32—34, *pt* 3·2. *Wings* well-saffroned at bases. Colour metallic bronze-green; *thorax* covered with brownish down; *notum* reddish brown. *Abd.* strongly clubbed; without markings. *App. sup.* 2·8, blunt, divergent; *inf.* 1·8, very strongly bifurcated. ♀ similar to ♂, but *abd.* broader, not clubbed; *apps.* 2·5, leaf-like. Rare; Southern. May—July.

Genus VIII. *Somatochlora* Selys (fig. 173). Venation as in VII, but *hw* with a weak *st*, and *t* either crossed or free. *App. inf.* of ♂ simple.

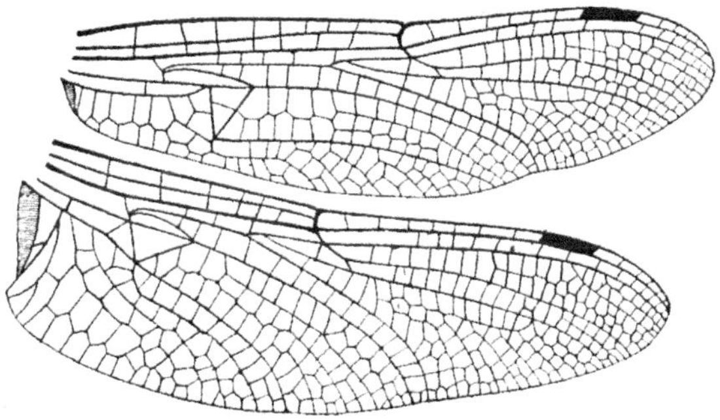

Fig. 173. Wings of *Somatochlora arctica* Zett., ♂ (*Hw* 31 mm.). Original.

13. *S. metallica* Vanderlinden. ♂ *abd* 33—38, *hw* 33—37, *pt* 2—2·5. A touch of saffroning at extreme bases of wings. Colour metallic bronze-green. *Abd.* much constricted at 3, 1—3 with small yellow spots. *Apps.* as in fig. 174 A. ♀ *abd* 37—41, *hw* 36—38, *pt* 2·5—3. Similar to ♂, but with broad unconstricted *abd.*; *apps.* 4, slender, lanceolate. Very rare. July—Aug.

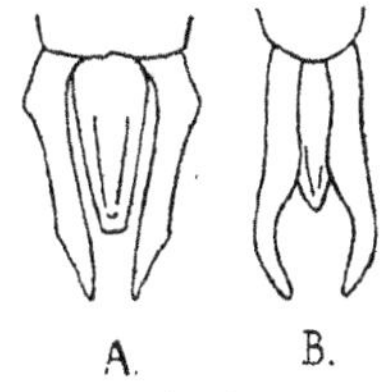

Fig. 174. Anal appendages of males of *Somatochlora*, dorsal view. A. *S. metallica* Vand. B. *S. arctica* Zett. Original.

14. *S. arctica* Zetterstedt. ♂ *abd* 34—35, *hw* 31—32, *pt* 2·5. *Wings* slightly saffroned at bases, and partly suffused with yellow near middle. Coloration bronze, coppery, or even steel black; sides of thorax metallic green. *Abd.* much constricted at 3; 1—3 with small yellow spots. *Apps.* as in fig. 174 B. ♀ *abd* 33—35, *hw* 30—32, *pt* 2·5. Similar to ♂, but wings yellower; *abd.* less constricted, wider, almost black; *apps.* 3, slightly lanceolate. Rare; Northern. June—July.

Subfamily **Libellulinae** (Characters, p. 269).

Tribe *Libellulini* (Characters, p. 269).

Genus IX. *Libellula* Linn. (fig. 131). *Fw* with about 12 *Ax*, the last complete. Sectors of *arc.* arising just separated; *t* crossed in both wings.

Wings coloured, at least by a dark basal patch on *hw*. *Abd.* short, considerably widened and depressed. Lobe of prothorax small.

15. *L. quadrimaculata* Linn. ♂ *abd* 28—29, *hw* 35—36, *pt* 4. *Wings* with a yellowish-brown band along *C*, broadest at base; a small dark patch at nodus; a large triangular dark brown patch at base of *hw*; tips of wings sometimes smoked; pattern variable. *Thorax* hairy, brown, with lateral yellow bands edged with black. *Abd.* brown, shaded with black; lateral yellow spots on 3—8. *App. sup.* 4, pointed, *inf.* 2 triangular, tip slightly bifid. ♀ *abd* 28—29, *hw* 33—36, *pt* 3·5. Similar to ♂, but *abd.* broader; *apps.* 3·5, leaf-like. Common. Apr.—Sept.

16. *L. fulva* Müller (fig. 131). ♂ *abd* 26—27, *hw* 33—34, *pt* 2·5. *Wings* hyaline except for a medium-sized triangular black patch at base of *hw*, and short brown lines in *cu* of *fw* and *sc* of *hw*; tips sometimes clouded. *Thorax* hairy, brown. *Abd.* of adult pale blue with terminal segments black. *App. sup.* 2·5, pointed; *inf.* 2, triangular. ♀ same size as ♂; wings suffused with yellowish-brown along *C*; tips smoky. *Abd.* brown with irregular dorsal black markings on 3—10; *apps.* 2, pointed. Local. June—Aug.

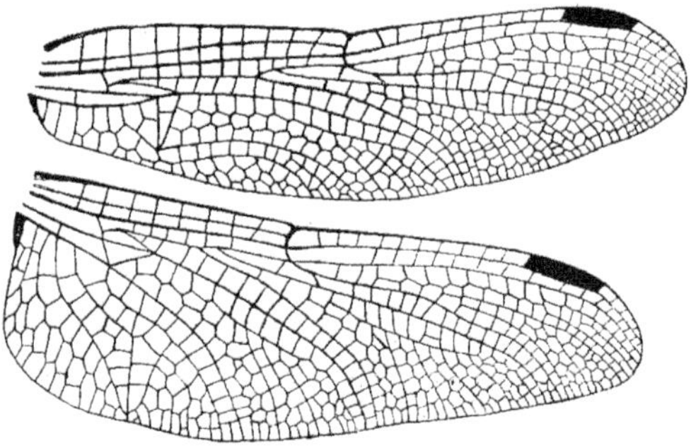

Fig. 175. Wings of *Orthetrum coerulescens* Fabr., ♂ (*Hw* 28 mm.). Original.

17. *L. depressa* Linn. ♂ *abd* 26—27, *hw* 35—37, *pt* 3·5. *Wings* hyaline, with an elongated dark brown patch at base of *fw*, and a larger triangular patch at base of *hw*. *Thorax* very hairy, brown, with bluish or yellowish dorsal stripes. *Abd.* excessively wide and depressed; blue in adult; 3—6 with yellow lateral spots. *App. sup.* 2, slightly toothed beneath; *inf.* 1·5, slightly notched. ♀ similar to ♂, but *abd.* brown; *apps.* very short. Very common. Apr.—Aug.

Genus X. *Orthetrum* Newman (fig. 175). *Fw* with about 12 *Ax* in *fw*, the last complete. Sectors of *arc.* united; *t* once crossed in *fw*, free in *hw*. *Wings* hyaline. *Abd.* only moderately widened. Lobe of prothorax large.

18. *O. coerulescens* Fabricius (fig. 175). ♂ *abd* 25—26, *hw* 28—29, *pt* 3·5, yellowish; *memb.* white. Only 1 cell-row between *Rs* and *Rspl.* *Thorax* blackish, with bluish or yellowish dorsal lines. *Abd.* of adult blue. *App. sup.*

2·5, black, pointed; *inf.* triangular. ♀ *abd* 26—28, *hw* 30—31, *pt* 3·5. *Wings* slightly suffused. *Thorax* brown; *abd.* brown or yellowish, with darker mid-dorsal line; *apps.* 1·5 pointed. Local; Southern. June—Sept.

19. *O. cancellatum* Linn. ♂ *abd* 30—31, *hw* 37—38, *pt* 2·5, dark; *memb.* greyish-black. Two cell-rows between *Rs* and *Rspl.* *Thorax* brownish; *abd.* blue in adult, with tip darkened; *abd.* broader than in No. 18. *App. sup.* 2·5, black; *inf.* triangular. ♀ *abd* 31—32, *hw* 39—40, *pt* 3. *Abd.* brown or yellow, with two black dorsal bands; *apps.* short, black. Local; Southern. May—July.

Tribe *Sympetrini* (Characters, p. 270).

Genus XI. *Sympetrum* Newman (fig. 133) (=*Diplax* Charpentier) (fig. 176). *Fw* with less than 10 *Ax*, the last incomplete. No dark patch at base of *hw*, but frequent saffroning. *Disc. field* of *fw* narrowed towards wing-border. Sectors of *arc.* united in both wings. *Abd.* not dilated. Important specific differentiation of hamuli on seg. 2 of ♂, and of *v.v.* of ♀.

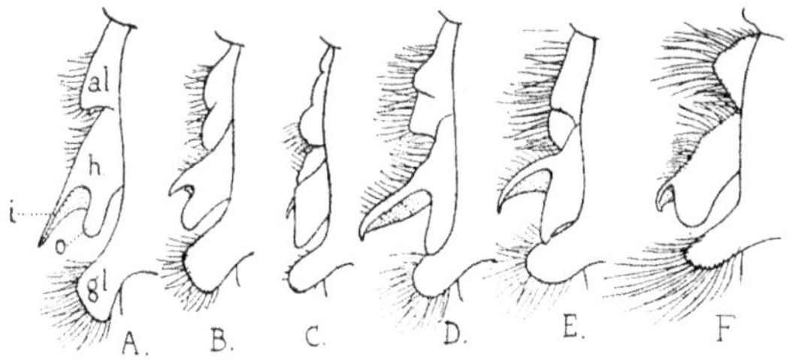

Fig. 176. Lateral view of appendages of second abdominal segment in the males of *Sympetrum*. A. *S. striolatum* Charp. B. *S. vulgatum* Linn. C. *S. fonscolombei* Selys. D. *S. sanguineum* Müller. E. *S. danae* Sülzer. F. *S. flaveolum* Linn. Original (except F. drawn by Mr H. Knight).

20. *S. striolatum* Charpentier (fig. 133). ♂ *abd* 27—29, *hw* 27—29, *pt* 2·5, reddish. *Wings* just tinged with saffron at bases; venation reddish-brown. A black line between vertex and frons, not extending to eyes. *Thorax* brown, hairy. *Abd.* red, 3—8 with small black spots posteriorly, 8—9 with mid-dorsal black spot. Hamulus with long straight inner branch (fig. 176 A). ♀ similar, but *abd.* brown with larger black markings. *V.v.* not prominent, notched. Very variable species. (*S. nigrescens* Lucas, 1912, is a variety superficially resembling No. 25.) Very common. June—Nov.

21. *S. vulgatum* Linn. ♂ *abd* 26—27, *hw* 27—28, *pt* 2·5. Closely resembles No. 20, but darker in colour. Black line between vertex and frons extends downwards alongside the eyes. Hamulus with short hooked inner branch, not so long as outer branch (fig. 176 B). *V.v.* of ♀ very prominent, not notched. ♀ *abd* 24—25, *hw* 27—28, *pt* 2·5. Rare; Southern. July—Sept.

22. *S. fonscolombei* Selys. ♂ *abd* 28—29, *hw* 29—30, *pt* 3, yellow or reddish. Venation red. *Hw* with a small saffron patch at base. Hamulus

very small (fig. 176 C). Colour scarlet all over. ♀ *hw* 30—31, *pt* yellow. Coloration yellowish brown. *V.v.* very small, not prominent. Very rare; Southern. May—June.

23. *S. flaveolum* Linn. ♂ *abd* 24—25, *hw* 28—29, *pt* 2, red. Venation dark, *C* yellowish. *Fw* with large basal patch of saffron; *hw* richly saffroned throughout basal half. *Thorax* dark red; *abd.* light red. Hamulus with short equal branches, the inner much bent (fig. 176 F). ♀ similar but olive yellow; *wings* even more saffroned; *v.v.* very small, triangular. Local; Southern. Aug.—Sept.

24. *S. sanguineum* Müller. ♂ *abd* 22—23, *hw* 27—29, *pt* 2·5, dark red. Venation reddish black. Bases of wings slightly saffroned. *Thorax* reddish brown. *Abd.* constricted at 4, deep red, except 1 and base of 2 black; 3—7 with two tiny black spots; 8—9 with mid-dorsal black streak. Hamulus large, with nearly straight inner branch longer than outer (fig. 176 D). ♀ *abd* 24—25, *hw* 29—30, *pt* 2·5, deep red. Coloration olive-yellow; *abd.* tapering posteriorly. *V.v.* small, triangular. Local; Southern. July—Oct.

25. *S. danae* Sülzer (= *S. scoticum* Don). ♂ *abd* 21—22, *hw* 24—26, *pt* 2, broad, black. *Wings* hyaline; venation black. Coloration black. *Abd.* much constricted at 4, dilated at 8. Hamulus large, with both inner and outer branches equally long (fig. 176 E). ♀ *abd* 22—23, *hw* 26—27, *pt* 2·5. Coloration orange-brown, reddish-brown or brownish-black when mature, yellowish with black markings when immature. *Abd.* tapering posteriorly. *V.v.* large, very prominent, pointed. Common. July—Oct.

Tribe *Leucorrhiniini* (Characters, p. 272).

Genus XII. *Leucorrhinia* Brittinger (fig. 134). *Fw* with less than 10 *Ax*, the last one complete. *Hw* with a dark patch at base. Sectors of *arc* slightly separated in *fw*, united in *hw*. *Disc. field* of *fw* widened towards wing-border.

26. *L. dubia* Vanderlinden. ♂ *abd* 26—27, *hw* 27—28, *pt* 1·5—2, dark brown. A dark brown triangular patch at base of *hw*; a small dark spot at base of *fw*. *Thorax* black with two crimson dorsal stripes, also some lateral stripes and lines. *Abd.* subcylindrical, black, with crimson spots. *App. sup.* 1·5, black; *inf.* truncated. ♀ *abd* 25—26, *hw* 27—28, *pt* 2—2·5. Conspicuous saffroning around basal mark of *hw*. Crimson of ♂ replaced by yellow. *V.v.* wide apart at bases, converging, blunt at tips. Local; Northern. July—Aug.

Suborder ZYGOPTERA (Characters, p. 273).

Family CALOPTERYGIDAE (Characters, p. 274).

Subfamily **Calopteryginae** (Characters, p. 275).

Genus XIII. *Calopteryx* Leach (fig. 177). *Wings* broad in middle. No regularly formed *pt*. No cross-veins in *ms*.

27. *C. virgo* Linn. ♂ *abd* 35—39, *hw* 27—36 by 10—11 wide, *pt* absent. *Wings* dark brown, with bright metallic blue reflections, except at extreme base and tip. Body metallic blue. ♀ *hw* about 1 mm. narrower. *Wings* uniformly dark brownish; a small white false *pt* present. Body metallic green; a median yellow line on *abd.* 8—10. Common. May—Sept.

28. *C. splendens* Harris (fig. 177). ♂ *abd* 33—39, *hw* 27—36, by 9—10 wide, *pt* absent. *Wings* hyaline, with a large dark patch extending from level of nodus to within 4 mm. of apex, shewing metallic blue reflections. Body metallic blue, tinged with green. ♀ *hw* about 1 mm. narrower. *Wings* uniformly pale translucent yellowish-green. Body metallic green, end of *abd.* coppery; 10 with a median yellow line. Common; Southern. May—Sept.

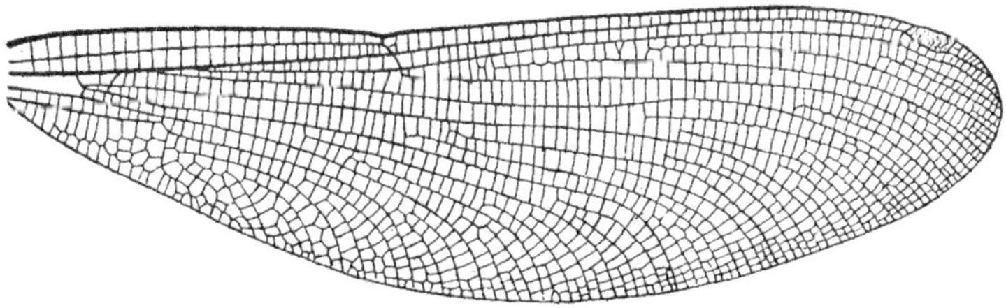

Fig. 177. Fore-wing (35 mm.) of *Calopteryx splendens* Harris, ♀. Original.

Family Lestidae (Characters, p. 276).

Subfamily **Lestinae** (Characters, p. 277).

Genus XIV. *Lestes* Leach (fig. 142). M_2 leaves M_1 about 3 cells distad from nodus. *Abd.* considerably longer than wings. ♂ with *app. sup.* forcipate. Metallic coloration.

29. *L. dryas* Kirby (=*L. nympha* Selys). ♂ *abd* 26—35, *hw* 19—26 *pt* 1—1·5, black. Robust build. Colour bronze-green; powdered blue on underside of thorax, and on *abd.* 1, part of 2, and 9—10. *Apps.* black, as in fig. 178 A. ♀ *pt* dark brown. Colour more coppery, especially on *abd.* 8—10. No blue powdering. Very rare; Fens. June—Aug.

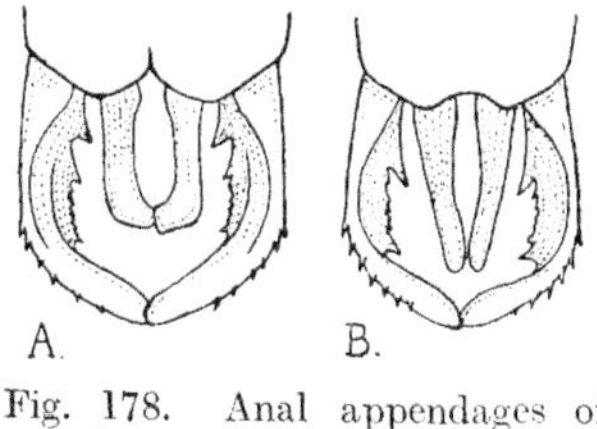

Fig. 178. Anal appendages of males of *Lestes*, dorsal view. A. *L. dryas* Kirby. B. *L. sponsa* Hans. Original.

30. *L. sponsa* Hansemann. ♂ *abd* 25—30, *hw* 18—21, *pt* 1—1·5, very narrow, dark brown. Slender build. Colour bronze-green; powdered blue on underside of thorax, and on *abd.* 1—2 and 9—10. *Apps.* black, as in fig. 178 B. ♀ with stouter *abd.*, more coppery colouring, and no blue powdering. Common. July—Oct.

Family Agrionidae (Characters, p. 277).

Subfamily **Platycneminae** (Characters, p. 279).

Genus XV. *Platycnemis* Charpentier (fig. 145). *Quad.* rectangular elongated. Tibia 2—3 dilated especially in ♂ (fig. 9 B).

31. *P. pennipes* Pallas. ♂ *abd* 27—28, *hw* 20—21, *pt* 0·8, rhomboidal, brown. *Thorax* black with green dorsal lines. *Abd.* blue with black markings; *apps.* very short. ♀ *abd* 26—30, *hw* 20—25, *pt* 1. *Abd.* stouter, blue replaced by greenish-yellow. Local; Southern. May—Sept.

Subfamily **Agrioninae** (Characters, p. 279).

Tribe *Agrionini* (Characters, p. 280).

Genus XVI. *Ischnura* Charpentier (fig. 18 J). *Quad.* with sharply acute distal angle. M_2 arises 4 cells distad from nodus in *fw*, 3 in *hw*. *Pt* in *fw* of ♂ bicolorous, proximally black, distally white. Postocular spots present. End of seg. 10 in ♂ projecting upwards; ♀ with a ventral spine under seg. 8. Coloration greenish-black, slightly metallic; a little blue on ♂. Dimorphic females.

32. *I. elegans* Vanderlinden. ♂ *abd* 22—26, *hw* 13—17, *pt* 0·6. Postocular spots, thoracic bands and *abd.* seg. 8 bright blue, 9—10 black. ♀ *abd* 22—26, *hw* 15—20, *pt* 0·8, unicolorous. *Normal form* of ♀ closely resembles ♂, but *abd.* stouter. *Heteromorphic form* has head, thorax, legs and *abd.* 1—2 orange, with black markings. Very common. May—Sept.

33. *I. pumilio* Charpentier. ♂ *abd* 18—24, *hw* 12—17, *pt* 0·5. Resembles No. 32, but *abd.* 8 blackish, 9—10 pale blue. ♀ *hw* 13—19. *Normal form* of ♀ has blue replaced by yellowish-green on head and thorax; *abd.* fairly stout, greenish-black. *Heteromorphic form* (var. *aurantiaca*) has head, thorax, legs, *abd.* 1—2 and part of 3 orange. Rare; Southern. June—Sept.

Genus XVII. *Agrion* Fabricius (fig. 148). *Quad.* with sharply acute distal angle. M_2 arises 5—6 cells distad from nodus in *fw*, 4—5 in *hw*. *Pt* unicolorous. Postocular spots present. ♂ with *abd.* seg. 10 not upturned. ♀ with a ventral spine under seg. 8. Colours bronze-black and bright blue.

34. *A. armatum* Charpentier. ♂ *abd* 29—32, *hw* 16—19, *pt* 0·6. *Abd.* 1—3 as in fig. 179, *a*. *App. inf.* large, longer than seg. 10. ♀ *abd* 27—32, *hw* 18—21, *pt* 0·6. *Abd.* stouter, blue colour less bright, or greenish; seg. 2 with black mark pointed basally. Very rare; Norfolk Broads. May—June.

35. *A. pulchellum* Vanderlinden. ♂ *abd* 23—32, *hw* 16—20, *pt* 0·6. Very slender build. *Abd.* 1—3 as in fig. 179, *b*. *App. inf.* small. ♀ *abd* 25—29, *hw* 18—22, *pt* 0·6; head and thorax with blue replaced by greenish; *abd.* a little stouter; seg. 2 with apical two-thirds or more black, the black mark concave or three-pointed basally. Local. Apr.—Sept.

36. *A. hastulatum* Charpentier. ♂ *abd* 28—34, *hw* 17—20, *pt* 0·7. Slender build. *Abd.* 1—3 as in fig. 179, *c*. *App. inf.* small. ♀ *abd* 29—32, *hw* 18—22, *pt* 0·7; *abd.* stouter, the blue replaced by greenish; seg. 2 with black mark along the whole length. Very rare; Inverness-shire. June—Aug.

37. *A. mercuriale* Charpentier. ♂ *abd* 21—26, *hw* 15—19, *pt* 0·5. More robust than No. 36. *Abd.* 1—3 as in fig. 179, *d*; *app. inf.* small. ♀ *abd* 21—27, *hw* 15—19, *pt* 0·5; closely resembles No. 36, but *abd.* stouter, *pt* shorter; blue replaced by greenish. Rare; New Forest. May—Aug.

38. *A. puella* Linn. ♂ *abd* 24—29, *hw* 17—20, *pt* 0·7. Very slender species, of larger size than Nos. 36, 37. *Abd.* 1—3 as in fig. 179, *e*. ♀ *abd* 23—29, *hw* 18—23, *pt* 0·8; *abd* slender, seg. 2 with black mark along its whole length; blue replaced by green, blue-green or yellowish. Very common; May—Sept.

Genus XVIII. *Erythromma* Charpentier. Venation as in XVII. No postocular spots. No ventral spine under seg. 8 in ♀. Coloration of ♂ bronze-black and pale blue. *Eyes* reddish in ♂, golden-brown in ♀.

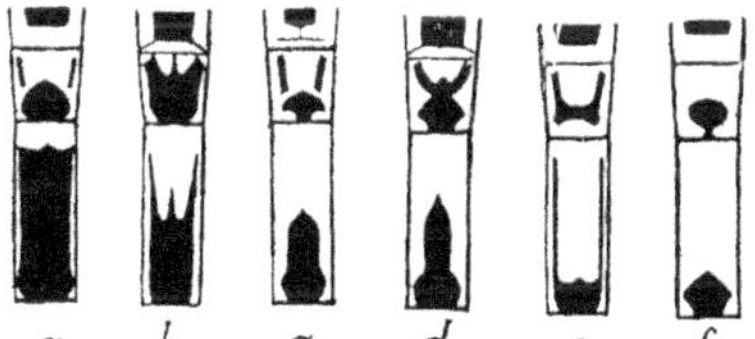

Fig. 179. Colour-pattern (bright blue and bronze-black) of first three abdominal segments in the males of *Agrion* and *Enallagma*. *a*, *Agrion armatum* Charp.; *b*, *A. pulchellum* Vand.; *c*, *A. hastulatum* Charp.; *d*, *A. mercuriale* Charp.; *e*, *A. puella* Linn.; *f*, *Enallagma cyathigerum* Charp. After Ris.

39. *E. naias* Hansemann. ♂ *abd* 25—28, *hw* 20—24, *pt* 0·9. *Thorax* black above, without dorsal stripes. *Abd.* pale blue on 1, sides of 2, and all 9—10. *App. sup.* stout. ♀ *abd* 26—29, *hw* 22—25, *pt* 0·9; *thorax* with two yellow dorsal lines; *abd.* 1 yellow with a black spot; rest bronze-black, sutures 2—6 yellow, 7—9 blue. Local; Southern. May—July.

Genus XIX. *Pyrrhosoma* Charpentier. Characters as in XVIII, but *abd.* red, or red and black.

40. *P. nymphula* Sülzer (=*P. minium* Harris). ♂ *abd* 26—30, *hw* 21—23, *pt* 0·9. Robust species. *Thorax* black with narrow red bands. *Abd.* bright red, 7 black, 8—9 partly black. ♀ *abd* 25—29, *hw* 22—25, *pt* 1; *abd.* stouter, with duller red and more black. Very common. Apr.—Aug.

41. *P. tenellum* de Villers. ♂ *abd* 22—25, *hw* 15—18, *pt* 0·5. Very slender species. *Thorax* dark bronze; *abd.* wholly red. ♀ *abd* 25—28, *hw* 17—20, *pt* 0·6; thorax darker; *abd.* 1—3 red, 4—10 bronze-black. Common; Southern. June—July.

Tribe *Pseudagrionini* (Characters, p. 280).

Genus XX. *Enallagma* Selys. Characters as in XVII (except for position of *Ac*, a tribal character).

42. *E. cyathigerum* Charpentier. ♂ *abd* 25—28, *hw* 18—21, *pt* 0·7, black. *Abd.* blue, with black pattern of 1—3 as in fig. 179, *f*, 6—7 considerably blacker,

8—9 blue with black points, 10 black. ♀ *abd* 22—28, *hw* 18—22, *pt* 0·8, yellow; blue parts replaced by grey-green or reddish-brown; *abd.* with elongated bronze-black markings on all segments. Very common. June—Sept.

The following reputed British species are not considered to have established a satisfactory claim to be included in the above list, either the determination or authenticity of the few captures recorded being open to doubt: *Gomphus flavipes* Charpentier, *Lindenia forcipata* Linn., *Sympetrum meridionale* Selys, *Leucorrhinia pectoralis* Charpentier, *Lestes viridis* Vanderlinden, *Lestes virens* Charpentier, and *Lestes barbarus* Fabricius.

LARVAE.

It is impossible to give a complete classification of the larvae of British species, partly because a number of them are still unknown, and partly because, in one or two genera, the similarity between the larval forms is so great that specific differences have not yet been definitely established. The following table should be used in conjunction with the characters given for the larvae in chap. XIV, and also with the figures of larval types and labial masks (figs. 29—42).

ANISOPTERA.

Subfamily *Gomphinae*. *Gomphus vulgatissimus*. Larva of type shewn in fig. 36 B. Mask of type in fig. 32 D. Found in débris on the bed of running streams.

Subfamily *Cordulegastrinae*. *Cordulegaster annulatus*. Larva of type in fig. 33, but with abdomen somewhat more elongated. Mask of type in fig. 32 L. In débris on the bed of running streams. (The only British larva with *divergent* wing-sheaths.)

Subfamily *Aeschninae*. Larva of type in fig. 29 A. Mask flat (figs. 29 D, 32 A)

Genus *Brachytron*. Eyes small, rounded, placed well forward on head. Mask in position of rest not reaching beyond second coxae; lateral lobe with narrow rounded apex. Anal pyramid short (length of seg. 10), the appendix dorsalis truncated at tip (*B. hafniense*).

Genus *Aeschna*. Eyes larger, hemispherical, occupying about half the length of the head. Mask in position of rest as in *Brachytron*; lateral lobe with broad squarish apex. Anal pyramid long (length of segs. 9 + 10), the appendix dorsalis notched at tip.

The five known larvae of this genus are best separated by the form of the supracoxal armature, as shewn in fig. 180. The larva of *Ae. coerulea* is unknown.

Genus *Anax*. Eyes very large and flat above, occupying two-thirds the length of the head. Mask reaching back to beginnings of third coxae; lateral lobe with apex somewhat narrowed and carrying a sharp inner tooth. Anal pyramid very long (longer than segs. 9 + 10) (*A. imperator*)[1].

Subfamily *Corduliinae*. Larva of type in fig. 38. Mask spoon-shaped (fig. 32 o), the teeth of the distal border in the form of large crenations.

Genus *Oxygastra*. Abdomen with no dorsal spines, but with fairly large lateral spines on segs. 8—9. Crenations of mask very large, one-third as high as wide, each carrying 14 or 15 short bristles (*O. curtisi*).

Genus *Cordulia*. Abdomen with small dorsal spines on segs. 5—9, rather small lateral spines on segs. 8—9. Legs very long; hind femur reaching to seg. 8. Crenations of mask medium, one-fourth as high as wide, each carrying 4 or 5 short bristles (*C. aenea*).

Fig. 180. Supracoxal armature of prothorax in larvae of *Aeschna*. *a*, *Ae. grandis* Linn.; *b*, *Ae. juncea* Linn.; *c*, *Ae. mixta* Latr.; *d*, *Ae. cyanea* Müller; *e*, *Ae. isosceles* Müller. After Lucas and Cabot.

Genus *Somatochlora*. *Somatochlora metallica*. Abdomen with very large dorsal spines on segs. 5—9, small lateral spines on seg. 8. Hind femur reaching to seg. 7. Crenations of mask medium, one-fourth as high as wide, each with 10 bristles.

S. arctica. Abdomen with neither dorsal nor lateral spines. Hind femur reaching only to middle of seg. 6. Crenations as in *S. metallica*, but with only 5 bristles on each.

Subfamily *Libellulinae*. Tribe *Libellulini*. Larva of type in fig. 36 F. Mask spoon-shaped (cf. fig. 32 N), the teeth of the distal border forming small flattened crenulations.

Species	Length	Lateral abdominal spines	Dorsal abdominal spines
Libellula quadrimaculata	23·5	8, moderate; 9, small	5—7, small; 8, moderate; 9, nil
,, *fulva*	21	8—9, large, bent	6—9, large
,, *depressa*	21	8, rudimentary; 9, nil	5—7, small; 8, moderate; 9, nil
Orthetrum coerulescens	16	very small, bent, on 8—9	7, rudimentary; 8, nil
,, *cancellatum*	23	small, on 8—9	,, ,,

In addition, *L. depressa* can be distinguished by having the crenulations of the lateral lobe of the mask about one-fifth as high as wide; in all the others they are excessively flattened down.

[1] Lucas describes and figures this larva as having the appendix dorsalis notched at tip, and nearly as long as the cerci. Ris, however, says that this organ is truncated ("gerade abgeschnitten") and barely half as long as the cerci!

Tribe *Sympetrini*. Larva of type in fig. 36 G. Mask much as in *Libellulini*, the crenulations excessively flattened, the lateral lobes very warty. Abdominal dorsal spines small.

Sympetrum striolatum. Length 16·5. Large lateral spines on segs. 8—9. Mask with 12—14 mental, 11 lateral setae. (The other species have not been definitely determined.)

Tribe *Leucorrhiniini*. Larva resembling that of *Sympetrini*, but with the occiput narrower. Strong lateral spines on segs. 8—9 of abdomen, small dorsal spines on segs. 4—5 (*Leucorrhinia dubia*).

ZYGOPTERA.

Family *Calopterygidae*. Larva of type in fig. 40. Mask with very deep median cleft (fig. 32 R). Caudal gills: lateral long, triquetral; median short, lamellar.

Calopteryx virgo. Pedicel of antenna 3·5 to 3·8 mm. long. Lateral gills 9 mm. (fig. 40).

Calopteryx splendens. Pedicel of antenna 4 to 4·2 mm., stouter. Lateral gills 12 mm.

Family *Lestidae*. Larva of type in fig. 41 B. Mask with setae on the movable hook (fig. 32 H). Caudal gills long, lamellar, tips rounded (fig. 84 B) (*Lestes*).

The two British species have not yet been determined with certainty.

Family *Agrionidae*. Larvae of type in fig. 42 D, E. Mask of type in fig. 32 F.

Species	Length + Gills	No. of setae on mask Mental	Lateral	Caudal Gills	General Remarks
Platycnemis pennipes	11·5 + 7	2	3	Long, opaque, very pointed	Stout, dark, long-legged, sluggish, muddy
Ischnura elegans	13·5 + 6·5	4	6	Subnodate, pointed, transparent	Variegated green or yellow, active, on weeds
Agrion pulchellum	13·5 + 5	4	5	Nodate, tip rounded	Rather opaque, dark brown
,, *mercuriale*	13·5 + 3·3	3	5	Denodate, short, pointed	Opaque ochreous, sluggish
,, *puella*	14 + 5·5	4	5	Nodate, very slightly pointed	Rather transparent, grey-green
Erythromma najas	18 + 7·5	3	6	Nodate, transparent, rich branching of tracheae	Rather transparent, greenish
Pyrrhosoma nymphula	11·5 + 4·8	3	6–7	Denodate, opaque, broad triangular tip	Broad, short, brownish, sluggish, muddy
Enallagma cyathigerum	14 + 6	4	6	Subnodate, transparent, bluntly pointed	Pale green or yellow, slightly marked

(The other species have not been satisfactorily determined.)

CHAPTER XIX

COLLECTING, REARING AND BIOLOGICAL METHODS

This chapter does not pretend to give a complete account of all the methods in use for the collecting, preserving, rearing and biological study of the Odonata. This would take up too much space. Ten years' experience in the field has convinced me that many of the more cumbersome methods are best discarded altogether. I shall therefore chiefly confine myself to the methods which I have finally come to employ, as being the best that I am at present aware of; though I shall not omit to mention other methods also, which are used by others, and appear to yield equally good results.

The Collection.

The collection must be made the basis of all work on Odonata, be it systematic or biological. Hence it is of the utmost importance that it should be well cared for, and arranged so that the specimens are easily accessible for study.

Field-Work. The apparatus required for the capture of Dragonflies is very simple. The only necessary articles are a *large* net and a tin full of paper triangles. A killing-bottle should not be used. For the net, I use the kite-shaped net, in four pieces and a Y; dimensions 2 feet by 18 inches. This is too large for most Lepidoptera, but is excellent for Odonata. It is also a useful net for "sweeping" bushes and long grass—a method which yields many Odonata in the early morning, and on dull days. In tropical countries, I find that the Y gets overheated, so that the joints melt. It is therefore advisable to have several Y's on hand, and have them well strengthened by a brazier.

To have really good specimens of Dragonflies, it seems to me to be essential that they should not be killed artificially, but allowed to die slowly. By this means, the whole of the contents of the alimentary canal is excreted in the form of small dry pellets, which can be emptied out of the paper every day, until the Dragonfly is dead. The result is that the colours keep far better than if the Dragonfly is killed with these impurities inside it. For storing the live Dragonflies, a number of paper triangles should be folded up in the manner indicated in the diagram (fig. 181), taking care that the *straight edge* of the flap comes to the left (for convenience in writing date and locality). These should be made of a moderately soft white paper (about the stiffness of newspaper) so as not to crush the eyes of the Dragonfly. For ordinary localities, three sizes are sufficient: *Agrionid*-size, $3\frac{1}{4} \times 2$ inches; *Libellulid*-size, $4\frac{1}{2} \times 3$ inches; and *Aeschnid*-size, 6×4 inches. These proportions give a reasonably broad flap for writing on. The dragonfly is placed in the paper with its wings folded back, and the flaps are carefully closed. It is convenient to have a flat tin in one pocket, full of empty papers, and a wider rectangular tin or small box in the other, for reception of the papered specimens. This prevents overcrowding, and consequent flattening of the eyes.

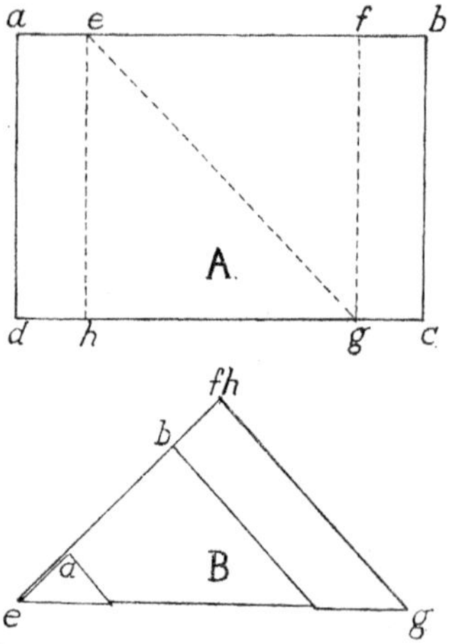

Fig. 181. Diagram to shew method of folding a paper triangle. In A, the dotted lines indicate creases. The paper should first be folded about *eg*; then the edges are folded over along *eh* and *fg*, the final result being shewn in B. Original.

Agility, except of the wrist, plays little part in the capture of Dragonflies. They are not to be caught by chasing. A swift back-hand stroke over one's head is the most effective way of capturing large species. With all fast or wary species a safe rule is to *strike from behind*, or, if the stroke must be from in front, then *downwards* upon the insect. A direct frontal stroke can be seen and dodged by most species. To get within striking distance, it is only necessary to watch the Dragonfly carefully, and note the places where he turns, or the twigs which he seems to favour. Most species will return often to the same spots. Also, in all movements of approach towards Dragonflies at rest, it is essential to move uniformly, without jerks of any kind. The collector who becomes expert in this will be surprised to find that he can usually rely upon getting within striking distance of even the largest species without alarming them. But if he must wink his eyelid, or make any sudden movement with his feet or hands, all his labour will be lost. . Nor should he wear any striking colours, such as a white hat or bright tie.

In choosing localities for collecting, *small* ponds or streams are always to be preferred to large ones. Even though the fauna be less numerous, the chances of catching them are increased in a much greater proportion. A brook of such a width that, by standing on a jutting promontory, the collector can command the whole passage of the stream, is the ideal place for catching Dragonflies.

Preservation. As soon as the specimen is dead, it should either be *set*, or permanently *papered*. Set Dragonflies take up a lot of room, but they are, to my mind, far preferable to papered specimens, and more easily looked after, since all the parts are spread out clearly to the view, and the pin is convenient for handling the insect without touching it with the forceps. In setting, a *flat* board must be used, with a groove deep enough to allow the body of the insect to be placed at least half-way up the pin. Uniformity of plan in setting is best obtained by placing the costa of the hind-wings at right-angles to the body, the fore-wings lying just free of it. The tips of the wings should be covered during setting, or they may turn up. Also, in the broad-winged *Libellulidae*, the anal field of the hind-wings should be raised from beneath by a small slip (fig. 182).

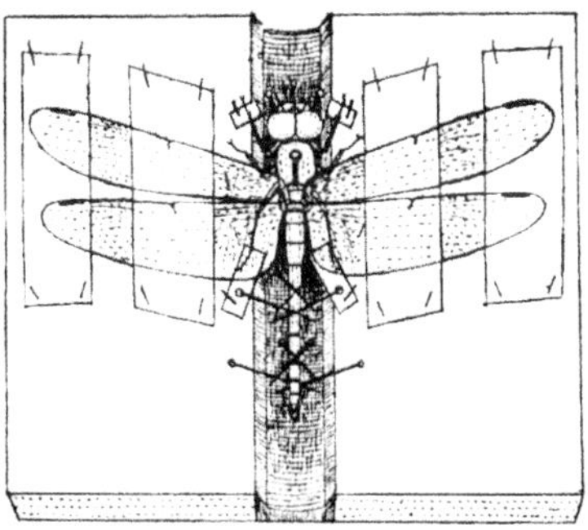

Fig. 182. A Dragonfly set on a flat board with double strips of transparent paper. Notice the pins supporting the head and abdomen, and the slips placed beneath the broad anal field of the hind-wings. Original.

The abdomen of a dried Dragonfly is very liable to break in pieces. This is best obviated by the process of "bristling" before setting (fig. 183). For Anisoptera, I use hog's bristles, such as can be obtained at all curriers and leather-sellers. For Zygoptera, a fine stiff horse-hair is the best. The bristle should be sharpened by a diagonal cut with the scissors. It is then inserted into the thorax above and between the mesocoxae, and run carefully down

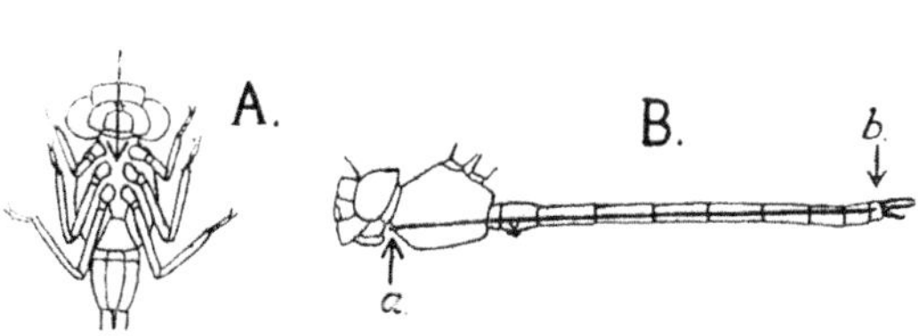

Fig. 183. Correct method of inserting a bristle. In A, the point of the arrow indicates the point of insertion. In B, the bristle is run through the abdomen as far as *b*, and is then cut off with sharp scissors at *a*. Original.

to the end of the abdomen, care being taken not to damage or displace the appendages, and a small allowance being made for shrinkage during the process of drying. The bristle is cut off close to the thorax with sharp scissors. The head and prothorax do not require bristling; if the head is accidentally knocked off, it can easily be replaced.

For the process of papering, the paper may be prepared by treatment with a solution of mercuric chloride, or a little dust of naphthaline or thymol may be placed with the insect. A small piece of pith gummed into the head-end of the triangle will serve to keep the two sides sufficiently apart to prevent damage to the head of the insect by pressure. The appendages of the second segment of the male are more easily studied in papered specimens than in set ones. The specimens should be kept in large tobacco-boxes in a dry place, and securely fastened down except when needed for examination.

Each specimen must of course be labelled with locality, date and name of collector. In set specimens, the label should be placed on the pin *upwards*, so that it can be read without moving the insect. In papered specimens, the label is either printed on the flap, or on a small slip placed inside with the Dragonfly.

Naphthaline, thymol, and an occasional dosing with carbon bisulphide will keep the collection clear of the usual insect pests. *Grease* is very rare on Dragonflies. To my knowledge, it only attacks the larger *Aeschninae*, particularly *Anax* and *Gynacantha*. It can be cured by immersion in powdered magnesia, or, better still, by careful evisceration of the specimen before setting, and re-filling the abdominal cavity with cotton-wool and powdered magnesia. In damp climates, a small white fungus is apt to attack the specimens. This can be removed by treatment with an alcoholic solution of mercuric chloride, or with carbolic acid.

Preservation of Colours. Much has been written on this subject. The *pattern*, though not the brilliance, of all dragonflies can be preserved, if the insect is allowed to die in the manner already described. For preservation of the bright hypoderm pigments, *rapid drying* is essential. A collection which I received from Central Australia, and which had travelled for three months per camel-mail, was found to be absolutely perfect in coloration, even down to bright reds and blues. Such was the effect of the hot dry climate! I find that a dry temperature of 105° Fahr. is the best. At this temperature, a Dragonfly will set in from eight to twelve hours, with all its colours perfect (except only the sky-blues of *Agrionidae* and the blues and greens of *Aeschninae*, which are unreliable). For this purpose I have designed the oven shewn in fig. 184, in which the setting-boards are placed along two sides of the oven, face to face vertically, and are held in place by the two ridges on the lid.

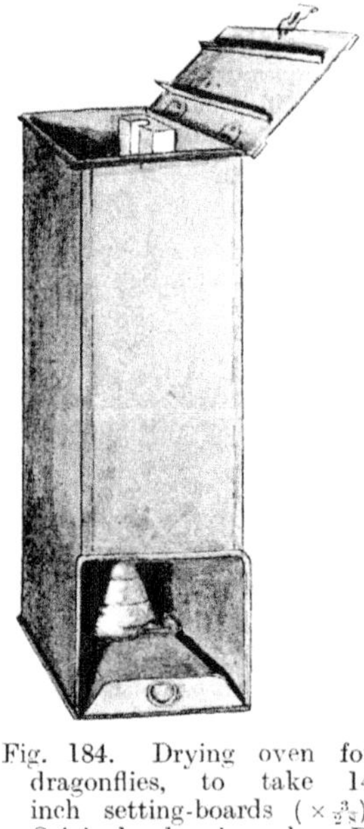

Fig. 184. Drying oven for dragonflies, to take 14 inch setting-boards ($\times \frac{3}{28}$). Original drawing by P. Tillyard.

The burner is an ordinary kerosene fount, such as is used in incubators. The boards should be turned every few hours, to equalize the heating. I have a series of the glorious red *Lathrecista festa*, taken at Cooktown in 1907, which are as brilliant to-day as when I caught them. This species turns dull brown if treated in any other way. Overheating (above 110° Fahr.) must be avoided, since it causes an unnatural iridescence of the wings.

The only safe way to preserve the brilliant blues of *Agrionidae* is to keep a few specimens always in 70 per cent. alcohol, in the dark.

Larvae. The best method of collecting live larvae is by using a small dredge-net. I use an ordinary 8 inch iron hoop, in one piece, with a strong ferrule and a bag of mosquito-netting, not too deep. With a stick four or five feet long, the water-weed, débris and bottom of any pond or river can be easily worked, and the larvae picked out. For some localities, a straight sharp edge is useful on the net, e.g. for running along a hard rocky bottom,

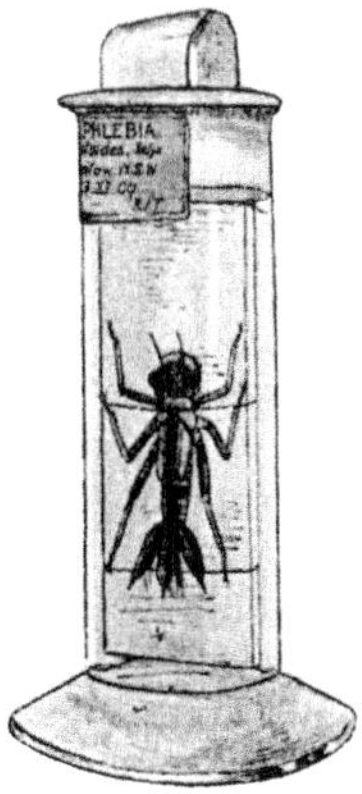

Fig. 185. Dragonfly larva mounted on opal glass in specimen jar containing 70 per cent. alcohol (× ½). Original drawing by P. Tillyard.

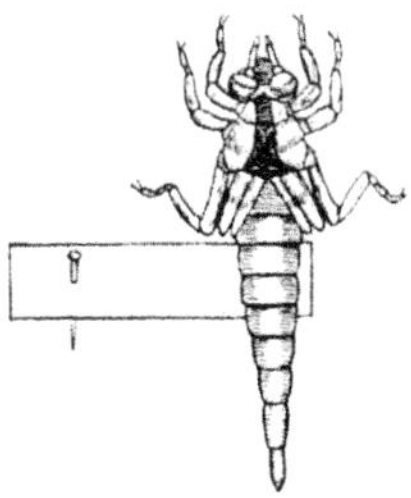

Fig. 186. Exuviae of *Austrogomphus manifestus* Tillyard (New South Wales), cleaned and mounted on a cardboard slip (× 1). Original.

or up smooth vertical rocks. Other ingenious devices are given by Needham [97].

Larvae can be dried and carded (like Coleoptera); but they make better specimens if kept in alcohol. The larva should be placed alive in alcohol (70 per cent., or 50–60 per cent. warm to hot). As soon as it is dead, a small incision should be made ventrally along segments 3–5 (mid-gut region), to allow the alcohol to penetrate there. If this is not done, the contents of this part of the abdomen will decay, and cause these segments to become unnaturally distended, often ruining the specimen. My series of larvae are mounted with silk on slips of opal glass, and placed in 4 × 1, or 4 × 1½ inch cylindrical specimen jars with strong bases, in 70 per cent. alcohol (fig. 185).

Exuviae. For external characters, at least in the Anisoptera, the exuviae are perhaps preferable to the actual larvae. They are much more easily collected, and much less difficult to deal with for the collection. Care must be taken in removing the exuviae from the reed-stems, to prevent damage to the legs. They should be prepared for the cabinet by being boiled in a weak solution of caustic potash for about half-a-minute. They can then be removed and thoroughly cleaned with a camel's-hair brush. When dry, the exuviae may be arranged in small boxes or trays, or mounted on a transverse slip of cardboard passing under the middle of the abdomen, with a little gum. The pin and label may be placed at the left-hand side, if the slip be made long enough (1 to 1½ inches). (Fig. 186.)

Rearing.

According to Needham [97], the best way to rear Dragonflies is to "let them rear themselves," by placing them in specially constructed cages which can be sunk in the water where they live. I cannot agree with this view, because (1) if the object of rearing be to get specimens, bred specimens are never so good as caught ones, since they can never be properly matured, (2) if the object is to study the life-history of the species, it can only be attained by having the specimens constantly under observation in the laboratory. Besides, very few of us live either on the edge of a lake or anywhere near a suitable locality for even occasional observation of such cages.

Obtaining the eggs. The rounded eggs of exophytic species are easily obtained by catching a female while ovipositing. If the tip of her abdomen be then placed in water in a small phial, she will usually continue to lay eggs in large numbers. It is necessary to have the water in the phial *dirty*, with mud, sand or small pieces of débris for the eggs to fall upon; otherwise the eggs will simply all stick together, and quickly go mouldy. The elongated eggs of endophytic species must be sought for in the leaves or stems of water-weeds. These may be dredged ashore in masses, and then patiently examined; or, if preferred, a female may be watched while ovipositing, and the reed-stems gathered when she has gone. The stems may be broken up into pieces and left in water, without harming the eggs.

In the laboratory, the eggs are best kept in flat Petri dishes, covered over to prevent evaporation of the water. Decaying vegetable matter seems to be *essential*. I have never succeeded in rearing eggs kept in pure water. Rounded eggs are the more difficult to hatch. I find, however, that, even if a strong growth of fungus appears on these eggs, they will usually hatch if mixed with mud and débris. Rounded eggs hatch in from five days to three weeks; elongated eggs take about three weeks or a little less.

Rearing the Larvae. The young larvae should be picked out with a camel's-hair brush and segregated (a few together) in slightly deeper Petri dishes, in water on which duck-weed has been floating for some time. They may be fed on *Paramoecium* and other Infusorians. They grow fairly rapidly.

At the fourth or fifth instar they should be transferred to larger "observation jars," and fed on water-fleas, or small mosquito-larvae. Only one species of larva should be kept in each jar. Later on they may be fed with larger mosquito-larvae, or small *Agrionid* larvae.

Glass cells, either rectangular or cylindrical, make excellent observation jars (fig. 187). These should have a clean sandy bottom, with water-weeds growing therein. I find *Vallisneria* and *Elodea* the best forms for aërating the water, particularly the former. The duck-weed (*Lemna*) and the pretty *Azolla* are also invaluable as surface weeds. A few sticks should be arranged so as to stand up some inches above water-level, for the larvae to climb up at metamorphosis.

For bottom-dwelling larvae I use a small circular tin pan (fig. 188), with

Fig. 187. Observation jar for rearing dragonfly-larvae (× $\frac{1}{4}$). Original drawing by P. Tillyard.

Fig. 188. Pan for rearing rock or bottom-dwelling larvae (× $\frac{2}{5}$). Original drawing by P. Tillyard.

sloping sides about three or four inches deep (a puppy's drinking pan is a useful kind). A sandy bottom, a small piece of rock or two, and a liberal supply of *Lemna* for the aëration of the water, are all that is required.

When the larvae are full-fed, a small cage, open below, and constructed of mosquito-netting on a wire framework, should be let down over the observation jar or pan. The larvae will crawl out up one of the sticks or on to the netting, and the imago will be found resting at the top of the cage. It may then be removed to a larger cage of similar design, to allow the wings and colours to mature. Bred specimens, however, are seldom at all satisfactory for the cabinet, as maturity depends not only on time and sunlight, but also on activity and feeding. Captive imagines, as far as my experience goes, cannot be persuaded to feed. I find it convenient to place all bred specimens, either set or pinned with folded wings, in a special box with their exuviae alongside them.

Biological Methods.

Chitin-Preparations. Cut off the selected organ or part, and place it in a strong solution of caustic potash for 24 hours or longer [or it may be boiled in the same solution for from a minute or two to half-an-hour, according to its nature]. Remove to physiological salt solution, and clean off all muscles, etc. left on. Wash in distilled water. Transfer to 70 per cent. alcohol, with at least one change. Run up through 90 per cent. to absolute alcohol, thence to clove oil. The preparation may be conveniently examined in this medium, and useless parts cut away. Mount in the usual manner with Canada Balsam. For staining chitin-preparations I find a solution of Orange G in water (24 hours or longer) gives the best results. Most chitin-preparations from Odonata do not, however, require any staining.

Cleared Larvae. Beautiful preparations of whole larvae, as well as of the head, thorax, abdomen, branchial basket or caudal gills separately, may be made as follows: Fix the larva (alive) in Carls' Fixative (see below); wash out in 70 per cent. alcohol with several changes extending over 24 hours or more; pass to 90 per cent. alcohol (6 hours), absolute alcohol (6 hours), absolute alcohol + cedar oil, equal parts (24 hours), and finally into cedar oil. Leave in cedar oil for a week or more (the longer the better). The whole larva will gradually become of a glassy transparency, until all the internal organs can be made out with great clearness. Keep the specimen always in cedar oil in a tube. The opened branchial basket, or the caudal gills, may be mounted in Canada Balsam within a raised ring.

Stained Whole-Mounts. This method may be used for studying the embryo, whole specimens of young larvae, various appendages (legs, caudal gills) and portions of internal organs. Fix as above, and wash well in 70 per cent. alcohol. Pass to 50 per cent. alcohol, thence to a solution of Grenacher's Borax Carmine in 35 per cent. alcohol. Leave in stain until thoroughly penetrated (24 hours to a week). Differentiate in 70 per cent. alcohol with 0·5 per cent. nitric acid added. Dehydrate, clear and mount in the usual manner.

Sections. Out of many methods which may be employed for this purpose, I shall only give in detail the two which I find to be most suitable for the Odonata in general. The processes are as follows:

A. *Single Embedding*:

1. Fix the animal, or part selected, in Carls' Fixative[1]; formula: absolute alcohol 15 parts, concentrated formol 6 parts, glacial acetic acid 2 parts, and distilled water 30 parts. For whole larvae, make a small slit for penetration of the mid-gut region (see above). Do not prolong fixation beyond 24 hours, as the fixative will then have a "swelling" action on the cell-layers.

[1] Used by Carls and Kurt-Bedau for studying the compound eyes of insects.

2. Wash thoroughly in 70 per cent. alcohol, with several changes during 24 hours or longer.

3. Pass into Soap Solution (soft soap in 70 per cent. alcohol) for 24 hours, or longer. This softens the chitin.

4. Wash again as in 2.

5. Pass into 90 per cent. alcohol; 6 hours.

6. Pass into absolute alcohol; 6 hours.

7. Pass into absolute alcohol + cedar oil (equal parts); 24 hours.

8. Pass into pure cedar oil; 24 hours.

9. Introduce fine parings of soft paraffin-wax (38° M.P.) into the cedar oil, until it has dissolved as much as possible.

10. Remove specimen to melted paraffin-wax (38° M.P.) at about 40°, for 2 hours, on top of bath.

11. Remove to melted paraffin-wax (52° M.P.) for 2 or 3 hours, within bath.

12. Embed in paraffin-wax (52° M.P.) in the usual way.

From this, the block may be cut out, trimmed up and sectioned in the usual way. The times given must be varied according to the size and character of the object. Care must be taken (*a*) to soften the chitin sufficiently, (*b*) not to leave the object too long in the higher grades of alcohol, as this causes hardening, (*c*) to extract all the cedar oil at stage 10 [I usually have two dishes of soft paraffin, and give 1 hour or more in each], (*d*) to keep the object in the higher paraffin only long enough to allow of complete penetration, and not "cooking."

The sections having been prepared, any of the usual methods of staining may be employed. I find the combination of Ehrlich's Haematoxylin with Eosin to be unequalled for general use. Another very good combination is Heidenhain's Iron Haematoxylin with Orange G.

B. *Double Embedding:* After carrying out processes 1–6 as in A, above, proceed as follows:

7. Pass into absolute alcohol + ether (equal parts); 24 hours.

8. Place in covered crystal-dish in ½ per cent. celloidin solution (dissolve the celloidin in equal parts of ether and alcohol); 24 hours.

9. Change to 2½ per cent. celloidin solution; 24 hours.

10. Change to 5 per cent. celloidin solution; 24 hours.

11. Harden in chloroform vapour under a bell-jar, until the celloidin block is of the consistency of cheese; 2–3 hours.

12. Shape and trim up block, with object correctly orientated; transfer to liquid chloroform for further hardening (half-an-hour or more).

13. Place hardened block in liquid chloroform with parings of paraffin-wax (58° M.P.) on top of bath (about 40° C.). Leave overnight. (The chloroform dissolves the wax, and does not wholly evaporate.)

14. Repeat processes 11, 12 as in A, above, treating the celloidin block as the object to be embedded. Section in the usual way.

Method B must be used for all the more delicate work, such as embryology, studies of the heart *in situ*, etc. It should also be used in sectioning those parts (e.g. caudal gills) in which the cuticle is very thick compared with the rest of the organ; for, otherwise, the cuticle may tear away and carry most of the tissues with it. Method A yields excellent results with the nervous-system, sense-organs, alimentary canal, rectal gills, etc., and also offers a less resistant cutting-surface to the razor. It is, also, of course, a much shorter method. Insects are, at the best, very difficult objects to section well, and the student must not expect to obtain first-rate results without long practice and many disappointments. The success of both methods depends chiefly on the softness of the chitin. There are two safe rules to be followed:

1. *Always use a specimen fresh from ecdysis or metamorphosis, if possible.*
2. *Avoid xylol, and substitute cedar oil, in all methods used.*

Tracheal Studies. The camera-lucida and a good photo-micrographic apparatus are, of course, invaluable adjuncts for which the methods of use need not be gone into here in detail. The latter may, however, be specially applied to the study of the tracheal system of the gills and wings of Odonate larvae. As the air passes out of the finer branches between one and two hours after death (or even sooner), it is clear that permanent preparations cannot be obtained. The part to be studied should be cut off immediately the larva is dead (a drop or two of chloroform shaken up in water is the best killing-agent) and placed on a slide, in pure water. It may then be quickly examined, and, if necessary, a special portion selected and cut off (in the case of gills). This should be floated on to a clear slide, with water, and a cover-glass lowered gently upon it. It should then be photographed to the desired magnification without loss of time. The tracheae appear black on a pale ground.

Wing-Venation Studies. For this purpose I employ a "triple-extension" camera with a total bellows-length of 2 feet, and a special magnifier (or micro-summar) to be used with the lens. A strong artificial light and a condenser are also needed. The wings may either be cut off and placed between two glass slides, or the set insect may be used (by sticking the pin into a small lump of putty placed on a piece of glass, or otherwise arranging it so that the wings are fixed in the desired plane). Slow plates must be used. The negative should be *under*-exposed and *over*-developed, to give a transparent venation on a black ground. The magnifications obtained by this camera are as follows:

(*a*) Without the magnifier: from one-half to a little over natural size.

(*b*) With the magnifier: from twice to four times natural size, according to the amount of extension of the bellows and the position of the object. The figures of wing-venation in this book were nearly all made from photographs taken by this method.

APPENDIX A

BIBLIOGRAPHY

[1] AMANS, P. "Recherches Anatomiques et Physiologiques sur la Larve de l'*Aeschna grandis*." Lab. Zool. Fac. Sci. Montpellier, 1881.

[2] ——. "Essai sur le vol des Insectes." Rev. Sci. nat. Montpellier, 3rd Sér., II et III, 1883—4.

[3] ASSMANN, A. "Beiträge zur Insecten-Fauna der Vorwelt." Breslau, 1870.

[4] BACKHOFF, P. "Die Entwicklung des Copulationsapparates von *Agrion*, etc." Zeit. wiss. Zool., 1910, XCV, pp. 647—706, 1 plate.

[5] BALFOUR-BROWNE, F. "The Life-History of the Agrionid dragonfly." Proc. Zool. Soc., London, 1909, pp. 253—285, 2 plates.

[6] BARLOW, W. F. "Observations on the Respiratory Movements of Insects." Phil. Trans. Roy. Soc., London, 1855, CXLV, pp. 139—148.

[7] BATHER, F. A. "The Florissant Exhibit." Proc. Geol. Ass., 1909, XXI, Part 3, pp. 159—165.

[8] BELLESME, J. de. "Phénomènes qui accompagnent la métamorphose chez la Libellule déprimé." C. R. Acad. Sci., Paris, 1877, LXXXV, pp. 448—450.

[9] BERGER, E. "Untersuchungen ueber den Bau des Gehirns und der Retina der Arthropoden." Arb. Zool. Inst. Univ. Wien, 1878, Hft II, pp. 1—48, 4 plates.

[10] BERLÈSE, A. "Gli Insetti, etc." Vol. I (Embryology and Morphology). Soc. Editrice Libraria, Milan, 1909, pp. 1—1004, figs. 1—1292.

[11] BOLTON, H. "On the Occurrence of a Giant Dragonfly in the Radstock Coal Measures." Quart. Journ. Geol., 1914, pp. 119—126, 2 plates.

[12] BORDAS, L. "Les glandes salivaires des Pseudoneuroptères et des Orthoptères." Arch. Zool. exp., 1897, V (3), pp. 345—384, 3 plates.

[13] BRANDT, A. "Beiträge zur Entwicklungsgeschichte der Libelluliden und Hemipteren." Mém. Acad. St Pétersbourg (Petrograd), 1869, Sér. 7, Tome XIII, pp. 1—33, 3 plates.

[14] BRANDT, E. "Ueber das Nervensystem der Wasserjungfer (Odonata)." Sitz. Horae Soc. Ent. Ross., Petrograd, 1879, XV, pp. xiii—xvi.

[15] BRANDT, J. F. "Bemerkungen ueber die Mundmagen- oder Eingeweidenerven (Nervus sympatheticus seu Nervi reproductorii) der Evertebraten."

Mém. Acad. St Pétersbourg (Petrograd), 1833, II. (French Trans. in Ann. Sci. Nat., Zool. (2) V, pp. 102—103, 1 plate, 1836.)

[16] BRODIE, P. B., Rev. "A History of the Fossil Insects in the Secondary Rocks of England." London, 1845.

[17] BRONGNIART, C. "Recherches pour servir à l'histoire des Insectes Fossiles des Temps Primaires." St Etienne, 1893.

[18] BRULLÉ, A. "Observations sur la bouche des Libellulines." Ann. Soc. ent. France, 1833, II.

[19] BUTLER, HORTENSE. "The Labium of the *Odonata*." Trans. Amer. Ent. Soc., 1904, XXX, pp. 111—134, 6 plates.

[20] BÜTSCHLI, O. "Nähere Mittheilungen über die Entwicklung und den Bau der Samenfäden der Insecten." Zeit. wiss. Zool., 1871, XXI, pp. 526—534, 2 plates.

[21] CABOT, L. "The Immature State of the Odonata." Mem. Mus. Comp. Zool., Cambridge, Mass. Part I (*Gomphinae*), II, 5, pp. 1—17, 3 plates, 1872; Part II (*Aeschninae*), VIII, 1, pp. 1—40, 5 plates, 1881; Part III (*Corduliinae*), XVII, 1, pp. 1—50, 6 plates, 1890.

[22] CALVERT, P. P. "North American Odonata." (Part I, General Account of Odonata.) Trans. Amer. Ent. Soc., 1893, XX, pp. 153—268, 3 plates.

[23] ——. "Preliminary Note on the Youngest Larval Stage of some Odonata." Ent. News, Philadelphia, 1895, pp. 181—182.

[24] ——. "The first filling of the larval tracheae with air." Ent. News, Philadelphia, 1898, p. 73.

[25] ——. "Odonata from Tepic, Mexico." Proc. Californian Acad. Sci., 1899, 3rd Ser., I, No. 12.

[26] ——. "The first moult." Entomologist, 1900, XXXIII, p. 350.

[27] ——. "Biologia Centrali-Americana. *Odonata*." London, 1901—8, pp. 17—420, 9 plates.

[28] ——. "Ganglia of *Odonata*." Proc. Acad. Nat. Sci., Philadelphia, 1903, p. 760.

[29] ——. "The Composition and Ecological Relations of the Odonate Fauna of Mexico and Central America." l.c., 1908, pp. 460—491, with Map.

[30] ——. "Some Comparisons between the eyes of Insects and of Man." Annals of Ophthalmology, 1909, pp. 1—10.

[31] ——. "Studies on Costa-Rican Odonata." Ent. News, Philadelphia, 1911, XXII.

1. "The larva of *Cora*," pp. 49—64, 2 plates.
2. "Habits of the plant-dwelling larva of *Mecistogaster modestus*," pp. 402—410.
3. "Structure and Transformation of *Mecistogaster modestus*," pp. 449—460, 3 plates.

[32] ——. "Progress in our knowledge of the Odonata from 1895 to 1912." Trans. Second Entom. Congress, Oxford, 1912.

[33] CALVERT, P. P. "The Fossil Odonate *Phenacolestes*, with a discussion of the Venation of the legion *Podagrion*, Selys." Proc. Acad. Nat. Sci., Philadelphia, 1913, pp. 225—272, 1 plate.

[34] CAMPION, F. W. and H. "Notes on Dragonfly Parasites (larval water-mites)." Entomologist, London, 1909, XLII, pp. 242—246.

[35] ——, H. "Some Dragonflies and their Prey." Ann. and Mag. Nat. Hist., London, 1914, XIII, pp. 496—504.

[36] ——, H. "The antenodal reticulation of the wings of Agrionine dragonflies." Proc. Acad. Nat. Sci., Philadelphia, 1913, LXV, pp. 220—224.

[37] CARPENTER, G. H. "The Geographical Distribution of Dragonflies." Sc. Proc. Roy. Dublin Soc., 1897, VIII (N.S.), Part V, No. 55, pp. 439—468, 1 plate.

[38] CARRADORI, G. "Esperienze ed 'osservazioni sopra la respirazione delle larve delle Libellule." Giorn. Fis. Chem. e Stor. nat., Briegnatelli, Tom VI, 1813, pp. 450—460.

[39] CHATIN, J. "Sur l'origine et la formation du revêtement chitineux chez les larves de Libellules." C. R. Acad. Sci., Paris, 1892, CXIV, pp. 1135—1138.

[40] CHILD, C. M. and YOUNG, A. N. "Regeneration of the Appendages in nymphs of the *Agrionidae*." Arch. Entwick. der Organismen, Leipzig, 1903, Bd XV, Hft 4, pp. 543—602, 3 plates.

[41] CHUN, C. "Ueber den Bau, die Entwicklung, und physiologische Bedeutung der Rectaldrüsen bei den Insekten." Abh. Senck. Naturf. Ges., 1875, Bd X.

[42] CIACCIO, G. V. "Del modo di unione dei bastoncelli ottici con i cristallini che sono negli occhi composti delle Sfinge delle Libellule e delle Squille." Ac. Sci. Ist., Bologna, 1885.

[43] COCKERELL, T. D. A. "Fossil Dragonflies from Florissant, Colorado." Bull. Amer. Mus. Nat. Hist., New York, 1907, XXIII, pp. 133—139.

[44] ——. "Fossil Insects from Florissant, Colorado." l.c., 1908, XXIV, pp. 59—69, 1 plate.

[45] ——. "Some Results of the Florissant Expedition of 1908." Amer. Nat., 1908, XLII, pp. 569—581.

[46] ——. "Descriptions of Tertiary Insects." Amer. Journ. Sci., 1908, XXVI, pp. 69—75.

[47] ——. "Two fossil Insects from Florissant, Colorado, with a discussion of the venation of the Aeshnine dragonflies." Proc. U.S. Nat. Mus., Washington, 1913, XLVI, pp. 61—64.

[48] COMSTOCK, J. H. and NEEDHAM, J. G. "The wings of Insects." Amer. Nat., 1898, XXXII, *Odonata*, pp. 903—911.

[49] DEWITZ, H. "Einige Beobachtungen, betreffend das geschlossene Tracheensystem bei Insektenlarven." Zool. Anz., 1890, XIII, pp. 500—504, 525—531.

[50] DUFOUR, L. "Sur la respiration branchiale des larves dès grandes

Libellules comparée à celle des Poissons." C. R. Acad. Sci., Paris, 1848, xxvi.

[51] Dufour, L. "Etudes anatomiques et physiologiques, et observations sur les larves des Libellules." Ann. Sci. Nat. Zool., 3rd Series, 1852, xvii, pp. 65—110, 3 plates.

[52] Evans, W. F. "British Libellulinae or Dragonflies." London, 1845.

[53] Exner, S. "Die Physiologie der fazettierten Augen von Krebsen und Insekten." Vienna, 1891, pp. 1—206, 7 plates.

[54] Faussek, V. "Beiträge zur Histologie des Darmkanals der Insekten." Zeit. wiss. Zool., 1887, Bd xlv, Hft 4, pp. 694—712.

[55] Gerstaecker, A. "Zur Morphologie der Orthoptera amphibiotica." Festschrift 100 Jahr. best. Ges. Naturfor. Freunde, Berlin, 1873, pp. 39—59, 1 plate.

[56] Giard, A. "La ponte des Libellules du genre *Lestes*." Feuille Jeun. Nat., 1894, Ann. 33 (4), pp. 189—192.

[57] Giebel, C. G. "Fauna der Vorwelt." Leipzig, 1856, Bd 2, Insekten u. Spinnen.

[58] Goddard, Martha E. "On the second abdominal segment in a few *Libellulidae*." Proc. Amer. Phil. Soc., 1896, xxxv, pp. 205—212, 2 plates.

[59] Griffiths, A. B. "On the Malpighian Tubules of *Libellula depressa*." Proc. Roy. Soc., Edinburgh, 1888, xv, pp. 401—403.

[60] Hagen, H. A. "Die Neuroptera des Lithographischen Schiefers in Bayern." Cassel, 1866.

[61] ——. "Beitrag zur Kenntniss des Tracheensystems der Libellen-Larven." Zool. Anz., 1880, iii, pp. 157—161.

[62] ——. "Essai d'un Synopsis des Larves de Caloptérygines." C. R. Soc. Ent. Belg., 1880, xxiii, pp. lxv—lxvii.

[63] ——. "Monograph of the earlier stages of the Odonata (*Gomphinae* and *Cordulegastrinae*)." Trans. Amer. Ent. Soc., 1885, xii, pp. 249—291.

[64] ——. "Ueber die früheren Zustände der Libellen nebst Andeutung über die Bildung des Thorax." Ent. Zeit. Stettin, 1889, Jahr. 50, pp. 162—166.

[65] Handlirsch, A. "Zur Morphologie des Hinterleibes der Odonaten." Ann. Hofmus., Wien, 1903, xviii, pp. 117—122.

[66] ——. "Bemerkungen zu der Arbeit des Herrn Prof. Dr Heymons über die Abdominalanhänge der Libellen." l.c., 1904, xix, pp. 59—63.

[67] ——. "Die Fossilen Insekten." Leipzig, 1906—7, pp. 1—1120, 51 plates.

[68] ——. "New Palaeozoic Insects from the Vicinity of Mazon Creek, Illinois." Amer. Jour. Sci., 1911, xxxi, pp. 297—377.

[69] Haupt, H. "Ein Beitrag zur Metamorphose der Libellen." Woch. Aquarienkunde, Brunswick, 1906, iii, pp. 382—384.

[70] Hesse, R. "Untersuchungen über die Organe der Lichtempfindung bei niederen Thieren. 7. Von den Arthropoden-Augen." Zeit. wiss. Zool., 1901, lxx, pp. 347—473, 6 plates. (Odonata, pp. 380—384.)

[71] Heymons, R. "Grundzüge der Entwicklung und des Körperbaues von Odonaten und Ephemeriden." Abh. Akad. Wiss., Berlin, 1896.

[72] HEYMONS, R. "Die Hinterleibesanhänge der Libellen und ihrer Larven." Ann. Hofmus., Wien, 1904, XIX, pp. 21—58, 1 plate.

[73] HICKSON, S. J. "The Eye and Optic Tract of Insects." Q. J. M. S., 1885, 2nd Series, XXV.

[74] HIGGINS, HELEN T. "The Development and comparative structure of the gizzard in *Zygoptera*." Proc. Acad. Nat. Sci., Philadelphia, 1901, LIII, pp. 126—141, 3 plates.

[75] HÜBNER, O. "Neue Versuche aus dem Gebiet der Regeneration und ihre Beziehungen zu Anpassungserscheinungen." Zool. Jahrb. Syst., XV, 1902, pp. 461—498, 2 plates.

[76] IMHOF, O. E. "Antennen der *Odonata*." Biol. Centrbl., 1901, XXI, pp. 255—256.

[77] JANDA, V. "Experimentelle Untersuchungen über die Regeneration der *Odonaten*." Zool. Anz., 1910, XXXV, pp. 602—608.

[78] KAMMERER, P. "Symbiose zwischen *Oedogonium undulatum* und Wasserjungferlarva (*Aeschna cyanea*)." Wiener Fest. Schr., Vienna, 1908, pp. 238—252.

[79] KELLICOTT, D. S. "An *Odonate* nymph from a thermal spring." Journ. Cincinnati Soc., 1899, XIX, pp. 63—65.

[80] KOLBE, H. "Ueber den Zweck der Appendices anales und der entsprechenden vicariierenden Organe am Hinderleibsende der Libelluliden." Jahrsb. Westfäl. Prov. Ver. f. Wiss. u. Kunst, 1881, IX, pp. 52—56.

[81] KORSCHELT, E. und HEIDER, K. "Lehrbuch der vergleichenden Entwicklungsgeschichte der wirbellosen Tiere." Jena, 1891—1902 (*Odonata*, etc., pp. 380 et sqq.).

[82] LAIDLAW, F. F. "*Odonata*. Part 1." Fasc. Malayenses, Zool., pp. 165—200.

[83] LANDOIS, H. "Die Ton- und Stimmapparate der Insekten." Zeit. wiss. Zool., 1867, XVII.

[84] LENDENFELD, R. von. "Der Flug der Libellen." Sitzb. kais. Akad. Wiss., Vienna, 1881, LXXXIII., Abt. I, pp. 289—376, 7 plates.

[85] LESPÈS, C., "Mémoire sur l'apparcil auditif des Insectes." Ann. sci. Nat. Zool., 4th Series, 1858, IX.

[86] LOWNE, B. T. "On the Compound Vision and the Morphology of the Eye in Insects." Trans. Linn. Soc., London, 2nd Series, Zool., 1884, II.

[87] LUBBOCK, J. "On the senses, instincts and intelligence of animals, with special reference to Insects." London, 1888, pp. 1—292.

[88] LUCAS, W. J. "British Dragonflies." London, 1900, pp. 1—356, 27 plates.

[89] MARTIN, R. "*Cordulines*." Cat. Coll. Selys, Fasc. XVII, 1906, pp. 1—98, 3 plates.

[90] ——. "*Aeschnines*." Cat. Coll. Selys, Fasc. XVIII—XX, 1908—9, pp. 1—211, 6 plates.

[91] MCGILL, CAROLINE. "The behaviour of the nucleoli during oogenesis

of the dragonfly, with especial reference to synapsis." Zool. Jahrb., Jena, Abt. f. Anat., 1906, XXIII, pp. 207—230, 5 plates.

[92] MEUNIER, F. "Les Agrionides Fossiles des Musées de Munich et de Haarlem." Ann. Soc. Ent. France, 1896, pp. 30—35, 3 plates.

[93] ——. "Nouveaux Insectes des schistes houillers de Commentry." Ann. Soc. Scient., Brussels, 1908, XXXII, pp. 241—244.

[94] MIALL, L. C. "The Natural History of Aquatic Insects." London, 1895. (Dragonflies, chap. IX, pp. 328—345.)

[95] MÜLLER, J. "Zur Vergleichenden Physiologie des Gesichts-sinnes des Menschen und der Thiere, etc." Leipzig, 1826.

[96] NÉANDER, A. "Zur Morphologie der Stigmen bei Aeschniden und Libelluliden-Larven." Vorl. Mitt. Arkiv Zool., Stockholm, 1913, VIII, No. 14 (6).

[97] NEEDHAM, J. G. "On rearing dragonflies." Canad. Ent., 1897, XXIX, pp. 94—96.

[98] ——. "The Digestive Epithelium of Dragonfly Nymphs." Zool. Bull., 1898, I, pp. 103—113.

[99] ——. "The fruiting of the Blue Flag (*Iris versicolor* L.)." Amer. Nat., 1900, XXXIV, No. 401, pp. 361—386.

[100] ——. "Aquatic Insects in the Adirondacks." N.Y. State Mus., Albany Bull., XLVII, 1901, pp. 383—612, 36 plates.

[101] ——. "Life-Histories of *Odonata*." Bull. N.Y. Mus., 1903, LXVIII, pp. 218—279, 11 plates.

[102] ——. "A genealogic study of dragonfly wing-venation." Proc. U.S. Nat. Mus., Washington, 1903, XXVI, pp. 703—761, 24 plates (No. 1331).

[103] ——. "New Dragonfly nymphs in the U.S. National Museum." l.c., 1904, XXVII, pp. 685—720, 7 plates (No. 1371).

[104] ——. "Descriptions of dragonfly nymphs of the subfamily *Calopterygînae*." Ent. News, Philadelphia, 1911, XXII, pp. 145—154, 2 plates.

[105] ——. "Notes on a few Nymphs of *Agrioninae* of the Hagen Collection." l.c., pp. 341—345, 1 plate.

[106] ——. "Notes on some Nymphs of *Gomphinae* of the Hagen Collection." l.c., pp. 392—396, 1 plate.

[107] NEEDHAM, J. G. and ANTHONY, MAUDE H. "The Skewness of the thorax in *Odonata*." Jour. N.Y. Ent. Soc., 1903, XI, pp. 117—124, 1 plate.

[108] NEEDHAM, J. G. and COCKERELL, T. C. "Some hitherto unknown nymphs of *Odonata* from New Mexico." Psyche, 1903, X, pp. 134—139.

[109] OGUMA, K. "On the rectal tracheal gills of a Libellulid-nymph, and their fate during the course of metamorphosis." Berliner Ent. Zeit., 1913, LVIII, pp. 211—225, 2 plates.

[110] OSBURN, R. C. "Observations and Experiments on dragonflies in brackish water." Amer. Nat., 1906, XL, pp. 395—399.

[111] OUSTALET, M. E. "Note sur la Respiration chez les Nymphes des Libellules." Ann. Sci. Nat. Zool., 5th Series, 1869, XI, pp. 369—386, 3 plates.

[112] Packard, A. S. "On the development of a dragonfly (*Diplax*)." Proc. Boston Soc. Nat. Hist., 1868, xi, pp. 365—372.

[113] ——. "Embryological Studies on *Diplax*, *Perithemis*, and the Thysanurous genus *Isostoma*." Mem. Peabody Acad., 1871, i, 2, pp. 1—21, 3 plates.

[114] Perkins, R. C. L. "Notes on some Hawaiian Insects." Proc. Cambridge Soc., 1897, ix, pp. 373—380.

[115] ——. "Fauna Hawaiiensis, *Neuroptera*," 1899, ii, pp. 31—89, 3 plates.

[116] ——. "Fauna Hawaiiensis, Introduction, being a review of the land-fauna of Hawaii," 1913, i, Part vi, pp. xv—ccxxviii, 16 plates.

[117] Picard, F. "Sur les changements de coloration chez les mâles de quelques Libellules." Bull. Soc. Ent., Paris, 1906, pp. 166—167.

[118] Pierre, Abbé. "L'éclosion des œufs du *Lestes viridis* Van der Lind." Ann. Soc. Ent. France, 1904, lxxiii, pp. 477—484, 1 plate.

[119] Plateau, F. "Recherches Expérimentales sur les Mouvements Respiratoires des Insectes." Mém. Acad. Roy. Belg., 1884, xlv, pp. 111—123, 1 plate.

[120] Poletaieff, N. "Die Flügelmuskeln der Lepidopteren und Libelluliden." Zool. Anz., 1880.

[121] ——. "Du développement des muscles d'ailes chez les Odonates." Horae Soc. Ent. Ross., Petrograd, 1881, xvi, pp. 10—37, 5 plates.

[122] ——. "Speicheldrüsen bei den Odonaten." l.c., pp. 3—6, 1 plate.

[123] ——. "Des glandes salivaires chez les Odonates." C. R. Acad. Sci., Paris, 1888, xci.

[124] Poletaieff, Olga. "Quelques mots sur les organes respiratoires des larves des Odonates." Horae Soc. Ent. Ross., Petrograd, 1880, xv, pp. 436—452, 2 plates.

[125] Poulton, E. B. "Predaceous Insects and their Prey." Trans. Ent. Soc., London, 1906, pp. 323—409. (Odonata, pp. 398—401.)

[126] Purser, G. L. "Preliminary Notes on some problems connected with respiration in Insects generally, and in aquatic forms in particular." Camb. Phil. Trans., 1914, xviii, Part 2, pp. 63—70.

[127] Rambur, P. "Histoire Naturelle des Insectes. Névroptères. Suites à Buffon." Paris, 1842.

[128] Rathke, H. "De Libellarum partibus genitalibus." Königsberg, 1832, pp. 1—38, 3 plates.

[129] Réaumur, R. A. "Mémoires pour servir à l'histoire naturelle et à l'anatomie des Insectes." Paris, 1734—1742, 6 vols., 267 plates (see Vol. 6, memoir xi).

[130] Riley, C. F. C. "Observations on the ecology of dragonfly nymphs; reactions to light and contact." Ann. Ent. Soc. Amer., 1912, v, pp. 273—292.

[131] Ris, F. "Untersuchung über die Gestalt des Kaumagens bei den Libellen und ihren Larven." Zool. Jahrb., 1890, xix, pp. 596—624.

[132] Ris, F. "Oviposition in *Cordulegaster*." Ent. News, Philadelphia, 1905, pp. 113—114.

[133] ——. "Die Süsswasserfauna Deutschlands. Odonata." Jena, 1909, Hft 9, pp. 1—67.

[134] ——. "Libellulinen." Cat. Coll. Selys, Fasc. IX—XVI, 1909—1913, pp. 1—1042, 8 plates.

[135] ——. "Kopulationsmarken bei Libellen." D. ent. Nationalbibl., Berlin, 1910, I, pp. 70—71, 79—80.

[136] ——. "Die Fauna Südwest-Australiens (Michaelsen u. Hartmeyer). Odonata." 1910, II, pp. 415—450, Jena.

[137] ——. "Ueber *Odonaten* von Java und Krakatau." Tijdschr. Ent., 1912, LV, pp. 157—183.

[138] ——. "Die mitteleuropäischen Corduliinen-larven." Mitt. Schweiz. ent. Ges., Berne, 1911, Bd XII, Hft 2, pp. 25—41.

[139] ——. "Die Atmungs-Organe der anisopteren Libellenlarven." l.c., 1913, Bd XII, Hft 4, pp. 92—99.

[140] Roster, D. A. "Contributo all' anatomia ed alla biologia degli Odonati." Bull. Soc. Ent. Ital., Florence, 1885, XVII, pp. 256—268, 2 plates.

[141] ——. "Contributo allo studio delle forme larvali degli Odonati." l.c., 1888, XX, pp. 159—170, 4 plates.

[142] Rothswand, S. J. "Ueber Phagocytose und die Pericardialzellen der Libellenlarven." Arb. Lab. Zool. Cabin. Univ. Warschau, Warsaw, 1898, p. 120.

[143] Rousseau, E. "Contributions à la connaissance des métamorphoses des *Odonates* d'Europe." Ann. Soc. Ent. Belg., 1908, LII, pp. 272—291.

[144] ——. "Etude monographique des larves des *Odonates* d'Europe." Ann. Biol. lacustre, Brussels, 1909, III, pp. 300—366.

[145] Sadones, J. "L'appareil digestif et respiration larvaire des *Odonates*." La Cellule, Louvain, 1896, XI, pp. 273—324, 3 plates.

[146] Schindler, E. "Beiträge zur Kenntniss der Malpighi'schen Gefässe der Insecten." Zeit. wiss. Zool., 1878, XXX, pp. 587—660, 3 plates.

[147] Scott, G. G. "The Distribution of Tracheae in the Nymph of *Plathemis lydia*." Marine Biol. Lab. Bull., Woods' Holl, Mass., 1905, IX, pp. 341—354.

[148] Scudder, S. H. "The Tertiary Insects of North America." Repr. U.S. Geol. Survey, Terr. XIII, 1890.

[149] ——. "Some Insects of special interest from Florissant, Colorado, and other points in the Tertiaries of Colorado and Utah." Bull. U.S. Geol. Survey, 1892, No. 93, 25 pp., 3 plates.

[150] Sellards, E. H. "Types of Permian Insects. Part I, *Odonata*." Amer. Jour. Sci., New Haven, 1906, Ser. 4, XX, pp. 249—258.

[151] Selys-Longchamps, E. de and Hagen, H. A. "Monographie des Gomphines." Brussels, 1858, pp. 1—460, 23 plates.

[152] Shafer, G. D. "Histology and Development of the divided eyes of

certain insects." Proc. Acad. Sci., Washington, 1907, VIII, pp. 459—486, 4 plates.

[153] SHARP, D. "Cambridge Natural History. Insects, Part I." 1901. (Dragonflies, chap. XVIII, pp. 409—428.)

[154] SIEBOLD, C. T. von. "Ueber die Begattung der Libellen." Archiv für Naturg., 1838, Jahrg 4, Bd 1, pp. 375—376.

[155] ——. "Ueber die Fortpflanzungsweise der Libellen." Germar's Zeit. für die Ent., Danzig, 1840, II, Hft 2, pp. 421—435.

[156] ——. "Sur la mode de reproduction des Libellulines." Revue Zool., 1842, pp. 283—288, 324—328.

[157] SNODGRASS, R. E. "The Thorax of Insects and the Articulations of the Wings." Proc. U.S. Nat. Mus., Washington, 1909, XXXVI, pp. 511—595, 30 plates.

[158] SONDHEIM, MARIA. "Wahrnehmungsvermögen einer Libellenlarve." Biol. Centrbl., 1901, XXI, pp. 317—319.

[159] SUCKOW, M. de. "Respiration der Insecten, insbesondere über die Darmrespiration des *Aeschna grandis*." Zeit. für die organische Physik, 1828, Bd I.

[160] —— "Sur l'anatomie et la respiration des larves des Libellules." Heusinger Zeitschrift, 1840, Bd III.

[161] THOMPSON, O. S. "Appendages of the Second Abdominal Segment of male dragonflies (*Odonata*)." Albany, N.Y. St. Ed. Dep. Mus. Bull., 1908, CXXIV, pp. 249—263.

[162] TILLYARD, R. J. "On the supposed numerical preponderance of the males in *Odonata*." Proc. Linn. Soc. N.S.W., Sydney, 1905, XXX, pp. 344—349.

[163] ——. "Life-History of *Lestes leda* Selys." l.c., 1906, XXXI, pp. 409—423, 2 plates.

[164] ——. "On Dimorphism in the Females of Australian *Agrionidae*." l.c., 1907, XXXII, pp. 382—390.

[165] ——. "Studies in the Life-Histories of Australian *Odonata*":
1. *Petalura gigantea* Leach. l.c., 1909, XXXIV, pp. 256—267, 1 plate.

[166] ——. 2. *Diphlebia lestoïdes* Selys. l.c., 1909, XXXIV, pp. 370—383, 1 plate.

[167] ——. 3. Description and Life-History of *Phyllopetalia patricia*, n.sp. l.c., 1909 (1910), XXXIV, pp. 697—708, 1 plate.

[168] ——. 4. Further Notes on the Life-History of *Petalura gigantea* Leach. l.c., 1911, XXXVI, pp. 86—97, 1 plate.

[169] ——. "Monograph of the Genus *Synthemis*." l.c., 1910, XXXV, pp. 312—377, 5 plates.

[170] ——. "On some experiments with Dragonfly Larvae." l.c., 1910, XXXV, pp. 666—676.

[171] ——. "On the genus *Cordulephya*." l.c., 1911, XXXVI, pp. 388—422, 2 plates.

[172] Tillyard, R. J. "On the genus *Diphlebia*, with descriptions of new species, and life-histories." l.c., 1911, xxxvi, pp. 584—604, 2 plates.

[173] ——. "On some new and rare Australian *Agrionidae*." l.c., 1912 (1913), xxxvii, pp. 404—479, 6 plates.

[174] ——. "A Study of the Odonata of Tasmania in relation to the Bassian Isthmus." l.c., 1913 (1914), xxxviii, pp. 765—778.

[175] ——. "On the Study of Zoogeographical Regions by means of Specific Contours." l.c., 1914, xxxix, Part 1, pp. 21—43, 1 plate, 3 transparencies.

[176] ——. "On some Problems concerning the development of the Wing-Venation of *Odonata*." l.c., 1914, xxxix, Part 1, pp. 163—216, 3 plates.

[177] ——. "The Hatching of the Larva of *Anax papuensis* Burm." Rep. British Ass. Australian Meeting, Sydney, Aug. 1914, pp. 424—425.

[178] ——. "Life-Histories and Descriptions of Australian *Aeschninae*." Journ. Linn. Soc., London, Zool. xxxiii, 1916, No. 222, pp. 1—83, 9 plates.

[179] ——. "On the Rectal Breathing Apparatus of Anisopterid Larvae." l.c., 1916.

[180] ——. "On the Development of the Wing-venation in Zygopterous Dragonflies, with special reference to the *Calopterygidae*." Proc. Linn. Soc. N.S.W., Sydney, 1915, xl, Part 2, pp. 212—230, 3 plates.

[181] Tschuproff, H. "Ueber die Entwicklung der Keimblätter bei den Libellen." Zool. Anz., 1903.

[182] Tümpel, R. "Ueber das scheinbar seltene Vorkommen der Weibchen mancher Libellen-arten." Illustr. Zeit. Ent., 1899, iv, pp. 227—228.

[183] Uchida, S. "Preliminary Report on the Larvae of Japanese *Aeschnidae*." Nip. Konch. Kw., Tokio, 1907, i, Part 1, pp. 4—23, 1 plate.

[184] Uexküll, J. "Studien über den Tonus. V. Die Libellen." Zeit. Biol. München, Munich, 1907 (1908), l, pp. 168—202.

[185] Viallanes, H. "Anatomie et dissection de la larve de la Libellule." Feuill. Nat., 1884, xiv, pp. 81—87, 1 plate.

[186] ——. "Le ganglion optique de la Libellule (*Aeschna maculatissima*)." Ann. Sci. Nat. Zool., 6th Series, 1885, xviii, pp. 1—34, 3 plates.

[187] Voinov, D. N. "Recherches physiologiques sur l'appareil digestif et le tissu adipeux des larves des *Odonata*." Bull. Soc. Bucharest, 1898, vii, pp. 49—52, 473—493, 2 plates.

[188] Walker, E. M. "The North American dragonflies of the genus *Aeschna*." Toronto Stud. Univ. Biol., 1912, xi, pp. 1—213, 28 plates.

[189] ——. "New Nymphs of Canadian *Odonata*." Canadian Entomologist, 1913, xlv, No. 6, pp. 161—170, 2 plates.

[190] ——. "Mutual adaptation of the sexes in *Argia moesta putrida*." l.c., No. 9, pp. 277—279, 1 plate.

[191] Wallengren, H. "Physiologische-biologische Studien über die Atmung bei den Arthropoden. 1. Die Atmung der gehirnlosen *Aeschna*-arven." Lund Univ. Årsskr., N. F., 1913, ix, Afd. 2, No. 16, 30 pp., 1 plate.

[192] Weele, H. W. van der. "Morphologie und Entwicklung der Gonapophysen der *Odonaten*." Tijdschr. Ent., 1906, xlix, pp. 99—198, 3 plates.

[193] WILLIAMSON, E. B. "The copulation of *Odonata*." Ent. News, Philadelphia, 1899, x, p. 42.

[194] ——. "Preservation of colours in dragonflies." l.c., 1901, XII, pp. 131—133.

[195] ——. "On the manner of oviposition, and on the nymph of *Tachopteryx thoreyi*." l.c., 1901, XII, pp. 1—3, 1 plate.

[196] ——. "Spiders as enemies of dragonflies." l.c., 1903, XIV, pp. 160—161.

[197] ——. "A Hybrid Dragonfly of the Genus *Gomphus*." l.c., 1903, XIV, pp. 253—258, 1 plate.

[198] ——. "The Dragonflies (*Odonata*) of Burma and Lower Siam. 1. Sub-fam. *Calopteryginae*." Proc. U.S. Nat. Mus., Washington, 1904, XXVIII, No. 1389, pp. 165—187.

[199] ——. "Oviposition of *Tetragoneuria*." Ent. News, Philadelphia, 1905, XVI, pp. 255—256.

[200] ——. "The medio-anal link in *Agrioninae*." l.c., 1913, XXIV, pp. 258—262.

[201] WILLIAMSON, E. B. and CALVERT, P. P. "Copulation of *Odonata*." l.c., 1906, XVII, pp. 143—148, 1 plate.

[202] ZAWARSIN, A. "Zur Frage über den Bau des sensiblen Nervensystems bei den Insekten." Vorl. Mitt., St Petersburg (Petrograd), 1911, XLII, 1, pp. 200—208, 222—223.

[203] ——. "Histologische Studien über Insekten. 1. Das Herz der *Aeschna*-larven." Zeit. wiss. Zool., 1911, XCVII, Hft 3, pp. 481—510, 2 plates.

[204] ——. "Histologische Studien über Insekten. 2. Das sensibel Nervensystem der *Aeschna*-larva." l.c., 1912, C, pp. 245—286, 3 plates.

ADDITIONAL BIBLIOGRAPHY

[In the following list are included a few publications of which the author was unable to obtain copies or abstracts, and which, therefore, have not been made use of in this book, together with all those recently-published papers which have been received between the date of completion of the text (March 1915) and the date of completion of the first proof-reading (July 1916).]

BERVOETS, R. "Sur le Système Trachéen des Larves d'Odonates." Ann. Biol. lacustre, 1913, VI, pp. 15—52.

CALVERT, P. P. "Studies on Costa Rican Odonata, V—VII. The Waterfall Dwellers: *Thaumatoneura*." Ent. News, Philadelphia, 1914—5, XXV and XXVI. (V. Imagos and Possible Male Dimorphism, XXV, pp. 337—348, 1 plate, Oct., 1914. VI. Transformation External Features and Attached Diatoms of Larvae, XXVI, pp. 295—305, 1 plate, July, 1915.

VII. Internal Organs of Larva, and the Respiration and Rectal Tracheation of Zygopterous Larvae in General, XXVI, pp. 385—395, 435—447, 3 plates, Nov.—Dec., 1915.)

CARUS, C. G. "Entdeckung eines einfachen vom Herzen aus beschleunigten Blutkreislaufes in den Larven netzflüglicher Insecten." Leipzig, 1827, 4to, pp. 1—40, 3 plates.

CHATIN, J. "Sur le processus général de la cuticularisation tégumentaire chez les larves des Libellules." Bull. Soc. Philom., Paris, 1892, VIII, T. 4, Nr. 3, pp. 105—106.

DUFOUR, L. "Des divers modes de respiration aquatique dans les Insectes." C. R. Acad. Sci., Paris, 1849, XXIX, pp. 763—770.

HAGEN, H. A. "Léon Dufour über die Larven der Libellen mit Berücksichtigung der früheren Arbeiten." Stett. Ent. Zeit., 1853, XIV, pp. 98—106, 237—238, 260—270, 311—325, 334—346.

KENNEDY, C. H. "Notes on the Life-History and Ecology of the Dragonflies (Odonata) of Washington and Oregon." Proc. U.S. Nat. Mus., 1915, XLIX, pp. 259—345.

LAMBORN, W. A. "Notes on the Habits of *Glossina morsitans*." Bull. Ent. Research, 1916, VI, Part 3. (*Orthetrum chrysostigma* recorded as attacking and devouring Tse-tse Flies.)

MARSHALL, W. S. "On the Anatomy of the Dragonfly *Libellula quadrimaculata* Linné." Trans. Wisconsin Acad. Sci. etc., 1914, XVII, Part 2, No. 1, pp. 755—790, 4 plates.

MARTIN, J. "Sur la respiration des larves de Libellules." Bull. Soc. Philomath., Paris, 1892, VIII, T. 4, pp. 122—124.

MEUNIER, F. "Un nouveau grand fragment de Libellulide géante du houiller de Commentry." Ann. Soc. Sci., Bruxelles, 1914, XXXVIII, pp. 93—94.

OGUMA, K. "A case of Prolapsus recti in a Dragonfly." Annot. Zool. Japonenses, Tokyo, 1915, IX, Part 1.

PORTIER, P. "Recherches Physiologiques sur les Insectes Aquatiques." Arch. Zool. Exp. Gén., 1911 (5), VIII, pp. 89—379, 4 plates.

RIS, F. "The identity of two *Odonata* Fossils." Bull. Wisconsin Nat. His. Soc., Milwaukee, 1910, VIII, pp. 102—105.

ROBERTS, E. W. "Nature of the intromittent organ of the *Odonata*." Trans. Amer. Microsc. Soc., Decatur, Ill., 1912 (1913), XXXI, pp. 201—203.

ROSENHOF, ROESEL VON. "Der Wasser-Insecten zweyte Klasse. Der monatlich-herausgegebenen Insecten-Belustigung." Nürnberg, 1749, Part 2.

ROSTER, D. A. "Cenno Monographico degli Odonati dei Gruppo *Ischnura*." Bull. Soc. Ent. Ital., 1886, XVIII, pp. 239—258, 5 plates.

SCHMIDT, E. "Vergleichende Morphologie des 2 und 3 Abdominalsegments bei männlichen Libellen." Zool. Jahrb., Jena, 1915, XXXIX, Hft 1, pp. 87—200, 3 plates.

SCHMIDT-SCHWEDT, E. "Kerfe u. Kerflarven d. süssen Wassers, besonders der stehenden Gewässer." Die Tier- u. Pflanzenwelt d. Süsswassers, von Dr O. Zacharias, Leipzig, 1891, Bd II, pp. 51—122.

SELYS-LONGCHAMPS, E. de. "Le Progrès dans la Connaissance des Odonates." Third Internat. Congr. Zool., Leyden, 1895.

TILLYARD, R. J. "On the Physiology of the Rectal Gills in the Larvae of Anisopterid Dragonflies." Proc. Linn. Soc. N.S.W., Sydney, 1915, XL, Part 3, pp. 422—437, 1 plate.

——. "Mesozoic and Tertiary Insects of Queensland and New South Wales." Queensland Geol. Survey, Brisbane, 1916, No. 253, pp. 1—47, 9 plates. (Description of *Mesophlebia antinodalis* and discussion of phylogeny of nodus, pp. 25—29.)

——. "Further Observations on the Emergence of Dragonfly Larvae from the Egg, with Special Reference to the Problem of Respiration." Proc. Linn. Soc. N.S.W., Sydney, 1916, XLI.

TUMPEL, R. "Die Geradflügler Mitteleuropas." 1st ed., Eisenach, 1898—9; 2nd ed., Gotha, 1908.

WALKER, E. M. "The known nymphs of the Canadian species of *Lestes*." Canadian Ent., 1914, XLVI, pp. 189—200.

——. "New and little-known nymphs of Canadian Odonata." l.c., pp. 349—357, 369—377, 2 plates.

WALLENGREN, H. "Physiologische-biologische Studien über die Atmung bei den Arthropoden." Lund Univ. Årsskr. N.F., x, Afd. 2, 1914.

2. Die Mechanik der Atembewegungen bei *Aeschna*-larven. A. Das Chitinskelett. B. Muskelatur des Abdomens. (No. 4, 1 plate.)
3. Die Atmung der *Aeschna*-larven. Die Ventilations-grosse des respiratorischen Darmes. Ist der Rhytmus der Atembewegungen von Wasserströmungen bedingt? Die Notatmung. (No. 8.)

WARREN, A. "A Study of the Food Habits of the Hawaiian Dragonflies or Pinau." Coll. Hawaii, Bull. No. 3, 1915, pp. 3—45, plates 1—4.

WESENBERG-LUND, C. "Odonaten Studien." Intern. Rev. Hydrobiol., Leipzig, 1913—4, VI, pp. 155—228, 373—422.

APPENDIX B

GLOSSARY

Alveolus (Lat. =a small hollow), a small hollow space in the interior of an organ.

Amoebocytes (*Amoeba* + Gr. κύτος, cell), the larger corpuscles found in the blood of Insects; so called because they undergo amoeboid movements.

Amphibiotic (Gr. ἀμφί, on both sides; βιωτικός, fit to live), indicates, in the Insecta, that the imago is terrestrial or aërial, but the larva aquatic.

Ampulla (Lat. =a flask), a small flask-shaped growth or outgrowth.

Anastomosis (Gr. =an outlet), a vessel forming an outlet or connection between two larger vessels. Often used for a network of small vessels connecting two larger ones.

Apodeme (Gr. ἀποδέμω, to build out), a ridge or ingrowth of the cuticle, serving for the attachment of muscles.

Asthenogenesis (Gr. ἀσθενής, weak; γένεσις, begetting), the development of a successful line of descent by the adoption of weakness in structure (e.g. the *Agrionidae* amongst Dragonflies). Contrasted with *menogenesis* (Gr. μένος, strong).

Autotomy (Gr. αὐτός, self; τέμνω, to cut), self-mutilation; the power to cast off an appendage at will.

Bilateral symmetry, the possession of a single longitudinal axis or plane of symmetry, separating exactly similar right and left halves.

Bromeliad, any plant belonging to the Pine-apple family (*Bromeliaceae*).

Caecum (Lat. =blind), a blind tube, i.e. one closed at its free end.

Caenogenetic (Gr. καινός, new; γεννάω, to beget), a *comparative* term used in contrast with "archaic," and signifying, phylogenetically, a comparatively recent line of development, as contrasted with an older (*archaic*) line. It may be used either of races of animals, or of types of organs.

Callus (Lat. =hard skin), a hard lump or swelling of the cuticle (e.g. at the wing-base).

Campodeiform (from *Campodea*, a wingless insect of the Order Thysanura), the term applied to larvae which are active and provided with well-developed limbs, as opposed to *eruciform* or grub-like larvae.

Capillary (Lat. *capillus*, a hair), a fine hair-like tube (tracheal system).

Carina (Lat. =a keel), an external ridge of the cuticle.

Caudal (Lat. *caudalis*, of the tail), used of organs situated at the posterior end of the abdomen.

Cell, a minute portion of protoplasm containing a *nucleus*.

Cerci (Gr. = tails), the appendages of the last abdominal segment.

Chiasma (Gr. = a crossing, from the letter χ), a diagonal intercrossing of two sets of nerve-fibres.

Chitin (Gr. χιτών, a tunic), the transparent horn-like substance which forms the cuticle of insects.

Chromatin (Gr. χρῶμα, colour), the substance forming the intra-nuclear fibres of the cell-nucleus (so called owing to its affinity for basic stains, such as haematoxylin).

Coelome (Gr. κοίλωμα, a hollow), in the Arthropoda, the body-cavity formed by fusion of the cavities of the mesodermal somites of the embryo; in the Insecta, it is much reduced. In some other Phyla, the coelome may have a different derivation.

Commissure (Lat. *commissura*, a connection), a cord of nerve-fibres connecting two ganglia.

Convergence, the term applied to indicate resemblance between two forms (either of animals or structures) derived from two widely different ancestries. The convergence may be due to the adoption of similar habits, or to the elimination, by reduction, of the original differences.

Corduliform (Gr. κορδύλη, a club), shaped like an Indian club (e.g. abdomen of certain Dragonflies).

Cornea (Lat. = horny), the outer covering of the eye (in Insects, continuous with the chitinous cuticle).

Cuticle (Lat. = the skin), the hard outer non-cellular covering of Insects; formed of chitin.

Cytoplasm (Gr. κύτος, cell; πλάσμα, anything moulded), all the protoplasm of the cell, except the nucleus.

Dendritic (Gr. δένδρον, a tree), with complex branchings, like those of a tree.

Diaphragm (Gr. = a partition-wall), a membrane dividing the body-cavity into two.

Dichotomy (Gr. διχοτομέω, to cut in two), a phylogenetic term indicating the branching of a single ancestral line into two equal and diverging lines of descent.

Dioptic or *dioptric* (Gr. διοπτικός, fit to look through, or διοπτρικός, pertaining to an optical instrument), used of the part of the eye concerned with the passage and focussing of the light-rays.

Ecdysis (Gr. ἐκδύω, to strip off), the periodical casting of the larval skin.

Ectoderm, see *Germinal Layers*.

Efferent (Lat. = carrying outwards), a vessel conveying air or blood outwards from a gill.

Embryo (Gr. ἔμβρυον, the young before birth), the young animal within the egg.

Endoderm, see *Germinal Layers*.

Endoskeleton (Gr. ἔνδον, within), the internal skeleton.

Epiphytic (Gr. ἐπιφύω, to grow upon), used of plants growing non-parasitically upon others, as ferns, Bromeliads, etc., upon the branches of trees.

Epithelium (Gr. ἐπί, upon; θηλέω, to flourish), a cellular membrane covering any internal or external surface.

Evagination (Lat. *ex*, out of; *vagina*, sheath), the outgrowth or outpushing of a hollow organ from within.

Exoskeleton (Gr. ἔξω, outside of), the external skeleton.

Exuviae (Lat. = the slough or cast skin), the cast skin of the larva at metamorphosis. The word does not exist in the singular.

Follicle (Lat. = a little bag), a small pit or bag-like hollow; in particular, the hollow cellular investment of the developing ovum.

Foramen (Lat. = a hole), a passage through any portion of the skeleton.

Fossa (Lat. = a ditch), an external depression or hollow.

Ganglion (Gr. γάγγλιον, a tumour), a definite swelling in the course of a nerve-cord, caused by an aggregation of nerve-cells.

Germinal Layers. In the early development of the embryo the blastoderm becomes marked out into three layers of cells, of which the outermost is called the *ectoderm*, the middle one the *mesoderm*, and the innermost the *endoderm*. The term *lower layer* is used for the last two together, when they are not clearly differentiated from one another.

Ginglymus (Gr. γιγγλυμός), a ball-and-socket joint.

Gonads (Gr. γόνος, begetting), the organs which produce the reproductive cells (sperms or ova).

Gonapophyses (Gr. γόνος, and ἀπόφυσις, outgrowth), appendages surrounding the genital pore.

Haemocoele (Gr. αἷμα, blood; κοῖλος, hollow), a body-cavity formed by secondary enlargement of the blood-spaces.

Hamule (Lat. *hamulus*, a little hook), a small hook-like process.

Histology (Gr. ἱστός, a weaver's web), the science of the minute structure of organisms.

Homology (Gr. ὁμολογέω, to be in accordance with), a correspondence in type of structure between parts or organs of different organisms, due to evolutionary differentiation from the same or a corresponding part or organ of some remote ancestor. In segmented animals, an organ of one segment is *serially homologous* with another of another segment if the two are derived from corresponding parts (e.g. thoracic legs and lateral abdominal gills).

Hyaline (Gr. ὑάλινος, glassy), colourless and transparent, like glass.

Hypertrophy (Gr. ὑπέρ, over; τροφή, nourishment), excessive development or overgrowth of a part.

Hypodermis (Gr. ὑπό, under; δέρμα, skin), in Insecta, the layer of cells which lies just under the cuticle and gives rise to it. These cells are *of ectodermal origin*. (Not to be confused with *hypoblast*, an alternative term for *endoderm*.)

Imago (Lat. = a likeness), the perfect insect.

Instar (Lat. = an image), the stage between two successive ecdyses.

Intima (Lat. = innermost), an internal cuticle or internal epithelium.

Invagination (Lat. *in*, and *vagina*, sheath), the ingrowth or inpushing of a hollow organ from without.

Lamella (Lat. = a small blade), a small flat blade-like organ.

Lamina (Lat. = a plate or blade), a thin flat firm layer.

Larva (Lat. = a ghost), the term applied biologically to any stage in the life of an animal which differs from the final or mature form to such an extent that a complicated change, or *metamorphosis*, is necessary to pass from the one to the other.

Leucocytes (Gr. λευκός, white; κύτος, cell), colourless or white blood corpuscles. Their function is to seize and carry off impurities; hence they are also called *phagocytes* (Gr. φαγεῖν, to devour).

Lower Layer, see *Germinal Layers*.

Lumen (Lat. = an opening), the internal canal of a hollow organ.

Mesenteron (Gr. μέσος, middle; ἔντερον, the gut), that part of the gut formed from endoderm—the *mid-gut*.

Mesoderm, see *Germinal Layers*.

Metamorphosis, see *Larva*.

Micron (Gr. μικρός, small), one-thousandth part of a millimetre. Denoted by μ.

Miocytes (Gr. μείων, smaller; κύτος, cell), the smaller (oat-shaped) corpuscles found in the blood of Insects.

Moniliform (Lat. *monile*, a necklace), like a string of beads.

Nucleus, see *Cell*.

Nymph (Gr. νύμφη, maiden, nymph, pupa of a moth), the term very unsuitably applied to the aquatic larvae of certain Insects, including Dragonflies.

Ocellus (Lat. = a small eye), a simple eye.

Oesophagus, the gullet.

Ommatidium (Gr. ὀμμάτιον, a little eye), one of the elongated separate elements of a Compound Eye.

Ontogeny (Gr. ὄντος, a being), the study of the development and growth of the individual animal.

Palaeontology (Gr. παλαιός, ancient), the study of extinct animals.

Papilla (Lat. = a teat), a small elevation like a teat or the finger of a glove.

Phylogeny (Gr. φῦλον, the race, tribe), the study of the descent or pedigree of the various tribes of animals.

Plasma (Gr. πλάσμα, anything moulded), what remains of the blood when the corpuscles are taken away—the *serum*, or liquid part, + the *fibrin*.

Pleura or *pleurites* (Gr. πλευρά, ribs, side), lateral sclerites or parts of the exoskeleton.

Podites (Gr. πούς, foot), the separate pieces of a jointed appendage of the Arthropoda. Of these, the basal piece is called the *protopodite*, the inner distal piece *endopodite*, and the outer distal piece *exopodite*.

Proctodaeum (Gr. πρωκτός, anus; ὁδός, entrance), a posterior invagination of the ectoderm, giving rise to the anus and hind-gut.

Retina (Lat. *retineo*, to retain), that part of the eye which retains or receives the image conveyed by the dioptic part. In the compound eye, the retina of each ommatidium is termed a *retinula*.

Saccus (Lat. =a sack or bag), a bag-like organ.

Sarcolemma (Gr. σάρξ, flesh, muscle; λέμμα, husk, peel), the elastic sheath enclosing a muscle-fibre.

Sclerite (Gr. σκληρός, hard), a separate hardened piece of the exoskeleton.

Sections, thin slices through an animal or organ; usually either *transverse* (in a plane at right angles to the axis of symmetry), *horizontal* (parallel to the ventral or dorsal surface), or *sagittal* (in a vertical plane parallel to the axis of symmetry). Most of the sections figured in this book are 10μ in thickness. T.S. =transverse section, L.S. =longitudinal section.

Sensilla (Lat., dim. of *sensus*, perception), a minute sense-organ.

Seta, a bristle or stiff hair.

Sphincter (Gr. σφιγκτήρ, a tight band), a circular band of muscle for closing or constricting the lumen or orifice of a tube.

Spiracle (Lat. *spiraculum*, air-hole) or *Stigma* (Gr. =spot), a breathing-aperture; one of the openings of the tracheal system.

Squame (Lat. =a scale), a flat piece attached to an organ.

Sterna or *Sternites* (Gr. στέρνον, breast), ventral sclerites or parts of the exoskeleton.

Stomodaeum (Gr. στόμα, mouth; ὁδός, entrance), an anterior invagination of the ectoderm, giving rise to the mouth and fore-gut.

Suture (Lat. *sutura*, a seam), the soft connecting portion between two sclerites.

Syncytium (Gr. σύν, together; κύτος, a cell), a layer of cells so fused together that their boundaries are obliterated and their nuclei lie in a single continuous protoplasmic mass.

Synthorax, the closely fused meso- and metathorax of the Dragonfly. The same two segments not so closely fused, in other Insects, are often termed the *pterothorax*, since they bear the wings.

Teneral (Lat. *tener*, tender), the term applied to the freshly-emerged or immature imago.

Terga or *Tergites* (Lat. *tergum*, the back), dorsal sclerites or parts of the exoskeleton.

Trachea (Gr. τραχεῖα, windpipe), an air-tube.

APPENDIX C

SOME IMPORTANT SYNONYMS

	Subfamily	Synonym	Name adopted here
ANISOPTERA	Gomphinae	*Aeshna* Fabr. et Kirby	*Gomphus* auct.[1]
	Aeschninae	*Aeshna* (recent American authors)	*Aeschna* auct.[1]
	,,	*Fonscolombea* Selys	*Boyeria* McLach.
	,,	*Oligoaeschna* Selys	*Jagoria* Karsch.
	Corduliinae	*Chlorosoma* Charp.	*Somatochlora* Selys
	,,	*Metathemis* Tillyard	*Eusynthemis* Foerster
	Libellulinae	*Neophlebia* Selys	*Tetrathemis* Br.
	,,	*Platetrum* Newman	
	,,	*Leptetrum* Newman	
	,,	*Plathemis* Hagen	*Libellula* Linn. (pars)
	,,	*Belonia* Kirby	
	,,	*Ladona* Needham	
	,,	*Libella* Br.	*Orthetrum* Newman
	,,	*Hydronympha* Newman	
	,,	*Nannodythemis* Br.	*Nannophya* Ramb.
	,,	*Diplax* Charp.	*Sympetrum* Newman
ZYGOPTERA	Epallaginae	*Euphaea* Ramb.	*Pseudophaea* Kirby
	,,	*Anisoneura* Selys	*Philoganga* Kirby
	,,	*Tetraneura* Selys	*Devadatta* Kirby
	,,	*Dineura* Selys	*Diphlebia* Selys
	Calopteryginae	*Agrion* Fabr.	*Calopteryx* Leach et auct.[1]
	,,	*Cleïs* Selys	*Umma* Kirby
	Megapodagrioninae	*Podagrion* Selys	*Megapodagrion* Selys
	,,	*Amphilestes* Selys	*Rhinagrion* Calvert
	Platycneminae	*Hypocnemis* Selys	*Prionocnemis* Selys
	,,	*Hemicnemis* Selys	*Leptocnemis* Selys
	,,	*Trichocnemis* Selys	*Coeliccia* Kirby
	,,	*Psilocnemis* Selys	*Copera* Kirby
	Protoneurinae	*Alloneura* Selys	*Caconeura* Kirby
	,,	*Neoneura* Selys	*Caenoneura* Kirby
	Agrioninae	*Coenagrion* Kirby	*Agrion* auct.[1]
	,,	*Micronympha* Kirby	*Ischnura* Charp.
	,,	*Erythragrion* Selys	*Telebasis* Selys (1865)
	,,	*Telebasis* Selys (1877)	*Teinobasis* Kirby
	,,	*Stenobasis* Selys	*Archibasis* Kirby

[1] Regarded as a *nomen conservandum*.

INDEX OF ILLUSTRATIONS

(The numbers are those of figures, not pages)

INDEX OF TEXT

References given in heavy type are principal references. † indicates fossils. Sub-orders of Odonata are printed in heavy capitals, families in capitals, sub-families in Italic capitals, and tribes in heavy type. All names of genera and species are printed in Italics.

CAMBRIDGE: PRINTED BY J. B. PEACE, M.A., AT THE UNIVERSITY PRESS.

Printed in Dunstable, United Kingdom

91231531R00238